Nutritional Aspects of Aging

Volume I

Editor

Linda H. Chen, Ph.D.

Professor and Chairman
Department of Nutrition
and Food Science
University of Kentucky
Lexington, Kentucky

CRC Press
Taylor & Francis Group
Boca Raton London New York

CRC Press is an imprint of the
Taylor & Francis Group, an informa business

First published 1986 by CRC Press
Taylor & Francis Group
6000 Broken Sound Parkway NW, Suite 300
Boca Raton, FL 33487-2742

Reissued 2018 by CRC Press

Library of Congress Cataloging in Publication Data

Main entry under title:

Nutritional aspects of aging.

   Includes index and bibliography.
   1. Aging--Nutritional aspects. 2. Aged--Nutrition.
3. Nutritionally induced diseases. I. Chen, Linda H.
QP86.N853   1986   613.2'0880565   85-9720
ISBN 0-8493-5737-3 (v. 1)
ISBN 0-8493-5738-1 (v. 2)

A Library of Congress record exists under LC control number: 85009720

ISBN 13: 978-1-315-89603-8 (hbk)
ISBN 13: 978-1-351-07513-8 (ebk)

Visit the Taylor & Francis Web site at http://www.taylorandfrancis.com and the
CRC Press Web site at http://www.crcpress.com

PREFACE

The phenomena of aging are so obvious that most of us assume that we know what the term means. However, one will be surprised to find that people have a hard time agreeing on an objective scientific definition of aging. The aging process involves progressive changes with age after maturity in various organs leading to decreased functional ability. It reflects changes in many molecular, cellular, and systemic processes that take place with time. Often, aging is associated with chronic diseases although aging itself is not a disease.

Proper nutrition throughout the life span is of importance in maintaining mental and physical health. We believe that nutrition in early and middle life may have an important impact in determining the rate of aging. Therefore, the study of nutrition and aging, especially the study of nutrition and life span, cannot exclude nutrition at early life and middle years. In addition, some nutrition-related diseases which occur more frequently in later years may be prevented by adequate nutrition or can be managed by nutrition intervention. Nutrition is one of the environmental factors influencing biological aging which can be controlled.

Although we should deal with biological age rather than chronological age in gerontology, it is not easy to define biological age due to the lack of clear measures of biological age. The elderly, or the aged, are generally accepted as people who are at or after retirement age which is usually 65. As far as the Title IIIb National Nutrition Program for Older Americans is concerned, people at age 60 or above are qualified to participate. The number and the proportion of the elderly population in the U.S. are increasing rapidly. The Bureau of the Census reported that in 1977 there were 23 million people over 65 years of age which was about 10.6% of the total population. It is predicted that by the year 2050 there may be about 30 million people or 20% of the total U.S. population over the age of 65. The life expectancy at birth in 1977 was 76.5 years for women and 68.7 years for men, whereas the life expectancy at birth in the year 2050 is predicted to be 81 years for women and 71.8 years for men. With increasing life expectancy and increasing size of the elderly population, the quality of later life is of concern. The elderly have special health requirements but their socioeconomic status often produces difficulties for them to meet these requirements; thus they are at increasing risk of health problems. Nutritional inadequacy and the role of diet in pathogenesis of diseases are important components in health care of the elderly. The health care of the elderly is not limited to the treatment of diseases but also includes the prevention of diseases. The need for information on nutrition and aging is increasing.

The first section in Volume I describes the fundamentals of nutrition and aging which include research strategies for the study of nutrition and aging. The nutritional modulation of the aging process which has provided a major breakthrough in the field of nutrition and longevity is also discussed. In the second section, factors affecting the nutritional status of the elderly are discussed. These include biomedical influences, and social and psychological aspects. Section 3 includes dietary characteristics of the elderly population and methods for the assessment of nutritional status. The nutritional status of the elderly with respect to individual nutrients as determined by dietary survey and by biochemical methods is described in Section 4. Section 4 also includes discussion on nutrient metabolism, requirements, nutritional imbalances, and deficiencies of nutrients. Energy metabolism and obesity as a factor in pathogenesis of diseases are also discussed.

In the first section of Volume II, toxicological factors affecting nutritional status are discussed. Medications and alcohol may affect nutritional status. Section 2 provides a discussion of nutrition-related diseases which occur more frequently among the elderly. Cardiovascular diseases including coronary heart disease and cerebrovascular disease are the leading causes of death in the U.S. The relative importance of cardiovascular diseases, in terms of all deaths for the given age group, rises steadily with age. The death rate from these diseases is 28% for the middle age group (age 35 to 44), and is 69% for the old age group (age 75 and above). This reflects the continued progression of artherosclerosis with aging. Cancer is the second leading cause of death in the U.S. The death rate from cancer also rises steadily with age. The death rate from diabetes mellitus increases progressively with age and more rapidly after the age of 45. The incidence of diabetes mellitus is 0.23% under age 25 and 6.2% over age 45.

Osteoporosis is found in men and women over the age of 50. The incidence is especially high among women: about one third of women above the age of 60 have osteoporosis. Periodontal diseases are a major cause of tooth loss after the age of 35. The incidence in the U.S. is 24 to 69% among the age group of 19 to 26, and 100% above the age of 45.

Hypertension is found in persons of all ages. The incidence is 13 to 20% in the U.S. This disease is an important risk factor in the cardiovascular diseases. Diverticular disease is found in more than one third of the population over 60 years old in Western nations. Nutritional anemia including Fe deficiency anemia, megaloblastic anemia, and/or pernicious anemia due to deficiencies of Fe, folic acid, and/or vitamin B_{12} are also found among the elderly. The etiology of anemia may also be secondary to diseases.

This book is designed to provide pertinent and useful information on the basic and applied aspects of nutrition and aging to health professionals. It may also be used as a reference book by researchers as well as professors and students for graduate courses at colleges and universities. It emphasizes recent advances and knowledge in the areas of basic, clinical, and community nutrition.

Linda H. Chen, Ph.D.

THE EDITOR

Linda Huang Chen, Ph.D., is currently Professor and Chairman of the Department of Nutrition and Food Science at the University of Kentucky.

Dr. Chen obtained her B.S. degree in Pharmacy in 1959 from the National Taiwan University in Taipei, Taiwan, Republic of China, and a Ph.D. in Biochemistry in 1964 from the University of Louisville. She received her postdoctoral training at the University of Louisville, School of Medicine, from 1964 to 1966. She served as an Assistant Professor from 1967 to 1972, an Associate Professor from 1972 to 1979, and has served as a Professor since 1979 at the University of Kentucky. It was in 1983 that she became the Chairman of the Department of Nutrition and Food Science.

Dr. Chen is a member of the American Institute of Nutrition, the American Society for Clinical Nutrition, the Gerontological Society of America, and the International Association of Vitamins and Nutritional Oncology. She served as a panelist for the National Program Standard in Gerontology for the Association for Gerontology in Higher Education, Washington, D.C. and the Gerontological Society of America. She also served a 4-year term as a member in the Cancer Research Manpower Review Committee of the National Cancer Institute. Her current research interests include vitamin interaction, nutrition and aging, nutrient-drug interaction, and nutrition and cancer.

CONTRIBUTORS

Thomas P. Almy, M.D.
Distinguished Physician
Veterans Administration
Professor of Medicine and of Community
 and Family Medicine
Dartmouth Medical School
Hanover, New Hampshire

Olav F. Alvares, B.D.S., Ph.D.
Associate Professor
Department of Periodontics
The University of Texas Health Science
 Center
San Antonio, Texas

Lynn B. Bailey, Ph.D.
Associate Professor
Department of Food Science and Human
 Nutrition
University of Florida
Gainesville, Florida

Charles H. Barrows, Sc.D.
Chief
Laboratory of Nutritional Biochemistry
National Institute on Aging
Baltimore City Hospital
Baltimore, Maryland

Rudy A. Bernard, Ph.D.
Professor
Department of Physiology
Michigan State University
East Lansing, Michigan

Judith Bond, Ph.D.
Professor
Department of Biochemistry
Medical College of Virginia
Virginia Commonwealth University
Richmond, Virginia

George A. Bray, M.D.
Division of Diabetes and Clinical
 Nutrition
Department of Medicine
University of Southern California
School of Medicine
Los Angeles, California

Toni M. Calasanti
Instructor
Department of Sociology
University of Kentucky
Lexington, Kentucky

Linda H. Chen, Ph.D.
Professor and Chairman
Department of Nutrition and Food
 Science
University of Kentucky
Lexington, Kentucky

Ching K. Chow, Ph.D.
Professor
Department of Nutrition and Food
 Science
University of Kentucky
Lexington, Kentucky

Harold H. Draper, Ph.D.
Professor and Chairman
Department of Nutrition
College of Biological Science
University of Guelph
Guelph, Ontario, Canada

Judy A. Driskell, Ph.D.
Professor and Head
Department of Human Nutrition and
 Foods
Virginia Polytechnic Institute and State
 University
Blacksburg, Virginia

Sandra E. Gibbs, Ph.D.
Senior Associate
Human Factors Department
IBM Corporation
Lexington, Kentucky

Janet L. Greger, Ph.D.
Professor
Department of Nutritional Sciences
College of Agricultural and Life Sciences
University of Wisconsin
Madison, Wisconsin

Jon Hendricks, Ph.D.
Professor
Department of Sociology
University of Kentucky
Lexington, Kentucky

Kang-Jey Ho, M.D., Ph.D.
Department of Pathology
University of Alabama
Veterans Administration Medical Center
Birmingham, Alabama

Jeng M. Hsu, Ph.D., D.V.M.
Research Chemist
Medical Research
Veterans Administration Medical Center
Bay Pines, Florida

Glenville Jones, Ph.D.
Associate Professor
Departments of Biochemistry and
  Medicine
Queen's University
Kingston, Ontario, Canada

Nancy L. Keim, Ph.D.
Research Nutrition Scientist
USDA, ARS
Western Human Nutrition Center
Presidio of San Francisco, California

Gertrude C. Kokkonen, B.A.
Laboratory of Nutritional Biochemistry
Gerontology Research Center
National Institute on Aging
Baltimore City Hospital
Baltimore, Maryland

Calvin A. Lang, Sc.D.
Professor
Department of Biochemistry
University of Louisville
School of Medicine
Louisville, Kentucky

Loren G. Lipson, M.D.
Chief
Division of Geriatric Medicine
Associate Professor of Medicine and
  Gerontology
University of Southern California School
  of Medicine
Los Angeles, California

Fudeko Maruyama, Ph.D.
Extension Professor
Department of Nutrition and Food
  Science
University of Kentucky
Lexington, Kentucky

Malcolm J. McKay, Ph.D.
Department of Biochemistry
Medical College of Virginia
Virginia Commonwealth University
Richmond, Virginia

Jaime Miquel, Ph.D.
Guest Research Scientist
NASA Ames Research Center
Moffett Field, California

Betty Jane Mills, Ph.D.
Assistant Professor
Department of Biochemistry
University of Louisville
School of Medicine
Louisville, Kentucky

Michael J. Monzel, M.D.
Central Maine Medical Center
Lewiston, Maine

Eve Reaven, Ph.D.
Research Physiologist
Geriatric Research, Education, and
  Clinical Center
Veterans Administration Medical Center
Palo Alto, California

Gerald M. Reaven, M.D.
Nora Eccles Harrison Professor
Head, Division of Gerontology
Stanford University School of Medicine
Director, Geriatric Research, Education
  and Clinical Center
Veterans Administration Medical Center
Palo Alto, California

Lora Rikans, Ph.D.
Associate Professor
Department of Pharmacology
College of Medicine
University of Oklahoma Health Sciences
  Center
Oklahoma City, Oklahoma

Howerde E. Sauberlich, Ph.D.
Department of Nutrition Sciences
University of Alabama
Birmingham, Alabama

Earl S. Shrago, M.D.
Professor
Departments of Medicine and Nutritional
 Sciences
Clinical Nutrition Center
University of Wisconsin
Madison, Wisconsin

Howard B. Turner, M.A.
Research Associate
Department of Rural Sociology
University of Kentucky
Lexington, Kentucky

Hans U. Weber, Ph.D.
Manager
The Alexander Medical Foundation
San Carlos, California

Aniece A. Yunice, Ph.D.
Associate Professor
Department of Physiology and Biophysics
College of Medicine
Oklahoma University Health Sciences
 Center
Veterans Administration Medical Center
Oklahoma City, Oklahoma

TABLE OF CONTENTS

Volume I

Volume II

SECTION I: TOXICOLOGICAL FACTORS AFFECTING NUTRITIONAL STATUS

SECTION II: NUTRITIONAL BASIS FOR DISEASES IN AGING

Section I
Fundamentals of Nutrition and Aging

Chapter 1

RESEARCH STRATEGIES FOR THE STUDY OF NUTRITION AND AGING

Calvin A. Lang

TABLE OF CONTENTS

I. INTRODUCTION

How do you study nutritional aspects of aging? Does one not simply use the same rationale and methodology as in any other nutritional research? Why study aging anyway? Is there not already complete information on nutrition of the elderly? Are there any challenging and new scientific problems to solve in this area?

These questions were raised many years ago when my dissertation advisor, Bacon F. Chow, suggested an extension of our studies on vitamin B_{12} metabolism to aging rats and eventually to humans. Little was known about a proper approach, so new and sometimes less than ideal strategies and methods were used. In the intervening decades of experience and errors, a number of basic criteria evolved which have facilitated our own research design and approach and also aided in the evaluation of other aging research.

Currently, nutrition and aging research is undergoing a renaissance due mainly to extensions of the discoveries of the gerontology pioneer, Clive McCay, of Cornell University during the 1930s. His basic finding was that restricted nutrition increases longevity. Today this is the best-known way to extend life span and also to maintain health in mammals.[1,2]

The recent increase in gerontological research was catalyzed by the establishment several years ago of the National Institute on Aging (NIA), a new unit of the National Institutes of Health. The advent of the NIA stimulated biomedical research and training in aging even though the economic climate for research funding had been poor. Since that time, however, this has changed, for in our affluent and healthier U.S. society there is an increasing number of elderly who will constitute a significantly larger proportion of the total population in the coming years. The political-economic impact of this population shift is gradually being recognized, and the elderly will have a greater influence in the future.

The area of health maintenance or preventive medicine has emerged in recent times because of the high cost of medical and nursing home care. Thus, the emphasis is now on *independent* living as long as possible, and one of the key health factors is to maintain an optimal nutritional status. However, there is a serious lack of definitive information, because almost all nutritional data are for growing or young adult subjects. For example, as indicated by others,[3-5] our ignorance of nutrient requirements for the elderly is illustrated by the lumping of 1980 Recommended Dietary Allowances (RDAs) into a 51 + year category rather than by intervals after age 50. This implies that there are no recognizable nutritional differences in any individuals older than 51 years! In the literature arbitrary ages such as 65 years have been used to separate "young" from "old", and these cutoffs were most likely based on political-economic or social security criteria and not on functional or biological status. The use of an age limit will mask the various aging changes that occur at different chronological ages in different individuals. The result will be that few aging-specific changes will be identified, so that the conclusions, recommendations, and treatments could be inaccurate. The practice of using an age limit or bracket is a common design error of many studies in the current literature — an error that must be corrected.

The objective of this chapter is to introduce some fundamental and unique considerations of aging research with special reference to nutrition. The scope is limited and includes definitions of aging, special criteria and methodologies for nutrition and aging studies, and the underlying rationale. The discussion represents what we believe are the best strategies and design for aging investigations. To the best of our knowledge there are few publications on such research strategies, and even these deal primarily with psychosocial and statistical aspects. Hopefully, this paper will guide and encourage newcomers to the field of nutrition in aging, and if the outcome provokes discussion, then a more important mission will have been accomplished.

II. HOW OLD IS OLD? FUNCTIONAL VS. CHRONOLOGICAL AGE

This fundamental question was a key point of early symposia on gerontological research many years ago. It is a critical and still timely subject which is not well understood and has resulted in studies of a wide range of chronological ages often giving conflicting and misleading conclusions.

Biological aging refers to changes that occur in the latter part of the life span. Hence, we use the term *aging* as a synonym for senescence. A frequent statement that aging begins at conception is erroneous and misleading, for it implies that any changes with time are part of the aging process. The objection is that earlier stages of the life span have been clearly delineated as growth, development, and maturity, periods with distinctly different characteristics from aging. For example, growth phenomena focus on net increases in cellular mass and function and should not be confused with senescent changes, which usually deal with losses in metabolic function and structure. Indeed, there has been no evidence so far that early growth affects senescence.

The best experimental aging approach is to study post-maturational changes during the mature and senescent periods of the life span. Thus, *aging-specific* changes which occur, especially during the period of decreasing survivorship, can be identified and characterized. Earlier growth and developmental changes *could* have subsequent effects on senescence, but such long-term relationships are too difficult to study at this time when few senescence-specific changes have been clearly identified and shown to be significant to the overall aging process.

A clear distinction should be made between the related areas of gerontology and geriatrics. Gerontology is the science of all old age phenomena whether normal or pathological and regardless of discipline. On the other hand, geriatrics is rigorously defined as the medical study of diseases of the aged. This distinction between normative aging processes and diseases of the elderly should be clearly appreciated, for their research objectives and findings are different, especially in studies of human subjects. It should be clear that aging is not a disease, for it occurs in *all* organisms who live long enough.

There are several ways to describe biological age. One of the most definite and accurate is *demographic age* — in other words, survivorship relative to a population (see Figure 1). From survivorship curves one can determine whether an organism falls in the mature, old, or very old categories. Arbitrarily, we have categorized adult animals in the 100 to 75% survivorship groups as mature, 75 to 25% as old, and 25 to 0% as very old. The rationale for using survivorship as an estimate of function or health is that survival depends on a certain functional level. During the "old" and "very old" stages there is a high probability that biological decrements will appear and lead eventually to death. A major advantage of this aging index is its applicability to different species with varying life spans, which allows comparison and integration of results. For example, our approach in studying mosquitoes and mice would have no meaning using chronologic age alone since their life spans differ by 30-fold (1 vs. 30 months), respectively. However, by using demographic age, they were compared on a common percent survivorship basis.

The most meaningful measurement of biological age is *physiological* or *functional age*, which also includes the underlying biochemical phenomena.[6] Functional age is expressed as percent of maximal function of the life span. The advantage is that functional level is probably the ultimate aging criterion. Other measurements or changes that occur with aging may be trivial.

A summary of some life span changes in different functions are shown in Figure 2. This is from a compilation of several human physiology studies and demonstrates several important life span changes.[6,33] First, there is a rapid growth increase of various physiological functions from birth to maturity, reaching a maximum which differs slightly for each function but

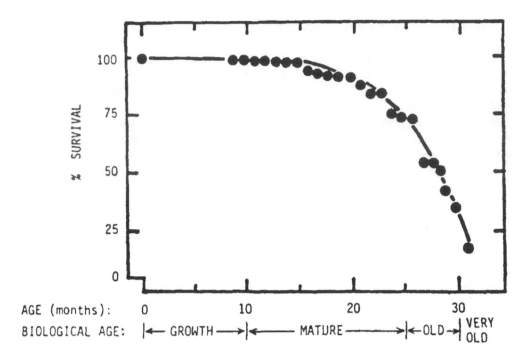

FIGURE 1.   Survival curve of the C57BL/6J mouse. The median life span is 28 to 29 months.

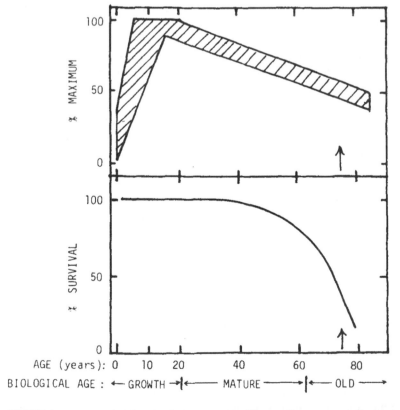

FIGURE 2.   Physiological function levels during the life span. Pulmonary, cardiac, and renal functions of human subjects are represented in the upper graph. A survivorship curve is in the lower graph. The arrows indicate the median life span at 75 years.

generally occurs by 20 to 25 years. This is followed by either a gradual decrease or no change during the remaining mature and senescent years. This life span profile of a relatively rapid rise to a peak and a gradual decline indicates the following: the age when the peak value is attained can be considered the beginning of senescence, and the decreases thereafter can be regarded as the aging process. Thus, the degree of function of an individual can be used as a measurement of functional age.

Also, it is of interest that none of these physiological functions decreased more than about 50% of maximum. In a reciprocal way this suggests that there is only a twofold reserve of function in view of the fact that these measurements were of *survivors* and assuming that the nonsurvivors had lower values. There is supporting evidence, as discussed later, that aging-specific biochemical or molecular functions also do not decrease much below 33 to 50% of maximum.

In these cross-sectional investigations, subjects of different ages were examined only once, so that the changes may reflect shifts in the surviving populations rather than aging changes in an individual. In this regard it has been pointed out in several studies that *apparent* increases in biological function may occur in very old organisms.[7] This is interpreted and confirmed by longitudinal studies in some cases to be due to the loss of a shorter-lived subpopulation with low values and thus the unmasking of a longer-lived subpopulation with high values. Thus, it is important to recognize that apparent aging changes in cross-sectional studies may not be "true" aging changes.

There has been great interest in finding an objective, quantitative, and simple measurement of functional age and the aging process. Obviously, demographic criteria are only appropriate for the study of relatively homogeneous inbred model organisms and not heterogeneous human populations. Also, the determination of physiological age is complex, according to the detailed discussion by Nathan Shock,[6] the leading pioneer in that field. In his Baltimore Longitudinal Study, no general aging index could be demonstrated from factor analyses of 14 physiological tests of 650 normal men (aged 20 to 96 years). It is of interest, however, that some of the changes which correlated well with chronological age include range of accommodation of the eye, pulmonary function (vital capacity), hearing loss, systolic blood pressure, and visual acuity.

The basic assumption for a biochemical aging index is that there is a key metabolite which can affect a variety of molecular and physiological functions and thus any aging decrease in this compound would have profound sequelae. This simple view seems unlikely. However, a prime candidate will be described later.

In summary, selection of ages for study should be based on functional changes that occur in old age, and not on chronological age alone. Thus, the results will be aging specific and have direct relevance to the aging process and senescence.

III. VALID ANIMAL MODELS ARE ESSENTIAL FOR GERONTOLOGICAL RESEARCH

Valid aging animal models are essential for aging research. Indeed, the use of *Escherichia coli* and related microorganisms were critical for the rapid advancements in molecular biology because the data from different laboratories could be utilized directly and extended without unnecessary duplication of effort.

The criteria for a valid aging model are as follows. First, the organism should be biologically similar to man, who will be the ultimate recipient of the information. It is realized that small morphologic, physiologic, or molecular differences will occur, but in general the model should be biologically as similar as possible. Second, rigorous control of genetics and environment of the model organism is necessary, otherwise, nonspecific changes can occur due to variations in these factors and may be misinterpreted as due to aging. Third, a short life span is desirable to obtain more rapid results. In fact, a primary deterrent to studying nutritional aspects of aging in human subjects is due to their long life span. Certainly there is no way a human investigator can follow the entire life span of a subject who probably will outlive the investigator. Fourth, the organism should be small enough to grow in large numbers for statistical purposes and yet large enough to manipulate individually without special apparatus. The ideal organism should be readily available and easily cultured and maintained. Finally, a backlog of information on the organism is helpful, but not critical.

The usual rodent models, the laboratory rat and mouse, have been used for aging research. The familiarity of these models and their background of available information has made them favorite and standard organisms for many studies. However, both the rat and mouse have some limitations that should be considered carefully before extending results to man. An excellent caveat by Bernard Oser[8] delineates the problems and marked differences between rat and man. The life spans are approximately the same for "normal" rat and mouse strains. The median life span is about 29 months, which is shorter than the current human life expectancy of approximately 75 years. Nevertheless, the rodent life span is still nonideal for obtaining rapid results.

The establishment and availability of aging colonies of mouse and rat strains of different ages throughout their life spans has been a major breakthrough. These colonies have been developed and subsidized by the NIA, and rodents of different ages can be obtained upon application. The enormous cost required to maintain aging colonies has deterred commercial and individual laboratories from this endeavor. Our own experience in maintaining an aging colony of C57BL/6J mice for over 10 years indicated that the cost for a 30-month C57BL/6J mouse was over $100 each.

IV. WHEN DO MATURITY AND AGING BEGIN?

In the mouse and rat, maturity or maximal body weight is attained by 10 to 12 months of age.[9,10] Also, organ weights do not reach maximal plateau levels until about 10 months.[9] A common misconception is that these rodents are mature by 90 to 100 days, when they become breedable. However, this may only be applicable for studies of reproduction, for other organs are still growing. For these reasons we always include rodents of 10 to 20 months of age as maturity reference points for comparison.

Aging in mouse and rat begins when the survival curve starts to drop. In our experience with the C57BL/6J mouse the median life span was about 29 months and hence, aging began about 25 months.[11]

Routinely we use age samples of about 12, 18, 24, 30, and 34 + months. Thus, mature, old, and very old (34 +) ages are included. Our rationale for studying ages around the period of decreasing survival (high mortality) is based on the assumption that aging changes will be amplified and identified more easily during that time.

It is recognized that tumors may occur with greater frequency in old age. Indeed, the use of rodents only up to 24 months of age has been justified on the basis that few deaths occur until after that age, and thus they were "normally aged". In fact, younger animals are known to have various pathological lesions, but the presence of these tumors may have little debilitating effect. Finally, one of the best, long-lived mouse strains, C57BL/6J, has one of the lowest tumor incidences of all strains, yet it is about 50%! Tumors or not, the animals may still be functionally young in age.

V. THE MOSQUITO IS A VALID AND USEFUL AGING MODEL

When we started aging research a major concern was the unavailability and high cost of maintaining aging rodents. To be sure, there was an important background of scientific information available for rats and mice, and they were accepted models for the study of human phenomena. Another possibility was the use of microorganisms, which were excellent nutritional models, because their rapid growth provided a quick way to discover and assay new growth factors and vitamins such as vitamin B_{12}. However, microorganisms were inappropriate for aging research because they have no comparable period of senescence as in metazoa.

By chance, I became exposed to insects, especially the mosquito, in my first job in a department of entomology. The most striking feature of the mosquito was its distinct stages of the life span, including larval growth, pupal metamorphosis, and the adult period. These clearcut stages suggested that it might be a suitable organism for aging studies without interference from growth phenomena. Unlike any mammal, the nutritional requirements of the mosquito and some related flies had been elucidated under axenic or germ-free conditions. As a result, we developed and standardized an axenic culture and a quantitative growth index for the yellow fever mosquito.[12] Its fastidious nutritional requirements were nearly identical to those of rat and man, suggesting that its intermediary metabolism was also very similar.

Direct evidence for the biochemical similarity of mosquito and man is the occurrence of the same enzyme systems and metabolic pathways. For example, the cytochrome system was first discovered in insects and has the same function as in man, differing only slightly in structural detail. There are many other biochemical similarities which far outnumber the differences.

At the beginning of our studies to develop and characterize the mosquito as a laboratory model, various biochemical parameters of growth and aging were determined to use as convenient, objective measurements of cell number and size. Thus, the first work described the RNA, DNA, and protein content of different stages of the mosquito throughout its life span.[13] The increase in body weight during larval growth was paralleled by concomitant increases in RNA and DNA. During metamorphosis, when there is no transport of any nutrients or metabolites, these parameters were stabilized except for losses due to molting. In the young adult immediately after emergence from the pupal stage there was relatively high metabolic activity such as DNA and protein synthesis as measured by our radioisotope studies in vivo.[14,15]

The first 8 to 10 days of adult life in the mosquito is actually a continuation of metamorphosis, a period which has been termed "adult development" in holometabolous insects like Lepidoptera and other Diptera. Thus, it is important in studies of insects, particularly the mosquito and other flies, that the entire adult period is not considered an aging stage but only the post-development period. For this reason we use the 8- to 10-day adult as a reference point. Thereafter, during the remaining adult life span there is no measurable change in fresh or dry body weight, protein, or RNA or DNA content. Any biochemical changes are due to aging and not growth. In brief, these biochemical measurements are useful, essential parameters in defining the various stages of the life span.

The mosquito is an ideal aging model that is stardardized and simple. We have used a strain of the yellow fever mosquito (*Aedes aegypti*) that has been inbred in our laboratory for over 25 years and 250 generations. Thus, it is relatively homogeneous from a genetic point of view. In addition, we developed standard culture procedures for routine and axenic rearing and aging.[16] With this genetic and environmental control the time schedule and biochemical parameters for growth, metamorphosis, and adult aging are highly reproducible and have not differed over a 25-year span even though many different individuals carried out the rearing and analyses.

Mosquito culture is simple, and large masses can be reared routinely on a bacterial infusion such as a rabbit feed pellet dispersed in water. For more controlled nutritional or radiolabeling experiments, an axenic or germ-free culture technique is used with a completely defined medium. In this way mosquito larvae can be pulse labeled or continuously labeled with various radioisotopic precursors of protein, nucleic acids, and other metabolites. These procedures were described in greater detail previously.[16]

The mosquito is unique among invertebrates, including other flies, because the adult female can be maintained on a blood-free diet of sucrose solution as it is in nature between blood meals. The result is that ovarian development is suppressed, and the ovaries remain teneral or thread-like. In effect, the females have been nutritionally ovariectomized. Other Diptera cannot be maintained without active, fully developed ovaries and, therefore, analysis of whole bodies will include metabolically active or dividing tissues which can confound the results. In the adult mosquito, which is post-mitotic, only turnover or replacement synthesis occurs as we have demonstrated by measuring radiolabeled DNA. Thus, any observed aging difference is due to changes from the homeostatic mature stage to senescence rather than from growth to maturity.

The only nutritional requirement for the *adult* mosquito is an energy source in the form of sucrose or fruit juices in nature. Indeed, the longevity of sucrose-maintained females is greater than those which are blood-fed. A blood meal is needed only for ovarian development and egg production. Since males cannot penetrate the skin and suck blood, they can subsist only on sugar solutions. This unusual and simple nutrition provides a well-defined and controlled system, for the metabolism is uncomplicated by nutrient intake or its variability. The organisms are healthy and not starved, for their weight, nucleic acid, and protein levels are unchanged as mentioned earlier.

Also, individual adult mosquitoes are injected routinely. The injection of a cold-anesthetized adult takes less than 1 min with ordinary equipment such as a low-power, stereodissecting scope and glass needles hand-drawn from disposable Pasteur pipettes. The technique is simple and as much as 1 $\mu\ell$ of solution is routinely injected into individual mosquitoes with no mortality for at least 48 hr. The injection volume was determined using radioactive solutions and was over 99% accurate.

VI. GLUTATHIONE LEVELS DECREASE IN AGING MOSQUITO, MOUSE, AND MAN

Evidence that findings in the mosquito can be extrapolated to mouse and man are our aging results on glutathione, the ubiquitous tripeptide. In the mosquito we demonstrated for the first time that reduced and total glutathione decreased in parallel with the survivorship curve.[17] In other words, glutathione content decreased specifically in senescence. This finding was extended to the C57BL/6J mouse, a long-lived and well-studied aging model. In various organs, the heart, liver, and kidney, the same aging decrease in glutathione was found.[9] Of special importance was the finding that the glutathione content of blood from these mice also decreased with old age.[18] This occurred both as a function of mouse age and as a function of red cell age. It should be pointed out that virtually all (97%) of the glutathione was present in erythrocytes and less than 3% was present in serum. These overall findings demonstrated that red cell glutathione levels paralleled and reflected other tissue levels in general. Thus, the results in mosquito and mouse led to our investigation of blood glutathione levels in man.

The glutathione content in healthy, human subjects decreased with adult age up to about 60 years of age in females[19] and 80 years of age in males.[20] Beyond those ages two distinct subpopulations were found: one with high levels of glutathione as found in young adults aged 20 to 40 years, and the other with very low levels of glutathione. This dichotomy or bimodal distribution in old subjects suggested that the subjects under study represented two different subpopulations of elderly, but there was no obvious explanation. Those with high glutathione levels were considered still biologically young, and those with low levels were biologically old. Proof of this may be obtained by our current longitudinal studies of individual subjects.

The difference in findings in mosquito and mouse compared to man may be due to the great heterogeneity of the human subjects compared to the relative genetic and environmental homogeneity of the animal models. Indeed, this distinction may be a reason for confusing or equivocal results from aging research on humans because their classification by chronological age is an inexact and highly variable indicator of their biological age.

We believe that the glutathione level is a prime candidate for an aging index because it is objective, quantitative, and is a regulator of a number of metabolic and detoxification reactions associated with senescence.[21]

The importance of glutathione as a key metabolite is its various roles which have been elucidated during the last 50 years. A major function is as a sulfhydryl buffer to maintain a reduced cellular environment for many synthetic reactions and to maintain the activity of various enzymes and proteins. Further, glutathione is an important membrane component as demonstrated by its role in red cell membrane stability. Finally, glutathione is a major detoxification mechanism for endogenous, deleterious compounds such as peroxides resulting from lipid metabolism and also from external toxic substances and xenobiotics such as drugs and environmental pollutants.[22] For example, in both rat and man glutathione is required to metabolize acetaminophen and prevent hepatotoxicity.

Recently we addressed the question of nutritional modification of this aging-specific decrease in glutathione in the mosquito. Can nutritional supplements of amino acid precursors of glutathione prevent, halt, or reverse this decrease? To this end various precursors were fed or injected into adult mosquitoes of different ages throughout the life span. One of the most striking findings is that administration of a very low concentration (0.001%) of thiazolidine-4-carboxylate in adult mosquitoes increased adult life span by 40%.[23] In addition, the glutathione content increased or maintained rather than decreased. These data demonstrated the close relationship of glutathione to longevity. Further work is under way to elucidate this mechanism. It is our belief that nutritional modification of this kind may play a significant role for a healthy, long life.

In brief, our "3M" approach with defined mosquito and mouse models has led to similar findings in man. This strategy was essential to carry out rapid, well-controlled aging studies at different levels of biological organization and to provide a rational basis for extending the results to human subjects.

VII. THE METABOLIC LIFE SPAN OF AN ORGANISM IS COMPRISED OF GROWTH, MATURE, AND SENESCENCE STAGES

Our initial effort in aging research was to consider changes in metabolism during different stages of the life span as shown in Figure 3.[24] This was based on the definition that metabolism is comprised of two opposing processes, namely anabolism or biosynthesis vs. catabolism or degradation. Further, these changes during the life span were expressed in terms of anabolism to catabolism ratios. During the growth stage this ratio would be greater than 1, to account for the net increase in tissue and tissue mass. When growth was complete we envisioned a period of maturity or homeostasis during which anabolism would equal catabolism, and the ratio would be 1.0. Finally, aging or senescence would occur with a decrease in the curve and ratios less than 1. Since this profile reflected changes in a ratio, the decrease in aging could be due to either a decrease in anabolism, an increase in catabolism, or a combination of both. Our aging studies since that time have been based on this idea. The evidence indicates that the decrease in anabolism to catabolism ratio in old age is due primarily, if not solely, to an impairment in anabolism and not to any change in catabolism. This was true in our in vivo studies of DNA and protein synthesis.

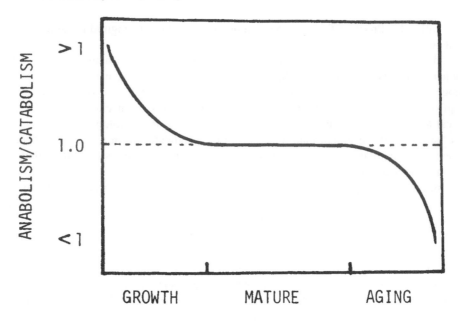

FIGURE 3.   Metabolic balance during the life span.

Our current nutritional practices for the elderly are based on requirements and recommendations that have been determined largely on growing or mature organisms and seldom on truly aged organisms. There is some information on nitrogen balance, but the data are scanty and do not describe biochemical details but rather the overall nitrogen intake and output. Very little work on a systematic basis has been carried out on elderly individuals who are past 70 and up to 90 years and over or on old animals with regard to free amino acid levels, their flux, and their relationship to protein turnover. Although these are difficult, relatively long-term experiments, the information is needed to provide a solid data base for nutritional recommendations.

A common misconception is that the protein and nutritional requirements of "old" rats have already been determined, but the data referred to were obtained with 3- to 5-month-old rats! With time and knowledge this kind of misunderstanding will hopefully be eliminated. Again this illustrates a need to educate the scientific community on a proper definition of aging.

VIII. "AN ADEQUATE DIET BY NO MEANS ASSURES ADEQUATE NUTRITION: NUTRITION INVOLVES MORE THAN THE NUTRIENT CONTENT OF THE DIET"*

These important statements were made by one of the pioneer nutritional biochemists who has studied the elderly for many decades. They summarize our view on the overemphasis of dietary aspects. The deluge of nutrition and aging books and articles which has appeared recently emphasize diet without considering other critical metabolic and utilization aspects of nutrition. Thus, the term "nutrition" is being misused to mean dietetics, which is only a part of the overall process.

Nutrition can be defined as "the fate of foodstuffs from the time of ingestion until the time of excretion". It is a complex, multidisciplinary area rather than a single discipline.

---

* A.A. Albanese, 1980.[25]

Nutrition is comprised of the psychological and sensory selection of diet, oral biology aspects, digestion, absorption, transport to target tissues, intermediary metabolism, and excretion via bile or the kidney. Indeed, many of these steps are impaired in old age. Thus, simply providing a "recommended" diet, which usually is based on healthy young adults, is an ineffective solution to the problem.

Nutritional surveys are a common means to assess nutritional status.[5,7] In the case of elderly as well as other subjects the answers may be inaccurate and misleading. Much has been said about the 24-hr recall procedure, but it is doubtful whether anyone can accurately give the amount of each nutrient or food, which is essential for evaluation. Thus, a survey approach is qualitative at best and is probably of value for only extremely malnourished subjects but not for chronic conditions as in the elderly. It should be recognized that answers to surveys may be questionable and fictitious because of the embarassment of many individuals, whether elderly or not, to admit they cannot remember. Assessments by clinical evaluation are better, but could be too late because such obvious nutritional defects may already have produced irreversible damage.

IX. THE BEST ASSESSMENT OF NUTRITIONAL STATUS IS BY BIOCHEMICAL MEASUREMENT

The important advantage of biochemical determinations is that subclinical deficiencies can be detected and provide earlier diagnosis and treatment than clinical findings. Some excellent summaries of nutritional assessment are available.[34,35] Many of these tests, however, have not been validated for specificity or completely established, except for some specific vitamins such as thiamin or vitamin B_6. Indeed, biochemical indexes can also uncover and pinpoint impairments of digestion, absorption, and excretion which are common in elderly. However, systematic and correlated analyses of blood, serum, urine, or other accessible body tissues need to be carried out to ensure they are indicators of tissue levels in general. Although detailed discussion is beyond the scope of this chapter, it should be emphasized that rigorous analytical techniques, including proportionality, recovery, and inhibitor-activator experiments, are required to determine the subtle quantitative changes that occur in senescence.

One of the standard areas of nutritional assessments of body fat and lean body mass are the anthropometric measurements of weight and height. In the past the reference standard tables were of questionable value since they included different and arbitrary categories of small, medium and large frames. More recently, a vogue is to determine body fat content via hydrostatic weighing through submerged and regular body weights. This indirect measurement is fraught with difficulties. From personal experience, I was surprised to learn that the hydrostatic weight tables apply only up to 52 years of age. Also, the algorithms or calculations for this procedure were developed with relatively scanty data using a limited number of cadavers. More sophisticated measurements of metabolic mass via potassium counting have been used but may not be practical for large-scale application. Probably the best new way of determining weight and fat content will be CAT scans (computed axial tomography), which enable the sequential "dissection" of a body by noninvasive X-ray techniques and a computerized measurement of intra- and extra-organ fat content.[26] This method should be invaluable to establish criteria based on quantitative measurements.

At the November 1983 national meeting of the Gerontological Society of America, Dr. Reuben Andres, the Clinical Director of the Gerontology Research Center, National Institute on Aging, presented a revised analysis of the Metropolitan Life Insurance height and weight tables issued earlier that year. The revision involved using the original raw data from over 4 million subjects, and computing the body mass index (BMI), which is the ratio of weight to height.[2] This ratio eliminates the arbitrariness in estimating frame size and also the sex differences. The BMI related to the lowest mortality rates for each age group was determined to find the ideal BMI. The result was a table of *age-specific* values for optimal height and weight. Minor limitations were that the original data were collected for only a 6.5-year period, and also the age range was somewhat narrow. In the BMI revision, persons older than about 35 years had higher recommended weights, and persons younger than 35 years had lower weights for any given height. This revision is an excellent demonstration and model of age-specific changes. Indeed, the measurement of height and weight is a simple, quantitative index of nutritional status, but more information is needed on older ages.

One of the greatest needs in nutrition and aging are *normal* reference values for essentially all biomedical measurements. This has been brought out and discussed by two groups in the recent Americal Society of Clinical Chemists symposium on "Aging: Its Chemistry, (1980)".[27,28] However, a distinction should be considered between *normative* aging changes, which refer to decreased physiological functions without any diagnosis of disease, and aging changes due to diseases of the elderly. Usually this distinction is difficult and not made because most elderly subjects are suffering from one or more diseases compared to the opposite situation in young human subjects, most of whom are healthy and free of disease. Thus existing data are often variable and inconclusive because of large differences in the health and disease status of elderly subjects. Normal reference values, therefore, depend on the study of subjects with carefully defined health status or with specific diseases.

Nutrition and aging information has been derived from major, nationwide surveys. In general, the ages are not old enough, and only a few surveys go as far as 75 or 85 years of age. Again, these data on older subjects probably do not rigorously discriminate whether the subjects are healthy or not.

A result of these surveys is that nonspecific and noncritical conclusions are made about the elderly. For example, the high incidence of iron-deficiency anemia that has been observed has resulted in a misconception that there is an "anemia of old age". In fact, this probably reflects an iron deficiency in these subjects which has nothing to do with aging, for this anemia can also be found in young persons. Our data on healthy men and women up to 102 years of age show no anemias of any kind.

Much of the nutritional information refers to enumeration or count data, and very little measurement data are presented to show the *degree* of change. Perhaps this has been due to the oversimplification tendency to set *critical levels* to discriminate between good and bad nutritional status. Establishment and use of these cutoff levels cannot be done with our present information. Indeed, these semiquantitative data have limited meaning in the study of aging where physiological changes are subtle and, therefore, require quantitative measurements of status.

Further, the absolute value of a tissue nutrient level may be less important than changes in that level which are determined repeatedly in the same individual at periodic intervals. This *trend* or *rate* of change in status is a critical procedure in the study of the elderly, for it provides a valid self-reference point that is more accurate and sensitive than any average pooled value.

The systematic sampling of several age groups is critical. Yet it is an uncommon procedure in nutritional research of aging, and especially in animal studies in which usually only two or three time points are determined. More points are required to determine the kinetics of the change, and these should include several points during the mature and senescent stages of the life span. With proper sampling, curves can be constructed to describe the changes accurately and provide necessary information on the onset and extent of the change.

Another reason for longitudinal studies and rate measurements is to determine the capability of an individual to adapt or cope with an apparently low intake of nutrients. Studies in other countries demonstrate that healthy, long-lived subjects can subsist and may be healthier on much lower nutrient allowances than in the U.S.

FIGURE 4. Nutritional control of physical, mental, and social health.

Indeed, the ability to adapt is dependent on health status which in turn may be regulated by physical activity. Recognition of the role of physical activity in determining the energy requirements of older subjects is well known but difficult to measure at this time using conventional methodology.

X. A HEALTHY PERSON IS A FUNCTIONALLY YOUNG OR MATURE PERSON

Recently we initiated the Louisville Longitudinal Longevity Program to determine and quantify the specific and unique characteristics of healthy, happy, long-lived persons. The rationale is that very old individuals are able to attain these ages by living differently from usual persons. This approach is shown in Figure 4 which integrates the well-known inter-relationship of physical, mental, and social health. However, we extended this by placing nutrition (N) in a central role and position. This figure shows that impairment in physical health can affect mental and also social health. In a similar manner, mental or social health changes can affect the other two components. We believe, however, that nutritional status is the common denominator and regulator. It is obvious that physical health is dependent on optimal nutrition. Further, new information indicates that various psychological functions like cognition and well being are influenced by good nutrition.[29,30] Finally, social health and the gathering of friends is frequently dependent upon and associated with eating meals together. In marked contrast are the elderly, who are alone tend to eat erratically and improperly with resultant deterioration of their physical health.

Our study focuses on laboratory and clinical assessments of physical health and also on established survey assessments of psychological and social health characteristics.[32] In our current first phase, baseline normative values were determined to establish reference standards. To reduce the experimental variables the study was restricted initially to white females between the ages of 60 to 102 years with an approximately equal distribution of subjects in each decade. The individuals were interviewed and self-assessed as healthy, ambulatory, and willing to participate, and the data were coded for double-blind design. All 58 women were found by direct physical evaluation to be in excellent physical health. Thus their laboratory measurements of blood and its biochemical components (SMAC 20) can serve as "normal" reference values of elderly.

There were two points of special interest. First there were no age differences in any of a battery of biochemical measurements whether a subject was 60 or 102 years old. Second, the values fell into the normal ranges from the Mayo Clinic for healthy subjects from ages 20 to about 55. Furthermore, values for these elderly had a much narrower variation compared to those from the Mayo Clinic. In summary, these data suggest that there is no difference between *healthy* adults who are *chronologically* young or very old. Thus, *healthy* adults may be considered "functionally young" regardless of chronological age. These results also suggest that *normative* aging may not exist, but instead decrements in function observed in elderly are due to disease or pre-disease conditions. Obviously, further research is required to establish these points.

Our current design and strategy are to determine baseline values in healthy subjects and then follow them to identify which of the various measurements are altered and which are maintained as the individuals age. This research is in its earliest phase, and in the future we plan to repeat analyses at short 8- to 12-month intervals, to extend the study to determine specific nutrients, and to investigate a more representative or random sample of the population.

XI. LONGITUDINAL DESIGNS ARE ESSENTIAL FOR THE STUDY OF NUTRITION OF THE ELDERLY

The use of a cross-sectional design in nutritional studies is common in animal nutrition. However, this is really a *pseudo-longitudinal* design because of the relatively homogeneous genetic and environmental background of the animal models, which minimizes individual variation. Thus, analyzing animals of different ages in the same experiment is equivalent to analyzing the same animal repeatedly. However, it is essential that the environmental background is consistent and well controlled, especially among different experiments. A common error is the application of survival data for Sprague-Dawley rats from one laboratory or a previous study to Sprague-Dawley rats obtained from a different source and reared under different conditions. A great deal depends on the consistency and high quality of long-term animal care, one of the most difficult aspects of aging research. The importance of longitudinal design is to obtain an accurate measure of the rate of change in the aging process. Determination of these rates also will indicate the time of onset when rapid aging occurs which can vary considerably in different individuals. The development of a valid biochemical index of functional age will obviate these variations of chronological age and also eliminate the necessity of long-term longitudinal measurements. At the moment the only definitive way to determine unequivocally the subtle, quantitative changes occurring in aging is by longitudinal studies. Detailed discussions on this and other designs and their advantages and limitations have been presented by Exton-Smith[7] and Elahi et al.[31]

XII. SUMMARY

The study of nutrition and aging is a doubly complex undertaking because each subject is a multidisciplinary area that includes psychosocial as well as biomedical aspects. Also, rapid changes are under way due to the new development of precise and specific biochemical methods. For example, the availability of high-performance liquid chromatography has greatly facilitated the rapid separation and analysis of tissue extracts. Also, the current availability of atomic absorption spectroscopy now enables the analyses of trace elements which could not have been done 20 years ago.

Finally, the recent interest in aging as a major societal problem has catalyzed many different research activities. Hopefully, some of the ideas that have been presented will be of value in planning future approaches and in the integration of diverse and multidisciplinary aspects of aging. If so, we have achieved our goal.

REFERENCES

1. **Barrows, C. H. and Kokkonen, G. C.,** Relationship between nutrition and aging, in *Advances in Nutritional Research,* Vol. 1, Draper, H. H., Ed., Plenum Press, New York, 1977.
2. **Young, V. R.,** Diet as a modulator of aging and longevity, *Fed. Proc. Fed. Am. Soc. Exp. Biol.,* 38, 1994, 1979.
3. **Exton-Smith, R. N.,** Nutritional status: diagnosis and prevention of malnutrition, in *Metabolic and Nutritional Disorders in the Elderly,* Exton-Smith, A. N. and Caird, F. I., Eds., Year Book, Medical Publishers, Chicago, 1980.
4. **Bowman, B. B. and Rosenberg, I. H.,** Assessment of the nutritional status of the elderly, *Am. J. Clin. Nutr.,* 35, 1142, 1982.
5. **Jensen, O. M.,** Dietary diaries and histories, in *Nutrition and Cancer: Etiology and Treatment,* Newell, G. R. and Ellison, N. M., Eds., Raven Press, New York, 1981.
6. **Shock, N.,** Physiological and chronological age, in *Aging — Its Chemistry,* Dietz, A. A., Ed., American Association for Clinical Chemistry, Washington, D.C., 1980, 3.
7. **Exton-Smith, A. N.,** Epidemiological studies in the elderly: methodological considerations, *Am. J. Clin. Nutr.,* 35, 1273, 1982.
8. **Oser, B. L.,** The rat as a model for human toxicological evaluation, *J. Toxicol. Environ. Health,* 8, 521, 1981.
9. **Hazelton, G. A. and Lang, C. A.,** Glutathione contents of tissues in the aging mouse, *Biochem. J.,* 188, 25, 1980.
10. **Ross, M. H.,** Aging, nutrition, and hepatic enzyme activity patterns in the rat, *J. Nutr. Suppl.,* 1, 97, 1969.
11. **Lang, C. A., Smith, D. M., and Sharp, J. B., Jr.,** The life span of the C57BL/6J mouse in the Louisville colony from 1963—1973, *Gerontologist,* 13, 40, 1973.
12. **Lang, C. A., Basch, K. J., and Storey, R. S.,** The growth, composition and longevity of the axenic mosquito, *J. Nutr.,* 102, 1057, 1972.
13. **Lang, C. A., Lau, H. Y., and Jefferson, D. J.,** Protein and nucleic acid changes during growth and aging in the mosquito, *Biochem. J.,* 95, 372, 1965.
14. **Mills, B. J. and Lang, C. A.,** The biosynthesis of DNA during life span of the mosquito, *Biochem. J.,* 154, 481, 1976.
15. **Lang, C. A and Smith, E. R.,** Protein metabolism in the aging adult mosquito, *Fed. Proc. Fed. Am. Soc. Exp. Biol.,* 31, 3762, 1972.
16. **Mills, B. J., Mastropaolo, W., and Lang, C. A.,** The biochemical analysis of insect DNA, in *Analytical Biochemistry of Insects,* Turner, R. B., Ed., Elsevier, Amsterdam, 1976, 37.
17. **Hazelton, G. A. and Lang, C. A.,** Glutathione levels during the mosquito life span with emphasis on senescence, *Proc. Soc. Exp. Biol. Med.,* 176, 249, 1984.
18. **Abraham, E. C., Taylor, J. F., and Lang, C. A.,** Influence of mouse age and erythrocyte age on glutathione metabolism. *Biochem. J.,* 174, 819, 1978.
19. **Schneider, D., Naryshkin, S., and Lang, C. A.,** Blood glutathione: a biochemical index of aging women, *Fed. Proc. Fed. Am. Soc. Exp. Biol.,* 41, 3570, 1982.
20. **Naryshkin, S., Miller, L., Lindeman, R., and Lang, C. A.,** Blood glutathione: a biochemical index of human aging, *Fed. Proc. Fed. Am. Soc. Exp. Biol.,* 40, 3179, 1981.
21. **Meister, A. and Anderson, M. D.,** Glutathione, *Ann. Rev. Biochem.,* 52, 711, 1983.
22. **Reed, D. and Beatty, P. W.,** Biosynthesis and regulation of glutathione: toxicological implications, in *Reviews of Biochemical Toxicology,* Vol. 2, Hodgson, E., Bend, T., and Philpot, R., Eds., Elsevier/North-Holland, New York, 1980, 213.
23. **Richie, J. P., Jr., Mills, B. J., and Lang, C. A.,** Magnesium thiazolidine-4-carboxylic acid increases both glutathione levels and longevity, *Gerontologist,* 23, 76, 1983.
24. **Lang, C. A.,** Macromolecular changes during the life span of the mosquito, *J. Gerontol.,* 22, 53, 1967.
25. **Albanese, A. A.,** Nutrition for the elderly, in *Current Topics in Nutrition and Disease,* Vol. 3, Alan R. Liss, New York, 1980.
26. **Borkan, G. A., Hults, D. E., Gerzof, S. G., Robbins, A. H., and Silbert, C. A.,** Age changes in body composition revealed by computed tomography, *J. Gerontol.,* 38, 673, 1983.
27. **Gillibrand, D., Grewal, D., and Blattler, D. P.,** Chemistry reference values as a function of age and sex, including pediatric and geriatric subjects, in *Aging — Its Chemistry,* Dietz, A. A., Ed., American Association for Clinical Chemistry, Washington, D. C., 1980.
28. **Hodgkinson, H. M.,** Biochemical diagnosis in the elderly, in *Aging — Its Chemistry,* Dietz, A. A., Ed., American Association for Clinical Chemistry, Washington, D. C., 1980.
29. **Goodwin, J. S., Goodwin, J. M., and Garry, P. S.,** Association between nutritional status and cognitive functioning in a healthy elderly population, *JAMA,* 249, 2917, 1983.

30. **Todhunter, E. N.,** Life style and nutrient intake in the elderly, in *Nutrition and Aging,* Winick, M., Ed., John Wiley & Sons, New York, 1976. ˙
31. **Elahi, V. K., Elahi, D., Andres, R., Tobin, J. D., Butler, M. G., and Norris, A. H.,** A longitudinal study of nutritional intake in men, *J. Gerontol.,* 38, 162, 1983.
32. **Lang, C. A., Richie, J. P., Jr., and Mills, B. J.,** High blood glutathione and top physical health in long-lived women, *Gerontologist,* 23, 92, 1983.
33. **Bafitis, H. and Sargent, F., II,** Human physiological adaptability through the life sequence, *J. Gerontol.,* 32, 402, 1977.
34. **Alpers, D. H., Clouse, R. E., and Stenson, W. F.,** *Manual of Nutritional Therapeutics,* Little, Brown, Boston, 1983.
35. **Sauberlich, H. E., Skala, J. H., and Dowdy, R. P.,** *Laboratory Tests for the Assessment of Nutritional Status,* CRC Press, Boca Raton, Fla., 1974.

Chapter 2

NUTRITIONAL MODULATION OF THE AGING PROCESS

Part I

DIETARY RESTRICTION AND LONGEVITY

Charles H. Barrows and Gertrude C. Kokkonen

TABLE OF CONTENTS

I. INTRODUCTION

Nutrition is probably the easiest intervention with which to affect the aging process. Nutritional manipulations are simple to bring about and unless there is severe modification of accepted dietary practices the overall physiological state of the organism is not disturbed. In addition, nutritional manipulations have been the only environmental variable shown to consistently increase the life span of laboratory animals. Life span extension has not always been considered a desirable goal of research. Gerontologists have suggested that efforts should be made to add life to years rather than years to life. Whether aging research will be successful in adding either life to years or years to life remains to be seen. However, experiments designed to increase life span through dietary manipulations are continuing. It is the aim of this review to discuss those nutritional manipulations which have been indicated to increase the life span of laboratory animals. Although the increase in life span associated with various methods of dietary restriction has been known for approximately 65 years, these studies have more frequently been looked upon as laboratory curiosities rather than providing information which may be applicable to man. Therefore, on the basis of available data, efforts will be made to determine whether these methods are practical for man and whether they may significantly add life to years as well as years to life.

II. DIETARY RESTRICTION AND LIFE SPAN

Dietary restriction has been shown to increase the life span of laboratory animals. In general, dietary restriction has been brought about by: (1) reducing the daily intake of a nutritionally adequate diet (one which supports maximal growth) (2) intermittently feeding a nutritionally adequate diet (e.g., feeding every 2nd, 3rd, or 4th day), and (3) feeding *ad libitum* a diet containing insufficient amounts of protein to support maximal growth. Any increase in life span associated with dietary manipulations is generally believed to be due to a restriction of dietary calories. However, most studies, in an attempt to accomplish caloric restriction, have restricted the intake of a nutritionally adequate diet so that not only has the caloric intake been reduced, but the protein and other dietary components as well. It must be recognized that it is experimentally difficult to hold all dietary components constant and reduce only calories. In order to achieve only caloric restriction under *ad libitum* conditions, there must be adjustments in the diets according to an animal's intake, which changes markedly with growth and is dependent upon dietary composition. The *ad libitum* feeding of a diet containing insufficient amounts of protein to support maximal growth has been shown to increase the life span of both young growing and adult animals.[1] However, the degree to which caloric restriction occurs under these experimental conditions is not clear. For example it has been reported that reducing dietary protein did not affect the caloric intake of adult rats.[2] However, Ross[3] reported that the caloric intake of young growing rats fed a synthetic diet containing 8% casein was reduced when compared to that of animals fed a commercial diet. Similar data has been reported by Barrows et al.[4] In contrast, Stoltzner[5] has reported a marked increase in the caloric intake of BALB/c mice fed *ad libitum* diets containing low amounts of protein. Therefore, on the basis of data presently available, it is not possible to conclude that calories are the sole dietary component which influence life span.

It has been generally believed that nutritional manipulations which increase life span had to be imposed during early growth. This concept originated as a result of the early work of Minot,[6,7] postulating that senescence follows the cessation of growth. In addition, McCay et al.[8,9] showed that increased life span of rats was associated with growth retardation. Furthermore, Lansing[10] indicated that aging in the rotifer involves a cytoplasmic factor the appearance of which coincides with the cessation of growth. However, more recent studies

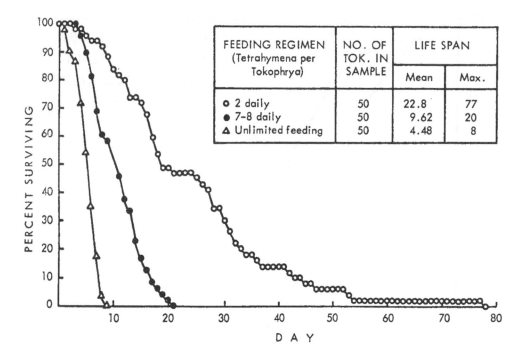

FIGURE 1.   The percent survivorship of *Tokophrya lemnarum.*

have indicated that dietary restrictions imposed in adult life were effective in increasing life span. Therefore, the results of the experiments reported here have been divided wherever possible into whether dietary restriction was imposed on young growing animals or on adult organisms.

A. Young Growing Animals
Increased life span associated with underfeeding has been reported in the following animal model systems: *Tokophrya*[11] (Figure 1), *Campanularia flexuosa*[12] (Figure 2), *Daphnia*[13] (Figure 3), rotifers[14] (Table 1), *Drosophila,*[15] and fish[16] (Figure 4). In addition, a number of laboratory experiments have been carried out on rodents. McCay et al.[17-19] carried out a series of three studies that supported the observation that nutritional deprivation increases life span. Since these early studies, the increased life span associated with underfeeding has been reported in rats by Berg and Simms[20] (Table 2), Ross[3] (Table 3), Yu et al.,[21] Leveille[22] (Figure 5), and Riesen et al.,[23] and in mice by Leto et al.[24] (Figure 6).

B. Adult Animals
The life expectancy of adult animals can be increased by dietary manipulations as can be seen in Figure 7[25] and Tables 4 to 9.[14,26-29] It is obvious from these data there are a number of inconsistencies regarding the effect of dietary restriction on the extension of life span in adult animals. Unfortunately, the limited amount of data does not allow for a critical evaluation as to the variables which may influence the ability of a mature organism to respond to dietary manipulation. However, the age at which dietary manipulations are initiated is important. This is clearly indicated by the data of Dunham[30] (Figure 8). *Daphnia* adequately fed up to the sixth instar, then subjected to dietary restriction, showed an increase in life span. However, a shortening in life span was observed if this dietary manipulation was imposed later in life. Indeed, there are a number of studies which indicate that the effective level of dietary restriction differs in rodents of different ages. This is referred to by Weindruch

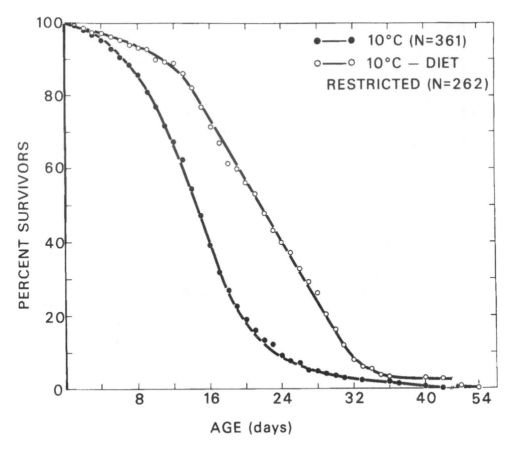

FIGURE 2.   The percent survivorship of *Campanularia flexuosa* fed artemia daily or every 3rd day.

et al.,[31] who indicated that adult animals were gradually restricted and employed different diets to bring this about. In addition, the data of Ross[3,27] (Tables 3 and 6) indicate that restricting the dietary intake of adult animals to that level imposed on younger animals resulted in early mortality. It is also apparent from the studies of Kopec[32] and David et al.[33] (Figure 9) that the degree of dietary restriction imposed on an adult organism may influence the life span of *Drosophila*.

The sex of an animal may also influence its response to dietary restriction in terms of life span. For example, the life-shortening effect of dietary restriction in adult *Drosophila* was more marked in males (33%) than in females (17%).[33] In addition, the life span of 445-day-old female, but not male rats, was increased by decreasing the dietary protein.[34] Thus, it is apparent that further studies must be carried out to define effective ways of consistently increasing the life span of adult organisms.

III. CHARACTERISTICS OF DIETARILY RESTRICTED ANIMALS

A. Biochemical Variables

Studies of the effect of dietary restriction on life span and biochemical variables have been limited to those measuring effects on collagen and tissue enzymes.

The breaking time of tail tendon collagen fibers of rats increased many-fold with age. Dietary restriction begun at 70 days markedly retarded the aging of collagen.[35] Harrison[36] has reported similar effects of food restriction in the genetically obese C57BL/6J-ob/ob female mouse.

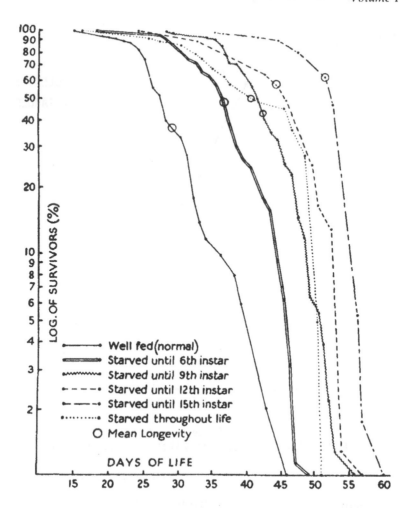

FIGURE 3.   Effect of restricted food upon the survivorship of *Daphnia longispina*.

When comparisons are made of intracellular proteins, such as enzymes, the only mean-ingful data are those based on some index of cell number. Two techniques have thus far been carried out in aging studies. Ross[37] counted the number of hepatocytes in his tissue preparations. Studies from our laboratory employed DNA per wet weight of tissue as an index of cell number. The validity of this was based on our finding of the constancy of nuclear DNA in the livers of rats of different ages.[38] Similar data have been described by Franks et al.[39] in the brains of mice.

Studies in which enzymatic activities of the tissues of normal and dietarily restricted mice and rats based on DNA[40] (Figures 10 to 12) or numbers of hepatocytes[37] (Figure 13) indicate a reduced cellular enzymatic activity in most but not all systems studied. These data led to the proposal that an increase in life span associated with dietary restriction was due to a reduced enzyme synthesis. However, recent studies do not support this concept. For example, in an effort to establish whether the various dietary manipulations which increase life span do so by a common biological mechanism, a recent study[41] measured biochemical variables in mice fed two dietary regimens reported to increase life span, namely low protein and intermittent feeding. The activities of succinoxidase, cholinesterase, and malic dehydrogen-ase were decreased as expected in the 4% protein groups. However, these enzymatic activities were essentially equal to the controls or even higher in the intermittent-fed animals. There-fore, these data do not indicate the existence of a common biochemical alteration to explain

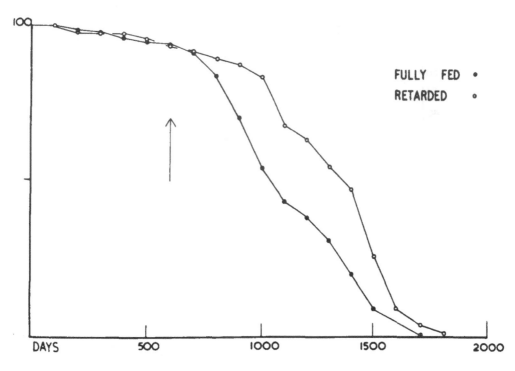

FIGURE 4.   Survivorship curve of female *Lebistes reticulatus* fed live Tubifex worms weekly (●) or biweekly (○). Arrow indicates realimentation of the restricted fish. The animals were maintained at 23°C.

Table 1
EFFECT OF NUTRITION ON LIFE SPAN OF ROTIFERS

Experiment		Mean life span (days)		
		Diet I[a]	Diet II[b]	Diet III[c]
1	Mean ± SEM	35.7 ± 2.1	43.8 ± 3.0	58.6 ± 1.9
2	Mean ± SEM	36.0 ± 1.2	45.6 ± 2.5	65.5 ± 2.2
3	Mean ± SEM	29.0 ± 2.8	46.2 ± 3.2	49.1 ± 2.2
Mean	Mean ± SEM	34.0 ± 1.1	45.3 ± 1.7	54.7 ± 1.3

[a]   Algae and fresh pond water daily.
[b]   Fresh pond water daily.
[c]   Fresh pond water: Monday, Wednesday, and Friday.

the increase in life span due to dietary restriction, nor do they support the concept that the benefits of dietary restriction are exclusively associated with low protein synthesis. These data may suggest that various dietary regimens which increase life span do so by independent mechanisms which result in differential effects on age-associated disease incidence and prevalence or physiological and biochemical alterations.

B. Physiological Variables

Animals whose life span has been increased by low-protein feeding (4%) have a lower rectal temperature than those fed the control diet (24% protein)[24] (Figure 14). Unfortunately, little information is available on the effect of body temperature on the life span of homeothermic animals. Nevertheless, the life span of poikilothermic animals increases with decreased environmental temperature.[42] It is generally assumed that this latter finding is a

Table 2
THE EFFECT OF REDUCED DIETARY
INTAKE ON LIFE SPAN OF MALE
SPRAGUE-DAWLEY RATS

Diet[a]	N[b]	Survivorship at 799 days (%)	Max. body weight (g)
Ad libitum	50	48	448
33% Restriction	48	87	342
46% Restriction	76	81	275

[a] Rockland "D free" pellets.
[b] Number of rats at start.

Table 3
THE EFFECT OF DIETARY INTAKES AND PROTEIN
LEVELS ON LIFE SPAN OF MALE SPRAGUE-DAWLEY RATS

Unrestricted dietary intake

Diets	Commercial	A	B	C	D
N[a]	150	25	25	25	25
Casein (%)	23	30	50.8	8	21.6
Caloric value (cal/g)	3.1	4.1	4.2	4.1	4.2
Food intake (g/day)	25	17.4	18.8	15	19.6
Max. body wt (g)	610		(Not available)		
Mean life span (days)	730	305	595	825	600
Max. life span (days)	1072	347	810	1251	895

Restricted dietary intake

N[a]		150	60	150	135
Casein (%)		30	50.8	8	21.6
Caloric value (cal/g)		4.1	4.2	4.1	4.2
Food intake (g/day)		14.3	8.5	14.3	5.3
Max. body wt (g)		420	287	390	162
Mean life span (days)		904	935	818	929

[a] Number of rats at start.

result of a decreased metabolic rate due to the lowering of the rates of biochemical reactions at the reduced temperature. However, the low body temperatures of these mice were associated with an increased oxygen consumption[24] (Figure 15). Furthermore, recent studies[43,44] in which poikilothermic animals have been exposed to different temperatures at various times in the life cycle, suggest a more complicated mechanism which may be independent of metabolic rate.

Masoro and associates reported that a marked increase in life span of Fischer 344 rats was brought about by reducing the daily allotment of food by 40%. These investigators[45] showed that the age-related decrease in lipolysis of adipocytes brought about by glucagon and epinephrine was delayed in the dietarily restricted animals. For example a 50% decrease in activity occurred in the former system at the age of 4 months in normal animals, but 18 months in dietarily restricted animals. In the latter system such decrements occurred at approximately 12 and 24 months in the normal and dietarily restricted animals, respectively. More recently, these authors have reported that this loss of responsiveness to glucagon can

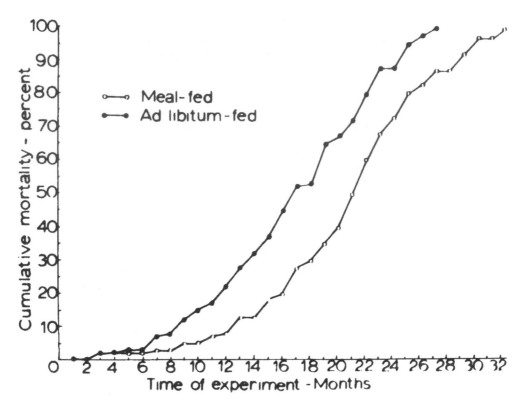

FIGURE 5.   Cumulative mortality for male Sprague-Dawley rats offered food for periods of 2 hr (O-O, meal fed) or 24 hr (●-●, *ad libitum*-fed) daily.

be slowly reversed so that animals dietarily restricted at 6 months of age have full responsiveness by 18 months of age.[46]

Tucker et al.[47] demonstrated a 4-fold increase in the amount of protein excreted by 21- as compared to 3-month-old normal Wistar rats. There was no increase in the protein excreted during this period in dietarily restricted animals.[47] This study also suggested that the age-associated increase in urinary protein excretion may be reversed by dietary restriction. For example, 12-month-old animals fed *ad libitum* excreted approximately 70 mg protein in 24 hr. Following 1 month of dietary restriction the excretion was 40 mg protein in 24 hr.

Slower rates of decrease in renal function in restricted as compared to normal mice have also been demonstrated by Harrison.[36] Urinary osmolality was found to decrease with age at a slower rate in restricted as compared to normal animals. The animals were restricted by reducing their intake by one third that of the controls.

Male Wistar rats subjected to dietary restriction by alternate days of feeding and fasting experienced not only a 40% increase in mean survival time, but also a substantial retardation of the normal age-associated loss of striatal dopamine receptors in brain. The authors[48] reported that the 24-month-old rats restricted since weaning had 50% higher dopamine receptor concentrations in striata than controls of the same age and were essentially comparable to 3- to 6-month-old control rats.

C. Immunological Variables

In general, the most frequently studied immune functions were mitogenic responses of lymphocytes and antibody responses to sheep red blood cells (SRBC). It seems fairly evident that the effect of low-protein feeding or intake restriction on the first function is an initial suppression during the first months of life.[49,50] As the animals mature and age the effect of

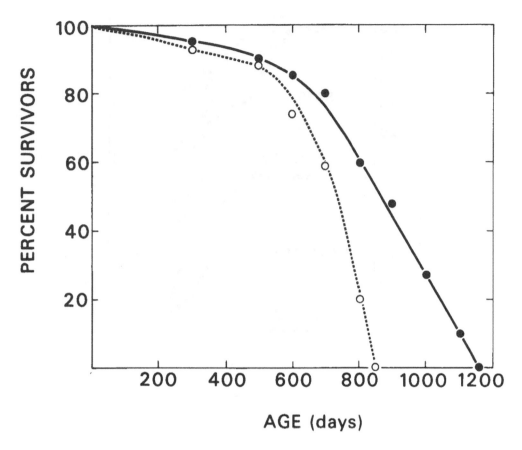

FIGURE 6. Survivorship curve of female C57BL/6J mice fed *ad libitum* either a 26% casein diet (○) or 4% casein diet (●); the mean life spans and SEM were 685 ± 22.8 and 852 ± 27.4 days, respectively.

low-protein feeding diminishes.[5] On the other hand the cells of animals with restricted intakes exhibit markedly higher responses than controls to a variety of mitogens.[50] Indeed, the data of Mann[49] clearly indicate that in mature animals the mitogenic response is markedly different in animals fed a low-protein diet as compared to those with restricted intake of an adequate diet. The data of Stoltzner[5] indicated a reduced primary response to SRBC in animals fed low-protein diets between the ages of 125 to 717 days. On the other hand, the data of Gerbase-DeLima et al,[50] indicate that although no differences were shown in young (18- to 20-week-old) and old (108- to 118-week-old) C57BL/6J mice, a markedly higher response to SRBC was found in middle-aged (52- to 55-week-old) mice whose dietary intake was reduced as compared to control animals. This again may demonstrate the differential effects of various dietary manipulations which increase life span.

Thus, it seems most likely that low-protein feeding results in immune suppression during early life with little effect in late life, whereas restricted dietary intake of an adequate diet results in early-life immune suppression followed by higher activities during middle life, which may continue into late life.

D. Diseases

Although the incidence of many diseases increases with age, the relationship between disease and aging remains unknown. Saxton[51] has reported that dietary restriction in adult as well as young growing animals delays the onset of a variety of diseases in the kidneys, heart, lung, and pituitary of the rat. Dietary restriction also delayed the onset of both benign

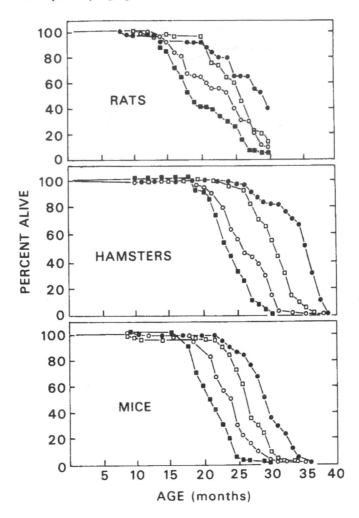

FIGURE 7.   Effect of various dietary regimens on the survivorship of rats, hamsters, and mice. Group 1 (■) was fed *ad libitum* throughout life; Group 2 (○) was fed one half the amount of food consumed by Group 1 throughout life; Group 3 (□) was fed *ad libitum* until 1 year of age and then restricted thereafter; Group 4 (●) was restricted until 1 year of age and then fed *ad libitum* thereafter.

and malignant tumors.[51] These findings have been repeatedly confirmed with greater detail added to our knowledge.

Renal disease was estimated by: (1) morphological examination and (2) urinary protein excretion. The data presented in Table 10 clearly show the reduction in renal disease resulting from dietary restriction.[52] In addition, Yu et al.[21] have shown that renal lesions occurred at an earlier age in *ad libitum*-fed rats than in restricted rats and progressed more rapidly. Tucker et al.[47] also indicate a reduced incidence and severity of renal lesions in dietarily restricted animals. It is interesting to note that the data of Bras[52] (Table 10) indicate a differential effect on the delay of renal lesions among the different dietary groups. For example, the animals consuming low calories (Groups B and D) had a significantly lower disease index than those offered higher caloric intakes (Groups A and C). Between the two latter groups of animals the beneficial effects of low protein feeding may be observed by the lower disease index in Group C as compared to Group A. Both the data of Tucker et al.[47] and Everitt et al.[35] indicate that age-associated proteinuria can essentially be eliminated

Table 4
THE EFFECT OF
INTERMITTENT FEEDING[a] ON
LIFE SPAN OF MALE WISTAR
RATS

Diet[b]	N[c]	Mean life span SEM (weeks)
Ad libitum[d]	25	133.1 ± 4.1
Ad libitum, restricted[e]	30	150 ± 4.6
Restricted, *ad libitum*[e]	30	149.2 ± 3.6
Restricted[d]	25	163.4 ± 3.9

[a]	Fed every other day.
[b]	Wayne Lab Blox Diet.
[c]	Number of rats at start.
[d]	Throughout life span.
[e]	Dietary regimens changed at 1 year of age.

Table 5
THE EFFECT OF CHANGES IN
NUTRITION FOLLOWING CESSATION
OF EGG PRODUCTION ON THE LIFE
SPAN OF ROTIFER (*PHILODINA
ACUTICORNIS*)

N[a]	Interval A—C[b]	Interval C—E[b]	Mean life span SEM (days)
30	I[c]	I	33.4 ± 1.27
28	I	III	41.8 ± 2.62
22	III	III	53.4 ± 3.65
29	III	I	57.7 ± 1.13

[a]	Number at start.
[b]	A = Day hatched; C = end of egg production, E = death.
[c]	I = Algae and fresh pond water daily; III = fresh pond water Monday, Wednesday, and Friday. Animals maintained at 25° C.

by protein restriction. In addition, both investigations indicate that the high levels of urinary protein observed in approximately 1-year-old animals can be lowered by dietary restriction and thus can reverse age changes.

Berg and Simms[20] and Yu et al.[21] report that dietary restriction delayed myocardial fibrosis and myocardial degeneration. Everitt[35] found that abnormal cardiac enlargement, which was taken as an estimate of cardiac pathology, was essentially eliminated by dietary restriction when imposed in early life.

Studies regarding the delay in onset of common pneumonia with bronchiectasis associated with underfeeding reported by Saxton[51] are not found in the literature. This is primarily due to the improvement in animal care which minimizes infectious diseases among the animals. Most studies in the current literature are carried out under barrier-specific pathogen-free (SPF) conditions or the animals are culled for respiratory diseases at very young ages. Therefore, little information is available regarding the effect of dietary restriction on infectious diseases.

Table 6
DAILY DIETARY ALLOTMENTS AND MORTALITY RISK AFTER 300 DAYS OF AGE

Diet (% casein)	Level of allotment (%)	Total food (g)	Casein (g)	Sucrose (g)	Corn oil (g)	Total cal (g)	Mortality[b] index (× 100)
Commercial	100	25				85.1	105
	80	20				68.4	106
	70	17.5				59.9	83
	60	15				51.3	79
A (30%)	100	18	5.40	10.98	0.90	73.6	5550
	90	16.2	4.86	9.88	0.81	66.3	1180
	80	14.4	4.32	8.78	0.72	58.9	1940
	71.5	12.9	3.86	7.85	0.64	52.6	723
B (50.9%)	100	19	9.66	6.44	1.61	78.9	178
	78	14.8	7.53	5.02	1.25	61.4	122
	55.9	10.6	5.40	3.60	0.90	44.1	115
	40	7.6	3.86	2.58	0.64	31.6	675
C (8%)	100	15	1.20	12.45	0.75	61.4	2103
	96	14.4	1.15	11.95	0.72	58.9	2882
	86.8	13	1.04	10.81	0.65	53.3	3195
	79.3	11.9	0.95	9.88	0.60	48.7	3195
D (2 1.6%)	100	20	4.32	10.81	2.70	84.8	200
	80.7	16.1	3.49	8.72	2.18	68.4	126
	59.6	11.9	2.58	6.44	1.61	50.6	99
	52	10.4	2.25	5.62	1.41	44.1	68

Note: Purina Lab Chow, 23% protein. Mortality ratio values less than 100 are indicative of the beneficial influences resulting from the change in dietary regimen; values more than 100, of the detrimental influences.

The delay in the onset of chromophobe adenomas of the pituitary gland reported by Saxton[51] has been confirmed by Ross and Bras.[53] However, the latter investigators also showed that feeding restricted amounts of diet for as little as 49 days following weaning had a marked influence on adenoma incidence as shown in Figure 16. This latter finding suggests that there may be time intervals, especially during early life, in which dietary manipulations may exert great impact on the future development and age-associated expression of an organism.

Ross and Bras[54] have extensively studied the relationship among spontaneous tumors, age, protein, and caloric restriction. Because of the high degree of specificity regarding the effectiveness of given diets on particular tumors, the investigations are difficult to summarize. Nevertheless, the beneficial effect of restrictive diets is evident.

In an experiment to study the effects of dietary restriction started at middle age on spontaneous cancer incidence, Weindruch and Walford[55] reported that male B10C3F1 mice on a restricted diet averaged increases in mean as well as maximal survival times of 10 and 20%, respectively, as compared to controls. In addition, spontaneous lymphoma was also inhibited. Cheney et al.[56] assessed tumor patterns in female B10C3F1 mice restricted throughout life or during only part of their life span (from weaning to mid-life or from mid-life to death). They reported the best mean and maximal survival and lowest late life mortality rate was found in those mice restricted throughout life, but noted that restriction during any part of the life span enhanced survival to some degree. The mean life span of tumor-bearing animals tended to be greater in restricted than in nonrestricted groups and thus corresponded to an age-decelerating effect.

Table 7
THE EFFECT OF REDUCED DIETARY
INTAKE ON THE MEAN SURVIVAL
TIME OF SPRAGUE-DAWLEY RATS

	Mean survival time (days)	
Diet[a]	Males[b]	Females[b]
Ad libitum	706	756
20% Restriction	856	872
40% Restriction	924	872
Ad libitum, 12 weeks, 20% restriction thereafter	801	871
Ad libitum, 12 weeks, 20% restriction thereafter	927	943
20% Restriction, 12 weeks, *ad libitum* thereafter	723	788
40% Restriction, 12 weeks, *ad libitum* thereafter	782	805

[a] Natural products diet: lipid, 18.5%; protein, 23%; ash, 6.2%; 4.4 kcal/g.
[b] 50 rats at start; diets started just after weaning.

Table 8
THE EFFECT OF VARIOUS DIETARY
REGIMENS ON LIFE SPAN OF FEMALE RATS

Dietary regimen[a]	Life expectancy ± SE (days)	Significant differences
A	763 ± 94	$p < 0.001$
B	980 ± 50	$p < 0.001$
C	828 ± 73	
D	282 ± 40	

[a] A: Stock diet throughout life; B: stock diet for 120 days, then 20% stock diet and 80% starch; C: 30% stock diet and 70% starch throughout life; D: protein-free diet. All diets were *ad libitum*.

IV. PERSPECTIVE

The characteristics of young growing animals subjected to dietary restriction leave doubt as to the applicability of this phenomenon to young growing human populations. For example, during early life, dietarily restricted animals are growth retarded, exhibit suppressed immunological function and cellular enzymatic activity, and have low body temperature and increased basal metabolic rate. Data obtained on invertebrates suggest that fertility may be adversely affected.[13-14] Therefore, it seems that if dietary restriction has any practical role in human populations it must be initiated during middle life. During the past 15 to 20 years a number of studies have shown that the life span can be increased when dietary restriction is initiated during the middle life of the rat (12 to 16 months). When the dietary protein

Table 9
THE EFFECT OF DIETARY
PROTEIN LEVELS ON THE
SURVIVAL OF 16-MONTH
FEMALE WISTAR RATS

Dietary protein levels (%)	N[a]	Survival (weeks)
24	44	29.5 ± 2.28[b]
12	44	37.0 ± 2.00[c]
8	44	30.0 ± 2.30
4	44	31.6 ± 1.70

[a]   Number of rats at start.
[b]   Mean ± SEM.
[c] $p = 0.001$.

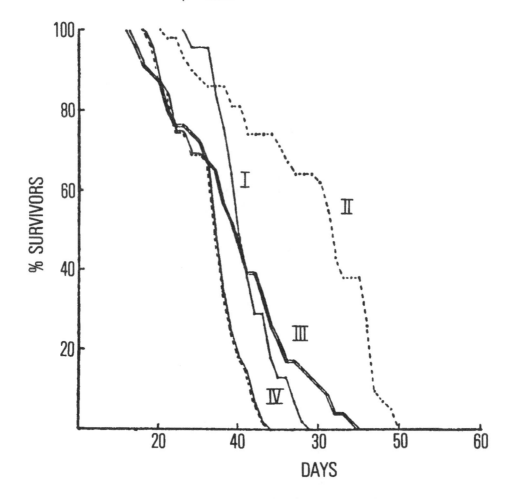

FIGURE 8.   The effect of dietary restriction on the survivorship curves of *D. longispira*. I, well-fed controls; II, semistarved controls; III, group well fed to 6th instar and then semistarved; IV, well fed to 12th instar and then semistarved. Semistarvation was brought about by diluting normal medium some 30 to 40 times with pond water.

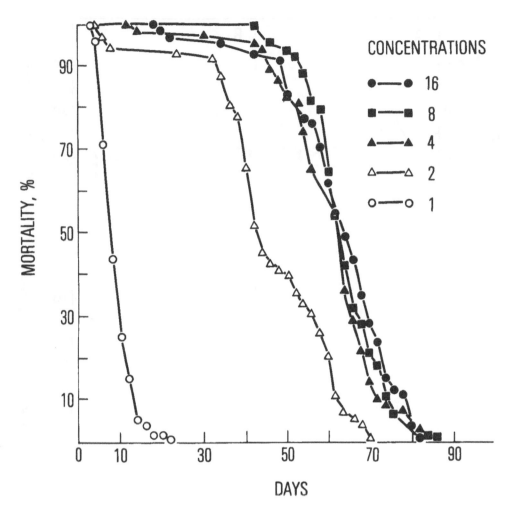

FIGURE 9. Survival curves of adult *Drosophila* (both sexes). The usual axenic medium 8% Brewer's yeast and 8% corn flour. This medium was diluted with an agar solution in order to produce concentrations ranging from 16 to 1%.

was reduced from 24 to 12% in 16-month-old female rats, the increased life span was accompanied by little if any change in the body weight of the animals. Beauchene and McCay have also indicated increases in life span may be affected when body weights are reduced only approximately 10%. Thus there are dietary manipulations which seem beneficial and which do not involve radical changes in physiological state as evidenced by body weight.

It may be recalled that there are a number of studies which indicated beneficial effects on physiological function when dietary restriction was initiated in middle life. For example, both the data of Tucker and co-workers and Everitt and associates indicated improvement in renal function associated with adult dietary restriction. Masoro reported that the loss of responsiveness of adipocytes to glucagon could be totally reversed in adult animals subjected to dietary restriction. Weindruch et al. showed improved mitogenic response of splenic lymphocytes to phytohemagglutinin (PHA) and concanavalin A (Con A) stimulation in animals dietarily restricted at 12, 17, and 22 months of age. Similarly, Barrows and Roeder[57] showed that the biochemical characteristics associated with dietary restriction were the same in 12-month-old animals as in weanling animals. Saxton[58] reported that dietary restriction imposed in middle life by McCay and co-workers reduced the frequency of a degenerative

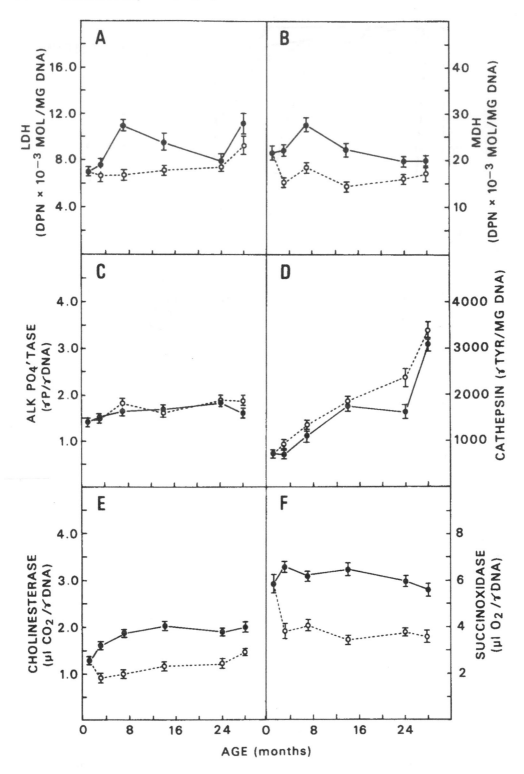

FIGURE 10.   Effect of age and diet on the enzymatic activities of liver of female C57BL/6J mice fed (●) 26% casein diet or (○) 4% casein diet. Vertical bars represent SEM. The mean life span and SEM of the animals fed either the low-protein or control diet was 852 ± 27.4 and 685 ± 22.8 days, respectively.

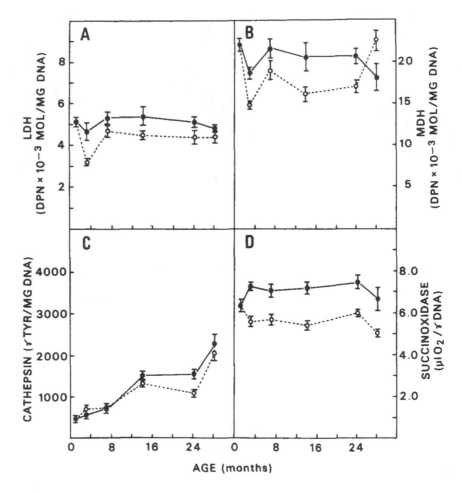

FIGURE 11. Effect of age and diet on the enzymatic activities of kidneys of female C57BL/6J mice fed (●) 26% casein diet or (○) 4% casein diet. Vertical bars represent SEM. The mean life span and SEM of the animals fed either the low-protein or control diet was 852 ± 27.4 and 685 ± 22.8 days, respectively.

disease of the kidney of rats. Dietary restriction imposed in 12- to 13-month-old mice by Weindruch and Walford reduced the overall incidence of cancer and lymphoma to levels which were of borderline significance ($p < 0.07$). These authors also indicated that dietary restriction imposed at the age of 12 months showed a reduced tumor incidence in animals killed between 19 and 25 months of age. These authors conclude that, apparently, restriction of the diet, even when started in middle-aged mice, can inhibit cancer. Thus, there are data which indicate that relatively moderate dietary restriction imposed in middle life can add years to life. This dietary manipulation also was found to reduce diseases as well as apparently reverse age-associated physiological and biochemical changes. Whether the latter findings qualify as adding life to years is uncertain.

V. SUMMARY

Dietary restriction has been shown to increase the life span of a variety of species. In young growing animals dietary restriction has been shown to reduce body temperature, increase basal metabolic rate, delay the onset of a variety of diseases, and delay age-associated changes in biochemical, immunological, and morphological variables. An increased life span

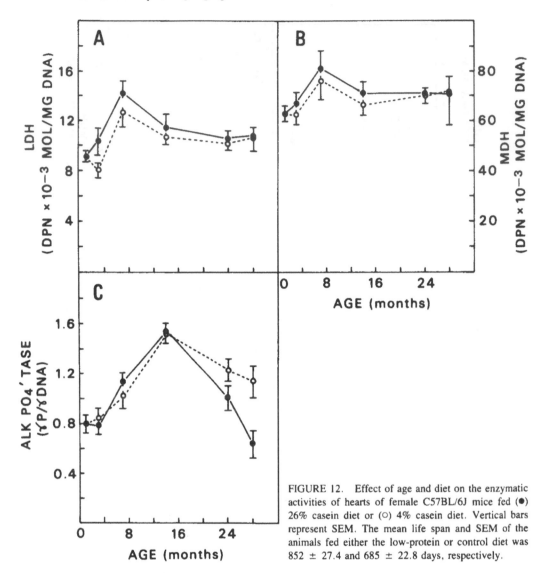

FIGURE 12.   Effect of age and diet on the enzymatic activities of hearts of female C57BL/6J mice fed (●) 26% casein diet or (○) 4% casein diet. Vertical bars represent SEM. The mean life span and SEM of the animals fed either the low-protein or control diet was 852 ± 27.4 and 685 ± 22.8 days, respectively.

associated with dietary restriction can be brought about when underfeeding is initiated in adult as well as young growing animals. Dietary restriction in adult animals is reported to improve renal and immune functions, reverse the age-associated decrease in responsiveness of adipocytes to glucagon, and reduce the frequency of disease. Finally, the data suggest that various dietary regimens which increase life span do so by independent mechanisms which result in differential effects on age-associated changes.

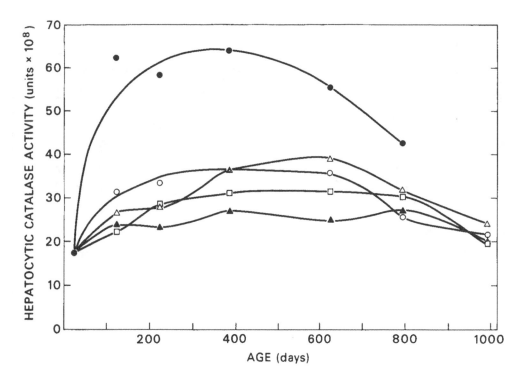

FIGURE 13. Effect of diet and age on the activity of hepatic catalase in male Sprague-Dawley rats. Rats maintained on commercial diet *ad libitum* (●); rats whose daily food allotment was restricted (see Table 3) (○) Diet A; (△) Diet B; (□) Diet C; (▲) Diet D.

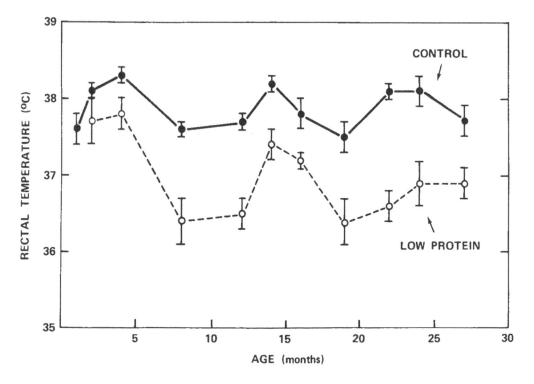

FIGURE 14. Effect of low-protein feeding on rectal temperature of female C57BL/6J mice. Vertical bars represent SEM. The mean life span and SEM of the animals fed either the low-protein or control diet was 852 ± 27.4 and 685 ± 22.8 days, respectively.

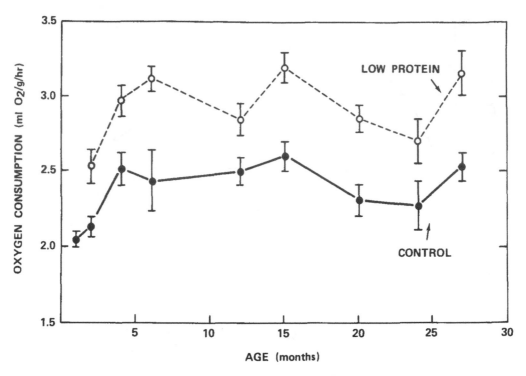

FIGURE 15.   Effect of low-protein feeding on oxygen consumption of C57BL/6J female mice. Vertical bars represent SEM. The mean life span and SEM of the animals fed either the low-protein or control diet was 852 ± 27.4 and 685 ± 22.8 days, respectively.

Table 10
PROGRESSIVE GLOMERULONEPHROSIS INDEX OF MALE SPRAGUE-DAWLEY RATS FED SEMISYNTHETIC DIETS

Dietary groups[a]	No. of cases		Disease index[b]
	Expected	Observed	
A	186.4	46	24.7
B	88.2	1	1.1
C	152.5	16	10.5
D	10.5	4	1.9

[a]   The intakes of animals fed diet A, B, C, or D were restricted (see Table 3).

[b]   Computed from rats dying from natural death only. Disease index expressed as percentage (computed as number of actual against expected cases). Expected cases equals disease rate at each age period of "control" population times exposure of experimental population. A value of the index of less than 100 indicates a beneficial effect of the experimental diet.

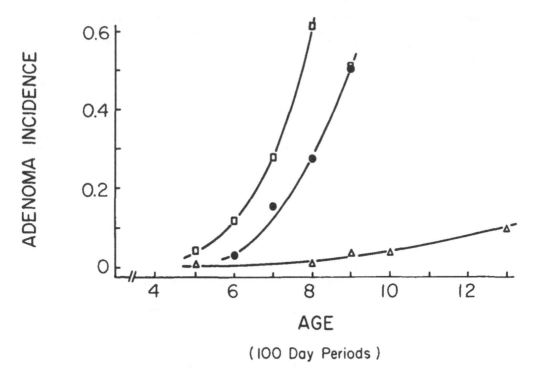

FIGURE 16.   Influence of dietary regimen on the incidence of adenomas in male COBS (Charles River) rats. (□) Rats fed *ad libitum* throughout post-weaning life; (△) rats fed a restricted amount of diet throughout post-weaning life; (●) rats fed a restricted amount of diet 21 to 70 days of age, and then fed *ad libitum*. Composition of diet: casein, 22.0%; sucrose, 58.5%; Mazola oil, 13.5%; salt mixture (USP XII), 6.0%; vitamins, and trace elements.

REFERENCES

1. **Barrows, C. H. and Kokkonen, G. C.,** Relationship between nutrition and aging, in *Advances in Nutritional Research,* Vol. 1, Draper, H. H., Ed., Plenum Press, New York, 1977, 253.
2. **Barrows, C. H. and Kokkonen, G. C.,** Protein synthesis, development, growth, and life span, *Growth,* 39, 525, 1975.
3. **Ross, M. H.,** Length of life and nutrition in the rat, *J. Nutr.,* 75, 197, 1961.
4. **Barrows, C. H., Roeder, L. M., and Fanestil, D. C.,** The effects of restriction of total dietary intake and protein intake, and of fasting interval on the biochemical composition of rat tissues, *J. Gerontol.,* 20, 374, 1965.
5. **Stoltzner, G.,** Effects of life-long dietary protein restriction on mortality, growth, organ weights, blood counts, liver aldolase, and kidney catalase, in BALB/c mice, *Growth,* 42, 337, 1977.
6. **Minot, C. S.,** *The Problem of Age, Growth, and Death; A Study of Cytomorphis Based on Lectures at the Lovell Institute, March 1907,* Lovell Institute, London, 1908.
7. **Minot, C. S.,** *Moderne Probleme der Biologie,* Jena, 1913.
8. **McCay, C. M., Ellis, G. H., Barnes, L. L., Smith, C. A. H., and Sperling, G.,** Chemical and pathological changes in aging and after retarded growth, *J. Nutr.,* 18, 15, 1931.
9. **McCay, C. M., Dilley, W. E., and Crowell, M. F.,** Growth rates of brook trout reared upon purified rations, upon dry skim milk diets, and upon feed combinations of cereal grains, *J. Nutr.,* 1, 233, 1929.
10. **Lansing, A.,** Evidence of aging as a consequence of growth cessation, *Proc. Natl. Acad. Sci. U.S.A.,* 34, 304, 1948.
11. **MacKeen, P. C. and Mitchell, R. B.,** Cytophotometric determination of cytoplasmic azure B RNA levels throughout the life span of *Tokphyrya lemnarum, Gerontologist,* 15(5), 27, 1975.
12. **Brock, M. A.,** Personal communication, 1976.

13. **Ingle, E., Wood, T. R., and Banta, A. M.,** A study of longevity, growth, reproduction, and heart rate in *Daphnia longispina* as influenced by limitations in quantity of food, *J. Exp. Zool.,* 76, 325, 1937.

14. **Fanestil, D. D. and Barrows, C. H.,** Aging in the rotifer, *J. Gerontol.,* 20, 462, 1965.

15. **Northrup, J. H.,** The effect of prolongation of the period of growth on the total duration of life, *J. Biol. Chem.,* 32, 123, 1917.

16. **Comfort, A.,** Effect of delayed and resumed growth on the longevity of a fish (*Lebistes reticulatus* Peters) in captivity, *Gerontologia,* 8, 150, 1963.

17. **McCay, C. M., Crowell, M. F., and Maynard, L. A.,** The effect of retarded growth upon the length of life span and upon the ultimate body size, *J. Nutr.,* 10, 63, 1935.

18. **McCay, C. M., Maynard, L.A., Sperling, G., and Barnes, L. L.,** Retarded growth, life span, ultimate body size, and age changes in the albino rat after feeding diets restricted in calories, *J. Nutr.,* 18, 1, 1939.

19. **McCay, C. M., Sperling, G., and Barnes, L. L.,** Growth, ageing, chronic diseases, and life span in rats, *Arch. Biochem.,* 2, 469, 1943.

20. **Berg, B. N. and Simms, H. S.,** Nutrition and longevity in the rat. II. Longevity and onset of disease with different levels of food intake, *J. Nutr.,* 71, 255, 1960.

21. **Yu, B. P., Masora, E. J., Murata, I., Bertrand, H. A., and Lynd, F. T.,** Life span study of SPF Fischer 344 male rats fed *ad libitum* or restricted diets; longevity, growth, lean body mass and disease, *J. Gerontol.,* 37, 2, 130, 1982.

22. **Leveille, G. A.,** The long-term effects of meal-eating on lipogenesis, enzyme activity, and longevity in the rat, *J. Nutr.,* 102, 549, 1972.

23. **Reisen, W. H., Herbst, E. J., Walliker, C., and Elvehjem, C. A.,** The effect of restricted caloric intake on the longevity of rats, *Am. J. Physiol.,* 148, 614, 1947.

24. **Leto, S., Kokkonen, G. C., and Barrows, C. H.,** Dietary protein, lifespan, and physiological variables in female mice, *J. Gerontol.,* 31, 149, 1976.

25. **Stuchlikova, E., Juricova-Horakova, M., and Deyl, Z.,** New aspects of the dietary effect of life prolongation in rodents. What is the role of obesity in aging? *Exp. Gerontol.,* 10, 141, 1975.

26. **Beauchene, R. E., Bales, C. W., Smith, C. A., Tucker, S. M., and Mason, R. L.,** The effect of food restriction on body composition and longevity of rats, *Physiologist,* 22, 4, 1979.

27. **Ross, M. H.,** Life expectancy modification by change in dietary regimen of the mature rat, *Proc. 7th Int. Cong. Nutr.,* 5, 35, 1967.

28. **Nolen, G. A.,** Effect of various restricted dietary regimes on the growth, health, and longevity of albino rats, *J. Nutr.,* 102, 1477, 1972.

29. **Miller, D. S. and Payne, P. R.,** Longevity and protein intake, *Exp. Gerontol.,* 3, 231, 1968.

30. **Dunham, H. H.** Abundant feeding followed by restricted feeding and longevity in Daphnia, *Physiol. Zool.,* 11, 399, 1938.

31. **Weindruch, R., Gottesman, S. R. S., and Walford, R. L.,** Modification of age-related immune decline in mice dietarily restricted from or after midadulthood, *Proc. Natl. Acad. Sci. U.S.A.,* 79, 898, 1982.

32. **Kopec, S.,** On the influence of intermittent starvation on the longevity of the imaginal stage of *Drosophila melanogaster, Br. J. Exp. Biol.,* 5, 204, 1928.

33. **David, J., Van Herrewege, J., and Fouillet, P.,** Quantitative underfeeding of Drosophila; effects on adult longevity and fecundity, *Exp. Gerontol.,* 6, 249, 1971.

34. **McCay, C., Maynard, L. A., Sperling, G., and Osgood, H.,** Nutritional requirements during the latter half of life, *J. Nutr.,* 21, 45, 1941.

35. **Everitt, A. V., Seedsman, N. J., and Jones, F.,** The effect of hypophysectomy and continuous food restriction, begun at ages 70 and 400 days, on collagen aging, proteinuria incidence of pathology and longevity in the male rat, *Mech. Aging Dev.,* 12, 161, 1980.

36. **Harrison, D. E.,** Personal communication, 1983.

37. **Ross, M. H.,** Aging, nutrition, and hepatic enzyme activity patterns in the rat, *J. Nutr.,* 97, 565, 1969.

38. **Falzone, J. A., Barrows, C. H., and Shock, N. W.,** Age and polyploidy of rat liver nuclei as measured by volume and DNA content, *J. Gerontol.,* 14, 2, 1959.

39. **Franks, L., Wilson, P., and Whelan, R.,** The effects of age on total DNA and cell number in the mouse brain, *Gerontologist,* 2, 79, 1967.

40. **Leto, S., Kokkonen, G. C., and Barrows, C. H.,** Dietary protein, lifespan, and biochemical variables in female mice, *J. Gerontol.,* 31, 144, 1976.

41. **Barrows, C. H. and Kokkonen, G. C.,** The effect of various dietary restricted regimes on biochemical variables in the mouse, *Growth,* 42, 71, 1978.

42. **Strehler, B. L.,** Studies on the comparative physiology of aging. II. On the mechanism of temperature life-shortening in *Drosophila melanogaster, J. Gerontol.,* 16, 2, 1961.

43. **Clarks, A. M. and Kidwell, R. N.,** Effects of developmental temperature on the adult lifespan of *Mormoniella vitripenuis* females, *Exp. Gerontol.,* 2, 79, 1967.

44. **Clarke, J. M. and Smith, J. M.,** Independence of temperature on the rate of aging in *Drosophila subobscura, Nature (London),* 190, 1027, 1961.

45. **Masoro, E. J., Yu, B. P.,** Bertrand, H. A., and Lynd, F. T., *Nutritional probe of the aging process,* *Fed. Proc. Fed. Am. Soc. Exp. Biol.,* 39, 3178, 1980.
46. **Bertrand, H. A., Masora, E. J., and Yu, B. P.,** Nutritional reversal of an age-related deficit, *Fed. Proc. Fed. Am. Soc. Exp. Biol.,* 41, 5, 1674, 1982.
47. **Tucker, S. M., Mason, R. L., and Beauchene, R. E.,** Influence of diet and feed restriction on kidney function of aging male rats, *J. Gerontol.,* 31, 354, 1976.
48. **Levin, P., Janda, J. K., Joseph, J. A., Ingram, D. K., and Roth, G. S.,** Dietary restriction retards the age-associated loss of rat striatal dopaminergic receptors, *Science,* 214, 561, 1981.
49. **Mann, P. L.,** The effect of various dietary restricted regimes on some immunological parameters of mice, *Growth,* 42, 87, 1978.
50. **Gerbase-DeLima, M., Lui, R. K., Cheney, K. E., Mickey, R., and Walford, R. L.,** Immune function and survival in a long-lived mouse strain subjected to undernutrition, *Gerontologia,* 21, 184, 1975.
51. **Saxton, J. A.,** Nutrition and growth and their influence on longevity in rats, *Biol. Symp.,* 11, 177, 1945.
52. **Bras, G.,** Age-associated kidney lesions in the rat, *J. Infect. Dis.,* 120, 131, 1969.
53. **Ross, M. H. and Bras, G.,** Lasting influence of early caloric restriction on prevalence of neoplasms in the rat, *J. Natl. Cancer Inst.,* 47, 1095, 1971.
54. **Ross, M. H. and Bras, G.,** Tumor incidence patterns and nutrition in the rat, *J. Nutr.,* 87, 3, 245, 1965.
55. **Weindruch, R. and Walford, R. L.,** Dietary restriction in mice beginning at 1 year of age: effect on life-span and spontaneous cancer incidence, *Science.* 215, 1415, 1982.
56. **Cheney, K. E., Lui, R. K., Smith, G. S., Meredith, P. J., Mickey, M. R., and Walford, R. L.,** The effect of dietary restriction of varying duration on survival, tumor patterns, immune function, and body temperature in B10C3F, female mice, *J. Gerontol.,* 38, 4, 420, 1983.
57. **Barrows, C. H. and Roeder, L. M.,** The effect of reduced dietary intake on enzymatic activities and life span in rats, *J. Gerontol.,* 20, 69, 1965.
58. **Saxton, J. A.,** Pathology of senescent animals, in *Conference on the Problems of Aging,* Shock, N. W., Ed., Josiah Macy, Jr. Foundation, New York, 1950, 135.

Chapter 2

NUTRITIONAL MODULATION OF THE AGING PROCESS

Part II

ANTIOXIDANT SUPPLEMENTATION AND LONGEVITY

Hans U. Weber and Jaime Miquel

TABLE OF CONTENTS

I. INTRODUCTION

The search for a means to prolong life and at the same time maintain vitality at a youthful level can be traced back to the roots of human culture. Popular myths and legends attest to the search for panaceas to retard aging or even defeat death and achieve immortality. Early attempts to attain these goals were necessarily empirical and difficult, if not impossible, to verify. Chinese herbal medicine, for example includes many preparations that have long been used to prevent or postpone some of the deterioration associated with aging. The efficacy of such preparations remains doubtful; knowledge and practice handed down through generations may assure safe use of medicinal preparations but does not prove any claims for its efficacy.

Aging, the universal phenomenon experienced by all, remains a puzzle to scientists. Many theories have been proposed, but none so far has been generally accepted by gerontologists. One of them, however, the free radical theory,[1-4] is steadily gaining acceptance as a plausible explanation of the primary chemical reactions involved in aging. This molecular theory postulates that oxygen-derived free radicals cause lipid peroxidation with damage to membranes and other structures in the cell.

Mitochondria, which transform energy released by oxidative reactions into chemical energy (mainly ATP), are the main source of free radicals and, therefore, the first target for free radical reactions. Harman[5] compared them to a "biologic clock". In a recent review, Fleming et al.[6] consider the inner membrane and the DNA of mitochondria as structures particularly prone to damage by free radicals. Lipid peroxidation in the mitochondrial membrane is the subject of an extensive review.[7] More recently, Lippman[8] reviewed the role of lipid peroxidation in aging.

Peroxidative damage is cumulative in post-mitotic cells and affects the structure and function of the mitochondria.[9,10] Loss of mitochondrial function or a decrease in mitochondrial cell density means a loss of energy transformation capacity in the cell.

Unsaturated fatty acids, essential components of cellular membranes, are particularly prone to peroxidation. Dormandy[11] compared aging with rancidification of autooxidative degradation of unsaturated lipids. More recently, Hegner,[12] described age-related changes in the properties of biological membranes and the role of protective systems and repair mechanisms. Hornsby and Crivello[13] reviewed the role of lipid peroxidation and biological antioxidants with particular reference to the adrenal cortex.

Cells which use oxygen had to evolve antioxidant defense systems to protect themselves against free radical damage. Any protective system, however, has a limited capacity of self-renewal and repair, and cannot prevent or repair all damage all the time. To quench oxygen-derived free radicals, mitochondria need a particularly effective system. In self-renewing cell populations, old cells with irreversibly damaged mitochondria are continuously replaced with young, undamaged cells. However, post-mitotic cells in many tissues cannot be replaced. They incur cumulative, irreversible damage which gradually impairs their functioning.

Tocopherols (vitamin E) appear to be among the most important antioxidants in biological membranes. In a study of lipid extracts from human red blood cells, Burton et al.[14] found that vitamin E, largely in the form of α-tocopherol, is the only significant lipid-soluble antioxidant which terminates free radical chain reactions in blood.

Exogenous antioxidants, e.g., vitamin E and A, ascorbate, sulfur amino acids, and selenium compounds, interact with endogenous antioxidants and free radical scavenging systems, including glutathione and various enzymes (superoxide dismutase, catalase, glutathione peroxidase, and other peroxidases). The rate of aging may be partially determined by the intake of dietary antioxidants, though a number of animal studies offer no convincing evidence that antioxidant supplements increase longevity.[15] Such a lack of proof is not surprising, however, in view of the complex interactions between nutrients and cellular antioxidant defense systems.[16] The use of only one antioxidant may increase the need for another and thus result in undesirable effects.[17]

Hoping to improve the natural, physiological antioxidant defense system, pharmaceutical companies have backed the search for "age retardants". Kormendy and Bender[18] discussed different types of potentially useful agents, including antioxidants.

II. ANIMAL STUDIES

Studies to test the theory that dietary antioxidant supplements would protect oxidation-prone structures in post-mitotic cells and thus retard aging and prolong the life span have been carried out with various synthetic compounds as well as the physiologic antioxidants, vitamins E and C, sulfur amino acids, etc. A wide variety of different organisms ranging from fungi to mammals have served as experimental models.

Some of the studies have not been designed carefully enough to exclude potential life-prolonging effects of reduced food intake in the antioxidant-supplemented group or to optimize the conditions for maximun life span in the control animals. Suboptimal life spans of the control animals could be caused by adverse environmental conditions, and a comparatively longer life span of the antioxidant-supplemented animals in such studies may be due to protective effects which are specific to the particular experimental conditions.

A. Rodents

In a series of experiments with different strains of mice, Harman[19-22] tested various antioxidants and free radical inhibitors for their effects on survival and tumor development. Some of the compounds, e.g., butylated hydroxytoluene (BHT) or 2-mercaptoethylamine (2-MEA) in concentrations of 0.5 or 1.0% (w/w) in the feed, significantly prolonged survival under certain conditions. However, these effects varied depending on the type of diet (commercial, semisynthetic, or synthetic) and strain and even sex of the animals. Male Swiss mice, for example, showed somewhat shorter survival times when their diet was supplemented with the hydrochlorides of 2-MEA, cysteine, of 2,2'-diaminodiethyl disulfide. Male LAF1 mice, which normally live for 26 to 30 months (16% survivors at 28 months) on a commercial diet, survived in significantly larger numbers (41% survivors at 28 months) on the same diet supplemented with 1% 2-MEA, but early mortality in the 2-MEA-supplemented group was slightly higher. In several experiments the antioxidant-supplemented animals weighed significantly less than the control animals. Since male LAF1 mice survived at most only about 20 months on a synthetic diet, the life-prolonging effects of antioxidant supplements under these conditions is not readily comparable to the effects obtained when a regular diet is used. Life span extensions achieved by feeding BHT (0.5% wt in a synthetic diet) or other antioxidants to male LAF1 mice still fall short of the life spans recorded for this strain on a regular commercial diet.

An elegant follow-up study by Harman and Eddy[23] indicates that antioxidant supplements given to female Swiss mice before mating until the offspring are weaned affects the life span of the offspring; 2-MEA in the maternal diet practically doubled the 26-month survivorship of both male and female offspring. Vitamin E in the form of α-tocopherol acetate (0.2 % wt) or sodium hypophosphite (1.0% wt) moderately increased survivorship of male but had hardly any effects on survival of female offspring. Santoquin (2-dihydro-6-ethoxy-2,2,4-thimethylquinoline) resulted in increased survivorship of female offspring but reduced that of male offspring.

Kohn[24] could not confirm the life-prolonging effects of 2-MEA or BHT in a study on weanling female mice and retired breeders. He also reported toxic effects such as weight loss and deaths with a BHT level of 1% wt in the feed. Clapp et al.,[25] however, reported reduced mortality in groups of BALB/c mice supplemented with BHT at 0.75% wt. The

BHT-supplemented animals lived on the average longer than the controls, but the maximum life span of about 1100 days was the same in all groups. The BHT-treated mice were generally heavier, had smoother coats, and looked healthier. The longest mean survival times were recorded in the group supplemented for life beginning at 11 weeks, but even the groups supplemented at an earlier age (8 weeks) fared better than the controls. All groups were kept in scrupulously clean conditions to optimize the environment.

Tappel et al.[26] compared a supplement consisting of D,L-α-tocopherol only with two antioxidant mixtures that contained BHT, ascorbic acid, D,L-methionine, and sodium selenite in addition to larger amounts of vitamin E. Male retired CD-1 breeder mice at an age of 9 months were used in this study, which continued for nearly 2 years. The antioxidant-supplemented groups showed less lipofuscin accumulation in testes and heart cells than the control group, but had mortality rates similar to those of the control animals.

Jones and Hughes[27] tested the flavonoid quercetin by feeding it as a dietary supplement (0.1 % wt) to LACA mice. The females of this strain can live as long as 35 months, about 2 months longer than the males. The quercetin supplement had a life-shortening effect, especially in the males. On the other hand, an extract of black currant juice, which contains quercetin, together with various other flavonoids significantly prolonged the life span in the 15 oldest surviving females. The results of this study are difficult to interpret since quercetin is not only a potent antioxidant but also acts as a chelating agent and has mutagenic activity. Quercetin and other flavonoids occur naturally in plant foods, and the average person may ingest anywhere from 50 to as much as 500 mg quercetin per day.

The effects of thiazolidine-4-carboxylic acid (TC) and its salts (Mg or Na) on mice as well as on fruit flies have been studied by our group[28-30] and are summarized in a recent review of TC.[31] In both species, Mg-TC not only reduced lipofuscin accumulation in post-mitotic cells, but also prolonged the life spans and improved vitality. Mg-TC was given to retired breeders (aged 23 months) at a level of 0.07% wt in the standard chow. The mean body weights of the supplemented animals remained similar to those of the control animals, but the serum cholesterol levels were significantly lower in the supplemented group. Mg-TC supplements also appeared to have a protective effect on the liver[32] and on liver mitochondria.[33]

We also studied tocopheryl-*p*-chlorophenoxyacetate (T-*p*-ClPh) as a dietary supplement for mice and fruit flies and reported effects similar to those of Mg-TC.[28,29,32,33]

Decreased tumor incidence and lower early mortality, but no life span extension, were reported by Blackett and Hall[34] in a study of C3H and LAF1 mice given supplements of D,L-α-tocopherol (0.25% wt). The authors suggest that these findings may indicate a low vitamin E content in the control diet. In a study by Ledvina and Hodanova,[35] a supplement of vitamin E plus sunflower oil slightly increased the maximum and mean but not the median life span of female mice.

The phenolic antioxidant, nordihydroguaiaretic acid (NDGA), given to Wistar rats at a concentration of 0.02% wt in the feed increased survival as well as number of litter, according to an early study.[36] NDGA supplements showed favorable effects in later experiments on Drosophila and mice,[28] but was not tested again in life span studies with rats.

Female rats were used in a study by Oeriu and Vochitu[37] of "Folcysteine", a combination of thiazolidine carboxylic acid and folic acid. Folcysteine was given by injection and did not increase the life span of rats, but appeared to increase survival of mice and guinea pigs. However, the relatively short life spans of the experimental animals used in this study indicate suboptimal conditions, and the results are dubious.

B. Insects

Fruit flies, (*Drosophila melanogaster*) were used in our studies of aging and the influence of antioxidants on aging and life span.[28,30,32,38-43] Adult fruit flies are made up almost entirely of post-mitotic cells which accumulate an increasing amount of lipofuscin as the flies age. Their rate of aging appears to be proportional to the metabolic rate and is increased at high ambient temperature. Regardless of differences in temperature and feed, an adult fruit fly consumes a total of 4 to 5 mℓ oxygen during its life. A low rate of oxygen consumption is conducive to a long life span and vice versa.[39]

A number of different antioxidants were tested for their effects on the life span of Drosophila. Propyl gallate at a level of 0.3 % wt increased the mean life span by 19.5% and the maximum life span by 14.5% according to the most recent study reported by Ruddle of Miquel's group.[43] The other three antioxidants tested in this study, namely NDGA, spermine phosphate, and 2,6-dimethyl-3,5-diethoxycarbonyl-1,4-dihydropyridine (Diludine) were also effective but less so than propyl gallate. Diludine had been studied as an antioxidant and free radical scavenger by Soviet researchers[44] who claimed dramatic life span extensions (mean and maximum!) in Drosophila as well as in rats given Diludine supplements. In a recent study,[43] the effects of Diludine on Drosophila were unspectacular.

Thiazolidine carboxylic acid (TC, 0.3% wt), vitamin E (0.4% wt), NDGA (0.5 % wt), or dinitrophenol (0.1 % wt) prolonged the mean life span of *D. melanogaster* by 31, 14, 12, and 12%, respectively.[39] This study also shows an inverse relationship between oxygen consumption and mean life span. The data suggest that these four antioxidants, and probably others as well, may prolong the life span of Drosophila by inhibiting the respiration rate. Previous experiments with the sodium or magnesium salt of TC have already been mentioned in connection with our studies on mice.[30]

The magnesium salt of TC also prolonged the median survival time of adult mosquitoes (*Aedes aegypti*) by about one third.[45] Mg-TC was fed in concentrations of 0.01 or 0.05% wt w/v in 10% sucrose solution. Determination of reduced glutathione (GSH) in mosquitoes after 1 week on Mg-TC showed about 40% more GSH than present in the nonsupplemented control group. This supports Lang's hypothesis that increased levels of tissue GSH are associated with slower aging and increased life span.

Sharma and Wadhwa[46] reported significantly increased median and maximum life spans in *D. bipectinata* supplemented with BHT at three levels, namely 0.002, 0.02, and 0.22% wt BHT (0.22% wt) also inhibited lipid peroxidation (measured with thiobarbituric acid).

C. Nematodes and Rotifers

Vitamin E added to cultures of nematodes increased both the average and maximum life span.[47-49] *Turbatrix aceti* in vitamin E-fortified cultures (0.01% v/v D,L-α-tocopherol) grew faster, though no larger, than the worms in the control group. Survival curves showed that vitamin E significantly increased survival in the first 18 days of life as well as at the end of life.[47,48] Epstein and Gershon[49] used the more stable α-tocopherolquinone (TQ, 0.04% w/v) in their study of *Caenorhabditis briggsae*. They found that increased survival in TQ medium was positively correlated with delayed lipofuscin accumulation in intestinal tissues of aging animals. These data support the hypothesis that vitamin E acts as a free radical scavenger blocking lipid peroxidation reactions resulting in age pigments.

Vitamin E at a level of 0.01% v/v significantly increased the average but not the maximum life span of the rotifer, Philodina.[50] Tween-80 was used to solubilize the water-insoluble D,L-α-tocopherol. Philodina in the vitamin E medium also had a significantly longer reproductive period and more offspring.

III. ANTIOXIDANTS

Antioxidants can be classified according to different criteria, e.g., chemical structure, polarity (solubility), or reducing power. Physiological antioxidants are usually assigned to one of two classes

1.   Water-soluble: ascorbic acid, glutathione, sulfur amino acids, selenium compounds, spermine and other polyamines, pyridoxine, etc.
2.   Lipid-soluble: tocopherols, carotenoids, retinol, ubiquinones

The antioxidant activity or reducing power with respect to a defined oxidant medium depends on the molecular structure. Active sites are hydroxyl, hepolic, sulfhydryl, or amino groups. Affinity of the molecule to water largely determines such processes as absorption, transport, and metabolic fate. Since biological systems are heterogeneous and consist of compartments differing from each other in phase (crystalline, liquid) and polarity, interaction of different types of antioxidants is essential for electron transport between different compartments.

The fate of the tested antioxidants and the antioxidant status of the organism have been largely disregarded in life span studies, and therefore, the interpretation of most findings remains speculative. Metabolites of certain synthetic antioxidants may have unforeseen pharmacologic effects or may interact with normal metabolic processes. High doses of physiological antioxidants may cause imbalances in other antioxidants. For example, Chen[17] showed that large supplements of vitamin C increased the vitamin E requirements of rats.

Under certain conditions different antioxidants may be used to prevent oxidant damage. For example, in a study of the effects of various antioxidants on the tissue antioxidant status in rats, Chen[51] reported that selenium compounds can act as a partial substitute for vitamin E. Her data suggest that seleno-amino acids and methionine have no direct antioxidant activity but are metabolized into compounds which improve the tissue antioxidant status.

IV. POSSIBLE MECHANISMS

A common criticism of life span studies concerns effects of antioxidants on food intake. Are the longer life spans really due to the tested agent(s) or merely the result of hidden dietary restrictions? Studies in which the food intake and the body weights of the animals are not monitored cannot answer these questions. Lower weights in the experimental group than in the control animals do not necessarily indicate dietary restriction; interference with food absorption or utilization or with the regulation of the metabolic rate is possible.

Reduction of the metabolic rate is associated with longer life spans in adult insects, e.g., fruit flies. Miquel and colleagues[39,40] reported decreased oxygen consumption by flies supplemented with antioxidants that prolonged the life span. Since the rate of chemical reactions is temperature dependent, the metabolic rate is reduced at lower body temperature. This explains why the life span of insects can also be manipulated by changes of the ambient temperature.

Reduction of oxygen consumption or of metabolic reaction rates in general is associated with a decrease in the generation of oxygen-derived free radicals. Agents which decrease the mitochondrial respiration rate and thus the production of free radicals may, therefore, simulate the effects of free radical scavengers. Certain agents may act both as depressants of the respiratory rate and as free radical scavengers.

The lack of fundamental data in most longevity studies, often the result of a compromise because of limited resources, precludes clearcut interpretations of the findings. Too many unaccounted interactive factors are operative even in apparently simple and strictly controlled experimental systems.

V. CONCLUSION

The available data on longevity studies show that a number of antioxidants are capable of extending the life span of experimental animals. However, the effects of any given antioxidant are not consistent for different animal species. Interactions with dietary constituents or differences in absorption and metabolism may be responsible for such inconsistencies and must be considered in dose-response relationships.

The hypothesis that antioxidants prolong life by protecting the organism from oxygen-derived free radicals is still tenable but insufficient for a full explanation.

Results of studies in which supplements of a single agent are used may not be predictive of studies in which the same agent is supplemented together with other agents. If we assume that antioxidants act by improving the efficiency of metabolic pathways either through better protection of oxidation-prone cellular structures or through reduction of the extent of uncontrolled oxidative reactions, it would follow that a combination of different antioxidants may have a better chance of success than any single agent.

The design of successful longevity experiments with antioxidant combinations requires extensive knowledge of cellular antioxidant systems and interactions. Bioavailability and metabolism must also be considered, particularly in the case of agents which are effective in in vitro but not in in vivo studies.

REFERENCES

1. **Harman, D.,** Free radical theory of aging: consequences of mitochondrial aging, *Age,* 6, 86, 1983.
2. **Harman, D.,** Free radical theory of aging: origin of life, evolution and aging, *Age,* 3, 100, 1980.
3. **Leibovitz, B. E. and Siegel, B. V.,** Aspects of free radical reactions in biological systems: aging, *J. Gerontol.,* 35, 45, 1980.
4. **Lippman, R. D., Ed.,** Developments in aging research. I. Free radicals — a primary cause of aging, Abstracts from Symposium at University of Uppsala, Sweden, *Uppsala J. Med. Sci.,* 86, 319, 1980.
5. **Harman, D.,** The biologic clock: the mitochondria, *J. Am. Geriatr. Soc.,* 20, 145, 1972.
6. **Fleming, J. E., Miquel, J., Cottrell, S. F., Yengoyan, L. S., and Economos, A. C.,** Is cell aging caused by respiration-dependent injury to the mitochondrial genome?, *Gerontology,* 28, 44, 1982.
7. **Vladimirov, Yu. A., Olenev, V. I., Suslova, T. B., and Cheremisina, Z. P.,** Lipid peroxidation in mitochondrial membrane, *Adv. Lipid Res.,* 17, 173, 1980.
8. **Lippman, R. D.,** Lipid peroxidation and metabolism in aging: a biological, chemical, and medical approach, *Rev. Biol. Res. Aging,* 1, 315, 1983.
9. **Vorbeck M. L. et al.,** Aging-dependent modification of lipid composition and lipid structural order parameter of hepatic mitochondria, *Arch. Biochem. Biophys.,* 217, 351, 1982.
10. **Clandinin, M. T. and Innis, S. M.,** Does mitochondrial ATP synthesis decline as a function of change in the membrane environment with aging? *Mech. Ageing Dev.* 22, 205, 1983.
11. **Dormandy, T. L.,** Biological rancidification, *Lancet,* 2, 684, 1969.
12. **Hegner, D.,** Age-dependence of molecular and functional changes in biological membrane properties, *Mech. Ageing Dev.,* 14, 101, 1980.
13. **Hornsby, P. J. and Crivello, J. F.,** The role of lipid peroxidation and biological antioxidants in the function of the adrenal cortex, *Mol. Cellular Endocrinol.,* 30, 1, 123, 1983.
14. **Burton, G. W., Joyce, A., and Ingold, K. U.,** Is vitamin E the only lipid-soluble, chain-breaking antioxidant in human blood plasma and erythrocyte membrane? *Arch. Biochem. Biophys.,* 221, 281, 1983.
15. **Barrows, C. H. and Kokkonen, G. C.,** Relationship between nutrition and aging, *Adv. Nutr. Res.,* 1, 253, 1977.
16. **Chow, Ch. K.,** Nutritional influence on cellular antioxidant defense systems, *Am. J. Clin. Nutr.,* 32, 1066, 1979.
17. **Chen, L. H.,** An increase in vitamin E requirement induced by high supplementation of vitamin C in rats, *Am. J. Clin. Nutr.,* 34, 1036, 1981.
18. **Kormendy, Ch. G. and Bender, A. D.,** Chemical interference with aging, *Gerontologia,* 17, 52, 1971.
19. **Harman, D.,** Prolongation of the normal lifespan by radiation protection chemicals, *J. Gerontol.,* 12, 257, 1957.
20. **Harman, D.,** Prolongation of the normal lifespan and inhibition of spontaneous cancer by antioxidants, *J. Gerontol.,* 16, 247, 1961.
21. **Harman, D.,** Free radical theory of aging: effect of free radical inhibitors on the mortality rate of male LAF1 mice, *J. Gerontol.,* 23, 476, 1968.

22. **Harman, D.,** Free radical theory of aging: beneficial effect of antioxidants on the lifespan of male NBZ mice: role of free radical reactions in the deterioration of the immune system with age and in the pathogenesis of systemic lupus erythematosus, *Age*, 3, 64, 1980.

23. **Harman, D. and Eddy, D. E.,** Free radical theory of aging: effect of adding antioxidants to maternal mouse diets on the lifespan of their offspring — second experiment, *Age*, 1, 162, (#40), 1978.

24. **Kohn, R. R.,** Effect of antioxidants on lifespan of C57BL mice, *J. Gerontol.*, 26, 378, 1971.

25. **Clapp, N. K., Satterfield, L. C., and Bowles, N. D.,** Effects of the antioxidant BHT on mortality in BALB/c mice, *J. Gerontol.*, 34, 497, 1979.

26. **Tappel, A., Fletcher, B., and Deamer, D.,** Effect of antioxidants and nutrients on lipid peroxidation fluorescent products and aging parameters in the mouse, *J. Gerontol.*, 28, 415, 1973.

27. **Jones, E. and Hughes, R. E.,** Quercetin, flavonoids and the lifespan of mice, *Exp. Gerontol.*, 17, 213, 1982.

28. **Miquel, J. and Johnson, J. E.,** Effects of various antioxidants and radiation protectants on the life span and lipofuscin of *Drosophila* and of C57BL/6J mice, *Gerontologist*, 15, 25, 1975.

29. **Patel, V., Miquel, J., Sharma, H. M., and Johnson, J. E.,** Hypocholesterolemic action of the antioxidants tocopherol *p*-chlorophenoxyacetate and magnesium thiazolidine carboxylate, *Age*, 2, 33, 1979.

30. **Miquel, J. and Economos, A. C.,** Favorable effects of the antioxidants sodium and magnesium thiazolidine carboxylate on the vitality and lifespan of *Drosophila* and mice, *Exp. Gerontol.*, 14, 279, 1979.

31. **Weber, H. U., Fleming, J. F., and Miquel, J.,** Thiazolidine-4-carboxylic acid, a physiologic sulfhydryl antioxidant with potential value in geriatric medicine, *Arch. Gerontol. Geriatr.*, 1, 299, 1982.

32. **Johnson, J. E., Jr., Miquel, J., Binnard, R., and Economas, A. C.,** Effects of antioxidants on the liver of aging mice, *Age*, 1, 162 (#41), 1978.

33. **Economos, A. C., Miquel, J., and Burns, M. K.,** Morphometrics of mouse liver mitochondria: effects of aging, alcohol and antioxidants, *Age*, 2, 134 (#51), 1979.

34. **Blackett, A. D. and Hall, D. A.,** The effects of vitamin E on mouse fitness and survival, *Gerontology*, 27, 133, 1981.

35. **Ledvina, M. and Hodanova, M.,** The effect of simultaneous administration of tocopherol and sunflower oil on the life span of female mice, *Exp. Gerontol.*, 15, 67, 1980.

36. **Buu-Hoi, N. P. and Ratsimamanga, A. R.,** Action retardante de l'acide nordihydroguaiaretique sur le vieillissement chez le Rat, *C. R. Acad. Sci. Paris*, 153, 1180, 1959.

37. **Oeriu, S. and Vochitu, E.,** The effects of the administration of compounds which contain -SH groups on the survival rate of mice, rats and guinea pigs, *J. Gerontol.*, 20, 417, 1965.

38. **Miquel, J., Economos, A. C., Bensch, K. G., Atlan, H., and Johnson, J. E., Jr.,** Review of cell aging in Drosophila and mouse. *Age*, 2, 78, 1979.

39. **Miquel, J., Fleming, J., and Economos, A. C.,** Antioxidants, metabolic rate and aging in Drosophila, *Arch. Gerontol. Geriatr.*, 1, 159, 1982.

40. **Miquel, J., Binnard, R., and Fleming, J. E.,** Role of metabolic rate and DNA repair in Drosophila aging: implications for the mitochondrial mutation theory of aging, *Exp. Gerontol.*, 18, 167, 1983.

41. **Economos, A. C., Ballard, R. C., Miquel, J., Binnard, R., and Philpott, D. E.,** Accelerated aging of fasted Drosophila. Preservation of physiological function and cellular fine structure by thiazolidine carboxylic acid (TCA), *Exp. Gerontol.*, 17, 105, 1982.

42. **Fleming, J. E., Leon, H. A., and Miquel, J.,** Effects of ethidium bromide on development and aging of Drosophila: implications for the free radical theory of aging, *Exp. Gerontol.*, 16, 287, 1981.

43. **Ruddle, D. L., Yengoyan, L. S., Fleming, J. E., and Miquel, J.,** Effects of structurally diverse antioxidants on the lifespan of Drosophila, *Gerontologist*, 23, 152A, 1983.

44. **Emanuel, N. M., Duburs, G., Obukhova, L. K., and Uldrikis, J.,** Drug for prophylaxis of aging and prolongation of lifetime, *PCT Int. Appl.*, 80, 02,373 (Cl. A61K311/44), 12 November 1980.

45. **Richie, J. P., Mills, B. J., and Lang, C. A.,** Magnesium thiazolidine-4-carboxylic acid increases both glutathione levels and longevity, *Gerontologist*, 23, 24A, 1983.

46. **Sharma, S. P. and Wadhwa, R.,** Effect of butylated hydroxytoluene on the life span of *Drosophila bipectinata*, *Mech. Ageing Dev.*, 23, 67, 1983.

47. **Kahn-Thomas, M. and Enesco, H. E.,** Relation between growth rate and lifespan in alpha-tocopherol cultured *Turbatrix acet*, *Age*, 5, 46, 1982.

48. **Kahn, M. and Enesco, H. E.,** Effect of alpha-tocopherol on the lifespan of *Turbatrix aceti*, *Age*, 4, 109, 1981.

49. **Epstein, J. and Gershon, D.,** Studies on ageing in nematodes. IV. The effect of antioxidants on cellular damage and life span, *Mech. Ageing Dev.*, 1, 257, 1972.

50. **Enesco, H. E. and Verdone-Smith, C.,** Alpha-tocopherol increases lifespan in the rotifer Philodina, *Exp. Gerontol.*, 15, 335, 1980.

51. **Chen, L. H.,** Effect of vitamin E and selenium on tissue antioxidant status of rats, *J. Nutr.*, 103, 503, 1973.

Section II
Impact of Aging on Nutritional Status

Chapter 3

BIOMEDICAL INFLUENCES ON NUTRITION OF THE ELDERLY

Linda H. Chen

TABLE OF CONTENTS

I. INTRODUCTION

When we refer to "age" we usually refer to "chronological age". Quite often, "chronological age" is not close to "biological age". Biological aging is characterized generally by a reduction in physiological activities and a reduced capacity to respond to environmental changes. With aging, there is a general decline in physiological functions, but the rate of decline differs from person to person. This leads to a wide range of "biological ages" among individuals within the same "chronological age" group. It is recognized that individual biological differences exist among people in the same age group. The degree of these individual differences is much greater among the elderly than among the younger population, i.e., the elderly are in many ways a more heterogeneous group than the young.

The extent of functional decrease due to aging differs among the various body systems. For example, the functional decrease is minimal in the liver but substantial in the kidney. Renal functions clearly deteriorate with aging. Age-associated decreases are noted in the following: numbers of glomeruli,[1] proximal tubular volume and length,[2] renal blood flow,[3] glomerular filtration rate,[3] maximum urine concentrating ability,[5] and maximum urine diluting ability.[6] The capacity of nephrons declines at a steady rate after the 4th decade. An accelerating decline in glomerular filtration rate occurs after the age of 40 at the rate of approximately 10% per decade.[4]

The functions of the cardiovascular system also decline with aging. The cardiac index and stroke volume decline as the mature adult ages.[7,8] The arterial distensibility decreases[9,10] and peripheral vascular resistance increases with aging.[7,8] Thus, the work demands on the heart are increased. Maximum oxygen consumption and maximum heart rate during exercise are diminished with aging in normal man.[8] Systolic pressure increases progressively with exercise and shows a greater change in older subjects.[8]

In the lung, maximum breathing capacity and vital capacity decrease with aging.[11-13] The diminished vital capacity is at the expense of all its component volumes. There is increased residual volume, decreased ventilation effectiveness, diminished bronchial potency, decreased blood filling of lung, reduced elasticity and increased rigidity of pulmonary vessels, and disturbed uniformity of pulmonary ventilation.[14] The residual volume[11] and physiological dead space[15] increase with advancing age. Therefore, the residual volume to total lung capacity ratio and the physiological deadspace to tidal volume ratio increase with age in adults.[11] Airway resistance is higher and conductance to volume ratios is lower in elderly men.[16] Both inspiratory and expiratory reserve volumes decrease with age in the adult,[11] as does the inspiratory capacity.[16] Both maximum voluntary capacity and the partial pressure of oxygen in the arterial blood decrease with age.[11,13]

It is generally believed that the functioning of the central nervous system declines with age. A loss of neurons with age occurs in some but not all areas of the cortex.[16] Nerve cells cannot reproduce themselves. The slowing in the response to environmental stimuli with age adversely affects human performance.[17] However, in the peripheral nervous system, age-related changes are minor.

In the neuromuscular system of man, motor function declines with age[18] as well as number of functional motor units.[19] The surviving motor units are often enlarged and tend to have relatively slow twitches. In aged humans, decreased performance of skeletal muscle is associated with a decrease in the number of motor units. The amount of muscle tissue decreases with age due to a decrease in both the total number and size of muscle cells.[20] Muscle cells, like nerve cells, cannot reproduce themselves. There is an increase in connective tissue and fat, as well as evidence of muscle wasting and weakness in the elderly.[19] Muscle contraction time and latency are both slightly prolonged with aging.

Many theories have been proposed to explain the biological changes associated with aging.[21] By reviewing and summarizing these theories, possible factors contributing to biological aging can be listed. They include genetic timetable, rate of living, increased abnormal chromosomes, deterioration of protein synthesis, age-dependent changes in enzyme systems, increased cross-linkage of collagen, breakdown in immunological processes, changes in membrane diffusion, slowed adaptation to stress, accumulation of free radicals, cellular mutation, accumulated results of ionizing radiation, waste product accumulation, and loss of cells.[22] The nutritional environment has important influences on some of the factors listed above. A lifestyle in which proper nutrition is practiced throughout life may decrease the rate of biological aging and increase the likelihood of effective longevity.

The nutritional status of the elderly is affected by multiple factors. The biomedical factors are discussed in this chapter.

II. BIOMEDICAL FACTORS AFFECTING FOOD INTAKE

A. Anorexia

Anorexia is common in the elderly and may lead to actual refusal of food. In addition to socio-behavioral factors (see next chapter), the causes of anorexia are multiple. Stomach distress and reduction of hunger contractions, which are common in the elderly, result in anorexia. Other causes of anorexia are discussed in the next six sections.

1. Decline in Taste and Smell Capacity

Decline in taste and smell capacity with aging does not occur abruptly but rather appears to develop gradually during the later years.

With advanced age, the number of taste buds decreases.[23-25] The gustatory papillae begin to atrophy in women at age 40 to 45 and in men between 50 and 60.[24] A decrease of taste buds has been noted in individuals aged 74 to 84 (less than 100 taste buds per papilla as compared to 250 taste buds in young adults).[25]

A human taste bud is comprised of 40 to 60 taste cells. Taste cells are of epithelial origin, have an average life span of 10 days, and are constantly renewed. Taste cell regeneration slows down with increasing age.[26]

With aging, some decline of taste sensitivity is found. Taste sensitivity is inversely related to taste thresholds. The thresholds for all four primary taste qualities (salt, sweet, bitter, and sour) increase with aging.[27] In one study, elderly subjects were found to have thresholds for sweet taste three times as great as young subjects.[28] The deterioration of salt taste may be due to the increased salivary sodium concentration in the elderly. However, no correlation has been seen between salivary sodium concentration and the taste threshold.[29]

A decrease in the number of taste buds is not indicative of a loss of taste in the absolute sense. More commonly, the ability to identify the four primary taste qualities remains, although the individual becomes progressively impaired in the capability to discriminate subtle blending of taste qualities (flavors) of foods.[30]

In the elderly, atrophy of the olfactory glomeruli and nerve fibers is partially due to the loss of sensory cells in the olfactory mucosa.[31,32] There is a general increase of the olfactory threshold, including food odor, with aging.[33] In one study, mean recognition thresholds for some food odors, including grape, cherry, lemon, orange, chocolate, mushroom, and bacon, were observed to be at least 11 times higher in elderly subjects than in young adults.[34]

Furthermore, there is a decrease in the sensitivity of the gustatory and olfactory nuclei in the brain. With age, the appetite centers decline so that appetite no longer operates to stimulate a desire for food.[35] Basic causes are the decreased blood supply to the brain and cell degeneration.

The general decline with aging in taste acuity and sensitivity to food odor may contribute to anorexia because taste and aroma are important in determining the palatability and acceptance of food.

2. Constipation

The incidence of constipation increases with aging. The rate for the above 65 age group is 9.6% as compared to 3.5% for the 45 to 64 age group.[36] Constipation and functional bowel distress affect the appetite.[37]

General causes of constipation include decreased muscle tone and motor function of the colon with aging, low fluid intake, low dietary fiber intake, low food intake due to poor dentition, and lack of physical exercise. Medications such as sedatives, tranquilizers, narcotics, calcium carbonate antacid, and some antihypertensives result in slowed colonic transit. Pathology of the colon such as chronic intestinal obstruction or inflammatory bowel disease also causes constipation.

3. Malnutrition

Deficiencies of thiamin, niacin, vitamin B_6, and folacin are associated with anorexia. Malnutrition may produce loss of taste acuity which, in turn, may result in anorexia. In addition to aging, nutrition is a major factor that affects the number of taste cells. Malnutrition slows down the rate of taste cell replication. Vitamin B_{12} and folic acid are important for cell regeneration.[38]

Vitamin A and/or zinc deficiency can cause hyperkeratinization of the epithelial cells. Zinc deficiency is associated with hypogeusia.[39]

4. Excessive Alcohol Consumption

Alcohol decreases food intake by displacing other dietary constituents and by suppressing appetite. The alcoholic may derive up to 50% or more of his caloric requirement from alcohol. Other causes of anorexia in elderly alcoholics are inebriation, gastritis, lactose intolerance, pancreatitis, hepatitis, cirrhosis, ketoacidosis, alcoholic brain syndromes, and withdrawal syndromes.[40] Anorexia in alcoholics may also be due to deficiencies of thiamin, niacin, folic acid, zinc, or protein.

5. Drugs and Therapy

Hypophagic drugs produce an anorexic effect. Examples of these drugs are anticonvulsants, antimetabolic agents, carbonic anhydrase inhibitors (e.g., diuretics), digitalis glucosides, estrogens, Flurazepam, indomethacin, levodopa, monoamine oxidase inhibitors, phenformin, tetracyclines, and thiazides.[41] Some cancer chemotherapeutic agents as well as radiation therapy for cancer induce anorexia associated with nausea and vomiting.[42]

Other drugs, such as D-penicillamine (a copper-chelating agent used in Wilson's disease) and ethambutol (a zinc-chelating agent used in tuberculosis) may produce zinc deficiency. They may cause a decline in taste acuity and affect appetite.[43]

Medications, commonly used by elderly, may produce an altered taste sensation (dysgeusia). Alkaloids and halogens transported back to the oral cavity by the circulatory, digestive, or respiratory system can result in bizarre taste sensations.[30] Similar taste aberrations are produced by clofibrate, penicillamine, uncoated penicillins, lincomycin, tetracyclines, chloramphenicol, salts of heavy metals, and even therapeutic doses of thiamin hydrochloride. Several tranquilizers can also produce dysgeusia.

6. Diseases

Acute diseases such as pneumonia, gastroenteritis, and cancer result in anorexia.[44] Chronic diseases, including gastrointestinal (GI) disease, alcoholic liver disease, cardiovascular diseases (e.g., congestive heart failure), respiratory diseases (e.g., emphysema with pulmonary decompensation), and chronic renal disease (e.g., chronic nephritis and uremia) also cause anorexia. In addition, elderly patients with partial or total gastrectomy are frequently anorectic.

B. Dental Problems

About 50% of people become endentulous by age 65, and 67% by age 75.[45] Elderly people suffering from tooth loss without dentures and those with dentures which do not fit have difficulty in chewing certain foods. Thus, some of these individuals avoid eating crisp and hard foods, such as meats, vegetables, and raw fruits, because of chewing difficulties. The diet is often changed to include large amounts of commercially prepared "convenience" foods and foods more easily masticated, high in carbohydrates and calories but deficient in protein, iron, calcium, riboflavin, and ascorbic acid.[30] In a study of the nutritional status of elderly residents in Missouri, 56% of males and 62% of females were edentulous; only 48% of the males and 30% of the females actually had dentures at the time of examination.[46] The dental health of the subjects was identified as the major nutrition problem in this study.

The biting force decreases from 300 lb/in^2 in young people to 50 lb/in^2 in the aged.[47] Masticatory muscles lose their force and undergo some fibrosis. With the use of artificial dentures, their efficiency is further lost. These physical changes may contribute to problems in the protein nourishment of the older age group.

A high percentage of prosthodontic patients might be classified as "nutritionally disadvantaged geriatric patients".[48] Nutrition deficiency signs, such as atrophic mucous membrane of the oral cavity, excessive resorption of alveolar bone, or slow healing response following injury, are found in these patients. Those who have systemic diseases or oral conditions for which inadequate nutrition is a contributing factor cannot be restored to health until their nutritional problems are corrected.

C. Diseases

Certain diseases not associated with anorexia can cause difficulties in food ingestion by the patient.[49] For example, cancer of the esophagus or oral cavity produces painful swallowing, congestive heart failure produces food-related breathlessness, and stroke produces unconsciousness.

D. Special Therapeutic Diets for Diseases

When an elderly person has a chronic disease he is placed on a special therapeutic diet. When different diseases coexist in the individual, special diets may be superimposed, resulting in low food intake and dietary inadequacy.[50] Special diets may be a major cause of malnutrition in the elderly for the following reasons: the food energy intake may be too low, the number of food sources of essential nutrients may be too limited, the diet may be too unpalatable to be consumed, and/or foods permitted do not include certain nutrients.[50]

E. Physical Disabilities

The incidence of physical disabilities increases with advanced age. Approximately 21% of persons between the age of 45 and 64 years have limitations in activity and 4.5% are unable to function in major duties. Contrasting rates for the age group above 64 are 43% with limitation in activity and 16% unable to carry on major activities.[51]

The progressive impairment of vision with aging makes it difficult to read labels. At the same time, the danger associated with the use of kitchen appliances increases. Decrease in pupil size, reduction in transparency of the lens, and increase in the thickening of the lens and capsule all combine to greatly diminish the light that reaches the retina.[52]

Other physical disabilities, such as poor coordination, paralysis, lack of mobility, dementia, and/or actual physical confinement affect the ability of the elderly to shop and select foods, to prepare foods, and sometimes, even to eat foods.

III. BIOMEDICAL FACTORS AFFECTING DIGESTION AND ABSORPTION

A. Aging-Related Changes in the Oral Cavity and GI System
1. Digestion

There is an age-related impairment in the general digestive capacity without any manifest GI disorder,[53] although it may not be clinically significant.

The diminished function of both major and minor salivary glands may result in an absolute or relative xerostomia or "dry mouth". Complaints of abnormal taste sensations and stomatodynia may be related to salivary changes.

Age-related atrophy and fibrous infiltration of the salivary glands occur in the elderly. With aging, the flow rate of salivary secretion (volume of saliva per unit time) is decreased, particularly in the parotid gland.[54,55] This may result from a loss of functional tissue. The acinar and duct cells of the salivary glands, particularly parotid, undergo a marked fatty degeneration. Decreased salivary flow in itself may result in diminished taste reception. The concentration of salivary amylase or ptyalin declines with age. In one study, less than one third of the amount in young adults was present in individuals above age 60.[54]

Dehydration and thinning of the gum tissue and shrinking of the connective tissue of the mouth occur with aging. The gingivae may undergo a reduction in vascularity. There is a loss of keratinization of the epithelium and decrease in connective tissue.[47] Elastic tissue of the periodontal ligament decreases.

Age-associated changes in the esophagus include tertiary contraction, delayed esophageal emptying, and dilatation of the esophagus.[56-58] There is a decreased ability to initiate relaxation of the lower esophageal sphincter and primary peristaltic contractions. The amplitude of esophageal contractions is diminished in men above the age of 80.[59] The nonperistaltic contractions are increased after the age of 70.[58] The incidence of failure of the lower sphincter to relax after swallowing is increased.[60]

Aging results in reduced basal gastric hydrochloric acid secretion in humans above the age of 60.[54,61,62] In a study of gastric secretion of normal persons over the age of 60, 15% had achlorhydria (complete lack of gastric hydrochloric acid).[62] In another study, 28% of women and 23% of men were achlorhydric.[63] The age-related decline in gastric juice secretion is almost entirely attributable to a concomitant decrease in the parietal component rather than the nonparietal component. The absorption of iron is decreased by achlorhydria. The rate of secretion of gastric acid is greater in men than in women. The acidity and the volume of gastric acid secreted following a suboptimal dose of histamine decline with age in both men and women, and the decline is more marked in women.[61,64,65] The volume aspirated from the human stomach at a fixed interval after ingestion of a test meal diminished with advancing age.[63] It may be due to reduction in the volume of gastric secretion.

Gastric secretion of pepsin declines with age after 70.[54,66] The secretion of gastric intrinsic factor is reduced with age and this decline is associated with a loss of parietal cells and, in some cases, the appearance of parietal cell autoantibodies.[67,68] The absorption of vitamin B_{12} is decreased because of the decline in the intrinsic factor.

Aging results in incomplete digestion of carbohydrates in the mouth and stomach of old individuals.[54] In one study, it was reported that the secretion of pancreatic amylase was only slightly decreased and pancreatic lipase concentration was unchanged with aging.[54] However, in another study it was reported that there was a 20% decrease in both the pancreatic lipase concentration of fasting duodenal content and stimulated pancreatic secretion.[66]

The pancreatic trypsin activity decreases sharply after the 4th decade of life.[54] The concentration of trypsin in stimulated pancreatic juice is similar in young and old individuals.[66] There is little or no age- or sex-related decline in the peak volume and bicarbonate responses of the pancreas to secretin stimulation.[69]

2. Absorption

Available information does not permit a solid conclusion about the effect of aging upon nutrient absorption under the normal (not diseased) condition.

In research studies with mice, it has been found that the cell cycle of the duodenal and jejunal epithelium increases with age due mainly to a longer crypt cell generation time.[70-73] In contrast, epithelial cells in the ileum of aged mice do not show any apparent change in crypt to villus transit time. Aging results in a decrease in the proliferation rate of cells in the small intestine.[74] This alteration of mucosal cell turnover in old age results in a preponderance of epithelial cells with a life span exceeding the 3-day renewal rate at a younger age.[30] With older cells relatively inactive in nutrient selection, especially if nutrients are present in small amounts, the reduced proliferation rate in the small intestine might contribute to intestinal dysfunction. The small intestine undergoes an age-related involution characterized by a proliferation of connective tissue and the formation of fibrous sclerotic tissue.[75,76]

In the duodenal mucosa of rats and mice, it has been found that the villus height and absorbing surface are reduced with advancing age when examined under the light microscope and in ultrastructure.[77,78] A biopsy study of aged human subjects showed a decrease in jejunal villus height but no change in the villus width and mucosal cell height.[79] In another study, a progressive reduction in mucosal surface area in intestinal biopsies of persons between the ages of 10 and 75 was found.[80] This decrease in the mucosal absorptive surface may be a major cause of intestinal absorptive insufficiency in old age.

The absorptive capacity associated with aging with regard to individual nutrients is discussed in Section IV of this volume and Chapter 10 of Volume II.

B. Diseases of the Oral Cavity and GI System

Although there may be some age-related impairment in the general digestive and absorptive capacity, the GI tract generally maintains an adequate level of functioning throughout life. Alteration or decrease in function and change in anatomical structure may cause some digestive and absorptive problems, but major problems are due to underlying diseases and drug use.[81]

With advanced age, masticatory efficiency may be impaired by loss of mobility of the mandibular joint due to osteoarthritis, periodontal disease, inflammation under a denture, or the edentulous or partially dentulous state.[82] Periodontal disease is common and is associated with loosening of the teeth, and dental abscesses. Angular cheilitis and/or stomatitis is common. The angular cheilitis may result from bite overclosure, drooling, or lack of riboflavin. More common problems of the tongue are glossodynia and glossopyrosis — painful and burning tongue.[47] Atrophic glossitis is common.

Presbyesophagus (disorganized and inefficient contractions), achalasia, and hiatus hernia are the most common GI disorders of the elderly.[81] The incidence of hiatus hernia is less than 10% under the age of 30, and increases to 60% above the age of 60.[57] The resting lower esophageal sphincter pressure decreases in elderly patients with symptomatic gastroesophageal reflux, suggesting that a deficiency in gastrin secretion may be responsible.[57,83]

Aging results in an increased incidence of chronic atrophic gastritis characterized by diffused lesions, thinning of the gastric glands, loss of parietal cells, and goblet cell metaplasia in individuals free of manifest gastric disease.[57,67,83-85] This atrophic gastritis results in decreased absorption of iron and vitamin B_{12}.[67]

The incidence of diverticulosis in the small intestine as well as the colon increases with aging.[86,87] Multiple jejunal diverticulosis may be accompanied by overt malnutrition.[88] Diverticulosis may affect normal intestinal mobility, and abnormal mobility is a factor in the intestinal bacteria overgrowth syndrome. Normal bacterial populations in the small intestine are given additional time to multiply. They reproduce on the host's ingested nutrients, especially amino acids, folic acid, and other vitamins needed by the host. The bacterial overgrowth interferes with the absorption of fat, iron, folic acid, and vitamin B_{12}. Nutritional deficiencies of the host become accentuated if the dietary intake is inadequate in quantity or quality. Other contributing factors involved in the intestinal bacteria overgrowth syndrome in the elderly are the progressive increase with aging in the incidence of hypochlorhydria and a tendency toward immunocompetence.

Diseases of the pancreas and diseases of the small intestine affect digestive juice secretion. Pancreatic insufficiency can result from pancreatitis, cancer of the pancreas, or resection of the pancreas. Insufficient secretion of bile salts, caused by liver disease, cystic duct obstruction, or bacterial overgrowth may result in steatorrhea. This condition reduces the absorption of calcium and fat-soluble vitamins. Diseases of the bile duct would affect the transport of bile into the small intestine. Cholelithiasis is not uncommon.

Dissacharidase deficiencies, the most common of which is lactase deficiency, are a cause of malabsorption. In lactose deficiency, only small amounts of lactose are tolerated without symptoms.

Surgical alteration of the GI tract affects digestion and absorption.[49] For example, total gastrectomy results in poor digestion of fat and protein and affects iron absorption. Gastrectomy also results in lack of intrinsic factor, which is secreted by the stomach and is needed for vitamin B_{12} absorption.

Some other diseases and disorders commonly found in the elderly which interfere with normal digestion and absorption include congestive heart failure, vascular insufficiency, endocrine problems, and radiation injury to the GI tract.[89]

Malnutrition also contributes to GI dysfunction. For example, protein malnutrition results in damage both to the mucosal ultrastructure and the absorptive function of the small intestine.

C. Drugs

Some medications commonly used by the elderly have the side effects of impairing digestion or absorption of nutrients. For example, neomycin, when administered orally to man in doses of 3 to 12 g/day for 3 to 7 days, may cause a reversible malabsorption syndrome and steatorrhea.[90,91] Malabsorption of fat, nitrogen, sodium, potassium, calcium, iron, lactose, sucrose, and vitamin B_{12} results in increased fecal loss of these nutrients. Neomycin can induce morphological and functional changes. Villi are shortened and the lamina propria of the intestinal mucosa is infiltrated with inflammatory cells and macrophages. In addition, neomycin forms a precipitate with bile salts[92] and decreases pancreatic lipase activity.[93] (See Chapter 1, Volume II for further discussion on the possible effects of drugs on nutrient digestion and/or absorption.)

D. Alcoholism

Alcoholism is conservatively estimated as affecting up to 10% of the population beyond the age of 60.[94] It affects digestion and/or absorption of nutrients by damaging the ultrastructure of GI mucosa. Alcoholism produces gastric inflammation, decreased gastric motility and emptying time, decreased villi and epithelial cell populations, and decreased villous to crypt cell ratio. The absorption of thiamin, riboflavin, pyridoxine, niacin, folic acid, vitamin B_{12}, vitamin A, zinc, magnesium, and iron may be impaired. Alcoholism also produces chronic or acute pancreatitis so that the production of pancreatic enzymes is impaired, resulting in maldigestion of protein and fats. Malabsorption of amino acids and fats is secondary to maldigestion.[95] Concurrent intake of some medications with prior alcoholism increases the risk of nutritional deficiency.[96] (See Chapter 2, Volume II for further discussion of the nutritional effect of alcoholism.)

Table 1
COMPARISON OF BODY
COMPOSITION OF
YOUNG AND OLD
REFERENCE MAN[98]

Age	25	70
Specific gravity	1.068	1.035
Water	61%	53%
Fat	14%	30%
Cell solids	19%	12%
Bone mineral	6%	5%

From Parizkova, J., *Proc. Nutr. Soc.*,
32, 181, 1973. With permission.

Table 2
PERCENT DIFFERENCES IN ANALYSIS OF HUMAN TISSUES,
70-YEAR-OLD VS. 30- TO 40-YEAR-OLD[104]

Constituent	Kidney	Liver	Spleen	Psoas muscle	Heart	Av. of tissues
Cl	+2	+18	+12	+56	+25	+23
Na	+5	+15	+24	+62	+0.3	+20
Ca	+60	+4	+14	+33	+31	+28
Total base	+3	+12	+4	+6	+7	+7
K	−19	+6	−13	−7	−9	−12
Mg	−9	+17	−10	−11	−2.5	−8
P	−13	−0.1	−8	−12	−9	−9
N	−9	+8.5	−13	−3	−4	−7
Ash	−11	+1	−8	−1	0	−5

From Simms, H. S. and Stolman, A., *Science*, 86, 269, 1937. With permission. Copyright 1937
by the AAAS.

IV. BIOMEDICAL FACTORS AFFECTING INTERMEDIARY METABOLISM

A. Body Composition

Body composition changes with increasing age. Table 1 compares body composition of
the young and old reference man.[97] Total body water as well as intracellular water decreases
with age. Replacing lean body mass, body fat content increases steadily with age until the
6th decade.[98-100] The specific gravity of the body decreases with aging until the 6th decade
of life.[101] This is mainly due to the increase in adipose tissue at the expense of lean body
mass. The decrease in bone mineral results in a high incidence of osteoporosis among the
elderly. Lean body mass declines with increasing age and the rate of decline increases in
the later years.[102,103] This decrease in lean body mass is reflected by a decrease in total body
potassium with age.[104,105] However, the decrease in potassium with age is more rapid than
the decrease in lean body mass, because the amount of potassium per kilogram of lean body
mass falls with age. Table 2 shows the percent differences in the content of some minerals
in tissues of humans 70 years old vs. those 30 to 40 years old. The amount of chloride,
sodium, calcium, and total base increases, while the amount of potassium, phosphorus,
magnesium, nitrogen, and ash decreases.[106]

The basal metabolic rate (BMR), the overall basal energy metabolism of the body in terms of oxygen consumption per square meter of body surface, declines gradually with age.[107] This age-related decline in BMR is mainly due to the decrease of lean body mass, because when based on the lean body mass the decrease of basal metabolism with age is no more than 1 to 2% per decade.[108]

Due to the change in body composition, a healthy 25-year-old man is not the same as a healthy 70-year-old man even though their body weights may be the same. The latter has a much higher adipose tissue content. Because of increasing adiposity with age, the ability to respond to a glucose challenge with adequate insulin secretion progressively declines.[109]

Increased adiposity with aging might alter lipid metabolism. A mechanism has been hypothesized whereby increased levels of triglyceride might result: obesity produces insulin antagonism in glucose-metabolizing tissues; this results in the pancreas producing more insulin, which increases endogenous triglyceride synthesis.[107] Cholesterol synthesis also increases with obesity.

Increased adiposity with aging is one of the contributing factors to the higher incidence of cardiovascular disease and diabetes mellitus associated with aging.

B. Protein

In aging tissues, including the kidney, liver, lung, muscle, tendon, and skin, the amount of metaplasmic proteins (fibrillar protein), especially collagen and elastin, increases.[108-110] The water-retaining power of the protein colloids is lowered. This irreversible dehydration of protein is due to changes in the structure of protein and a decrease in the number of free hydrophilic groups. Aging of cells is associated with an increase in the viscosity of protoplasm.

Collagen and elastin, the proteins of connective tissues, change in physicochemical properties with aging.[111] Collagen fiber becomes more crystalline, more rigid, and less soluble. Its elasticity and ability to shorten are also decreased. For example, collagen of human bone, cartilage,[112] and myocardium[113] becomes insoluble; collagen of tendon loses its swelling ability, and that of tendon and myocardium becomes resistant to enzymatic digestion.[114-117] In the collagen of human tendon, fluorescent material increases.[115] The resistance of collagen to solubilization increases with age. In the collagen of human bone and cartilage, Schiff bases decrease and hexitollysine increases.[110] Elastin fibers are thickened. Fragmentation of the ends takes place, an aggregation of fibers results in the formation of irregular masses. Sometimes this fragmentation is associated with pigmentation and calcification. In elastin of human artery and lung, isodesmosine and desmosine decrease, and carbohydrate content increases.[118]

The ground substance, glycosaminoglycans, also changes in physicochemical properties with increasing age. In human cartilage, chondroitin sulfate decreases, and keratin sulfate plus glucoprotein to chondroitin sulfate increases.[119,120] Glycosaminoglycans decrease and glycoproteins increase in the lung.[121] In the aorta and myocardium of man, the hexosamine to hydroxyproline ratio decreases.[122] In cartilage, the sulfate to hexauronic acid ratio increases, and the galactosamine to glucosamine ratio decreases.[120,123]

The susceptibility of the proteins of an organ as a whole (e.g., liver, heart, brain) to hydrolysis by proteolytic enzymes decreases with age. This may be indirect evidence of changes in the internal structure of proteins with age.[111]

The increase of fibrous connective tissue proteins and change of these proteins and glycosaminoglycans in physicochemical properties with aging affect the transport of oxygen and nutrients into the cells and the transport of catabolic products out of the cells because all the cells and organs are supported by connective tissues.[124]

C. Enzymes

The regulating factors of biochemical reactions in cells are enzymes and hormones. Substrate concentration is a factor in regulating the rate of enzyme reactions, and this supply of substrate available for enzyme reactions is regulated by hormones which influence the diverse reactions of intermediary metabolism.[125] Due to the lack of information regarding the factors which may affect the concentration of enzymes and the difficulties in distinguishing the enzyme levels and activities present in the tissues, the results of enzyme studies with aging cannot be explained systematically at the present time.

The level of many enzymes changes with aging. For example, the level of monoamine oxidase in human platelets and plasma increases progressively after age 40.[126] This enzyme catalyzes the degradation of biogenic amines such as norepinephrine and serotonin, the normal metabolism of which is essential to mental health as well as other physiological functions. Therefore, the level of this enzyme may play an important role in mental depression, to which the elderly are particularly susceptible. The graying of hair associated with aging is related to the gradual loss of activity of tyrosinase of the hair bulb melanocyte.[127] It is possible that graying may develop as a result of tyrosinase inhibition by metabolites which accumulate in the hair bulb or is caused by failure of enzyme synthesis during aging.

Studies of certain enzymes in rat or mouse liver indicate no definite pattern to age-dependent changes in hepatic enzyme activities. Studies by different researchers used different strains of animals at different ages, and enzyme levels were expressed on the basis of different measurements: wet weight, protein, DNA, or cell number. Some studies showed different results for male and female animals. With aging, activities of some enzymes decline, some increase, and others remain unchanged. Examples of the studies are shown in Tables 3 to 5.

No change as a function of age has been found in the overall respiratory capacity or the functional integrity of hepatic mitochondria in rats or mice.[136-138] Oxidative phosphorylation is not affected by aging.

Age-related changes in hepatic enzyme activities may vary depending on the animal strain. For example, hepatic glucose-6-phosphate dehydrogenase decreases about 50% between 3 and 24 months of age in Sprague-Dawley rats, while the activity of this enzyme doubles in Fischer 344 rats during the same age span.[139]

The enzyme levels of tissues other than the liver in rats of different ages have also been studied by many researchers. These studies have been reviewed by Finch[140] and Wilson.[141]

The induction period of some enzymes has been reported to be dependent on the age of rats. The duration in the lag period of glucokinase induction by insulin,[142] and that of NADPH-cytochrome reductase induction by phenobarbital are directly proportional to age.

The induction of some enzyme levels has been reported to decrease with age. For example, the dexamethasone induction of glucose-6-phosphatase and fructose-1,6-diphosphatase in the rat liver is decreased markedly in older animals.[143]

The turnover rate of some enzymes are reported to change with increasing age. For example, following administration of a standard dose of L-tryptophan to rats, hepatic tryptophan pyrrolase activity reaches maximum activity sooner and at a lower level, and begins to decline earlier in older rats than younger adult rats.[144]

D. Hormones

Age-related changes in some hormone functions are considered to be closely associated with the effect of age on physiological functions. The most apparent age-related changes occur in the sex hormones. The blood concentrations of testosterone in men[145] and of estrogen in women[146] decrease with aging. Figure 1 shows the changes of hormones during aging in humans. In many endocrine control systems, significant age-related alterations in resting hormone function and decreases in the ability of endocrine control systems to respond to stress or stimulation have been identified. Age-related changes of some hormones are discussed below.

Table 3
LIVER ENZYMES WHICH INCREASE WITH AGING

Species	Strain	Age compared (months)	Enzyme	Basis of measurement	Ref.
Rat	Sprague-Dawley	33/26	Alkaline phosphatase	Weight, cell	128
Mouse	C57	24/12	Alkaline phosphatase	Protein, weight	129
Mouse	N.M.R.I.	24/6	Acid phosphatase	Weight	130
Mouse	N.M.R.I.	24/6	β-Galactosidase	Weight	130
Mouse	C57	30/18	Acid phosphatase	DNA, protein, weight	131
Mouse	C57	30/18	ATPase	DNA, protein, weight	131
Mouse	C57	30/18	Phosphoglucomutase	DNA, protein, weight	131
Mouse	C57	30/18	β-Glucuronidase	DNA, protein, weight	131
Mouse	C57	30/18	Glucose-6-phosphate dehydrogenase	DNA, protein, weight	131
Mouse	C57	30/18	Lactic dehydrogenase	DNA, protein, weight	131
Rat	BN/BiRij	30—35/24	Cathepsin D	Protein	135
Rat	BN/BiRij	30—35/24	β-Galactosidase	Protein	135
Rat	Wild	34/14	Cathepsin	Weight	132

Table 4
LIVER ENZYMES WHICH DECREASE WITH AGING

Species	Strain	Age compared (months)	Enzyme	Basis of measurement	Ref.
Rat	Sprague-Dawley	33/26	Catalase	Weight, cell	128
Mouse	C57	30/18	Alkaline phosphatase	Protein, DNA, weight	131
Mouse	C57	30/18	Glucose-6-phosphatase	Protein, DNA, weight	131
Mouse	C57	30/18	5'-Nucleotidase	Protein, DNA, weight	131
Mouse	C57	30/18	Glycogen phosphatase	Protein, DNA, weight	131
Mouse	C57	30/18	NAD diaphorase	Protein, DNA, weight	131
Mouse	C57	30/18	Cytochrome oxidase	Protein, DNA, weight	131
Mouse	C57	30/18	Succinic dehydrogenase	Protein, DNA, weight	131
Mouse	C57	25/8	RNA polymerase I	DNA, protein	133
Mouse	C57	25/8	RNA polymerase II	DNA, protein	133
Mouse	CF-1	20/12	Alkaline DNase	Weight, cell	134
Rat	BN/BiRij	30—35/24	Acid phosphatase	Protein	135
Rat	BN/BiRij	30—35/24	Aryl sulfatase B	Protein	135
Mouse	C57	24/12	Acid phosphatase	Protein, weight	129

Table 5
LIVER ENZYMES WHICH DO NOT CHANGE WITH AGING

Species	Strain	Age compared (months)	Enzyme	Basis of measurement	Ref.
Mouse	C57/BL	30/6	Malate dehydrogenase	Protein	136
Mouse	C57/BL	30/6	Cytochrome oxidase	Protein	136
Mouse	N.M.R.I.	24/6	Cytochrome oxidase	Weight	133
Mouse	N.M.R.I.	24/6	Glucose-6-phosphatase	Weight	133
Mouse	CF-1	20/12	Acid DNase	Cell	134
Rat	Wild	34/14	D-Amino acid oxidase	Weight	132
Rat	Wild	34/14	Succinic dehydrogenase	Weight	132
Rat	Wild	34/14	Alkaline phosphatase	Weight	132

Hormone	Hormone concentration in blood	Response to physiologic or pharmacologic stimulation	Metabolism (disposal rate)	End-organ sensitivity
Growth hormone	↔	↓		↓
Gonadotropins	↑			
Thyrotropin (TSH)	↔	↓		↔
Thyroxine (T4)	↔	↔	↓	↑
Triiodothyronine (T3)	↓			
Parathyroid hormone	↓			↑
Cortisol	↔	↔	↓	
Adrenal androgens	↓	↓		
Aldosterone	↓		↓	
Insulin	↔	↓	↔	↔
Glucagon	↔	↔		
Testosterone	↓		↓	
Estrogens	↓ *		↓	

FIGURE 1.   Changes of plasma hormones during aging in man. Symbols: ↑, increase; ↓, decrease; ↔, no change; *, postmenopausal. (From Gregerman, R. I. and Bierman, E. L., *Textbook of Endocrinology*, Williams, R. H., Ed., W. B. Saunders, Philadelphia, 1974, 1069. With permission.)

1. Adrenal Cortex

Adrenals from several species show increased connective tissue and lipofuscin pigments with aging.[147,148] The increased formation of adrenal nodules with aging is associated with the development of hypertension.[149]

Studies have shown that in aging humans the glucocorticoid secretion rate is reduced from 25 to 40%.[147,150] The reduction in secretory rate is accompanied by an increase in circulatory half-life (slowed metabolic disposal), resulting in nearly stable blood cortisol concentrations in aged humans.[147] Adrenal response to ACTH stimulation and dexamethasone are normal in aged humans compared to young controls.[151]

Studies with rats suggested that the hypothalamic-pituitary control mechanism for ACTH release in the young responds differently than in the aged to elevated serum corticosterone.[152] The response was significantly higher in young than in old rats of each sex. In another study, since the adrenocortical control mechanisms of young rats were more responsive to restraint by stress-induced increases in corticosterone secretion, a decreased sensitivity of the adrenocortical control system to glucocorticoid negative feedback is thought to exist in aged rats.[153]

Aldosterone control systems may be affected by aging. The adrenocortical secretory rate and the metabolic clearance rate have been shown to decrease in the elderly.[154] The plasma concentration of aldosterone is decreased to levels about one half of those in young adults. Although older persons do increase aldosterone secretion with sodium restriction, those over age 60 show increases which are only 30 to 40% of those seen in young adults.[155] Serum renin concentration and plasma renin activity have also been shown to decrease in aged humans with parallel lowering of plasma aldosterone concentrations.[156] The decrease in secretion of renin parallels that of aldosterone with aging.

2. Pancreas

Decreased glucose tolerance has been well documented in aging human populations.[157]

However, the factors associated with the increased blood glucose after carbohydrate loading are not clear. Several studies reported a decreased secretion of insulin following pancreatic challenge.[158] Structural alterations in the pancreas such as fewer β cells and pancreatic vascular degeneration with increased age may contribute to β cell insensitivity in the aged.[159] However, other studies showed different results and indicate that insulin levels following glucose loading are unchanged with age.[160] Increased tissue resistance to insulin activity in the aged could be due to increased adiposity and the increased capillary basement membrane thickness in aged subjects.[161] (See Chapter 5, Volume II for further discussion.)

3. Thyroid

Some studies of the effect of aging on blood thyroid hormone levels showed a modest decline of about 20% in serum thyroxine or protein-bound iodide concentration.[162,163] However, other studies showed no change in serum thyroxine or protein-bound iodide concentration with advancing age.[164] The thyroxine clearance rate is reduced with aging[165,166] so that by age 80 it is about half that of young adults.[167] This was attributed largely to a progressive decline of the rate of cell degradation of the hormone. Thyroid secretion rate is also reduced with aging because serum thyroxine levels remain nearly constant. The concentration of thyroxine-binding globulin is minimally increased with aging.[168,169] The thyroxine distribution space decreases with age after the 6th decade of life.[166] This may be a result of the decrease of metabolic mass with age.

Triiodothyronine concentrations decrease with aging.[162,163,167] A decrease in extrathyroidal thyroxine deiodination may contribute to the reduced metabolism of thyroxine.

The effect of aging on thyrotropin response to thryotropin-releasing hormone in humans is not clear. Studies by one group of researchers reported a decline in old men but not in old women.[169,170] Another group of researchers found no age-related effects in a group of mixed sexes.[164] However, others found an increased response in the old age group of mixed sexes.[164] The levels of thyroid-stimulating hormone are found to be in the normal range in elderly subjects.[167]

4. Parathyroid

Serum parathyroid hormone concentration in men is increased from age 20 to 50, and then progressively decreased through the age of 90.[171,172] However, serum parathyroid hormone concentration in women is decreased from 20 to 49 years of age and then increased in older ages.[172] Estrogen appears to decrease the sensitivity of bone to parathyroid hormone.[173] The increased blood level of parathyroid hormone and increased sensitivity to parathyroid hormone in estrogen-lacking post-menopausal women may contribute to the high incidence of osteoporosis in elderly women.

E. Drugs

The functions of most micronutrients depend upon their involvement in various enzymatic processes required for normal metabolic pathways. Drugs modifying or interfering with the normal participation of nutrients in enzymatic reactions may have profound effects on intermediary metabolism. For example, aminopterin, like methotrexate, is a folic acid antagonist. It acts by preventing the formation of tetrahydrofolic acid from folic acid by inhibiting folic acid reductase.[174] Aminopterin is a useful chemotherapeutic agent in malignant diseases because it depresses DNA synthesis and inhibits cell division. The use of aminopterin results in hepatic, hematologic, and neurologic manifestations of folate deficiency. (See Chapter 1, Volume II for further discussion of the effect of drugs on intermediary metabolism.)

F. Alcoholism

Alcohol has a primary toxic effect on the liver. Alcoholism produces fatty liver, hepatitis, and eventually cirrhosis of the liver. Alterations of lipid, carbohydrate, and protein metabolism occur in the liver damaged by alcoholism. (See Chapter 2, Volume II for detailed discussion of the effect of alcoholism on intermediary metabolism.)

V. SUMMARY

Aging is accompanied by a general decline in physiological functions. The rate of functional decline differs among individuals and among the various body systems. Some of the possible factors contributing to biological aging are influenced by the nutritional status of the individual.

The nutritional status of the elderly is affected by many biomedical factors. The causes of lowered food intake are multiple. Anorexia is common in the elderly and can be caused by a decline in taste and smell capacity, constipation, malnutrition, excessive alcohol consumption, drugs and other therapy, chronic or acute diseases, and/or post-gastrectomy syndromes. Dental problems also affect food intake. Certain diseases can cause difficulties in food ingestion by the patient. Elderly people with chronic diseases are placed on special therapeutic diet(s) which may cause low food intake and dietary inadequacy. Physical disabilities such as poor vision, poor coordination, paralysis, lack of mobility, dementia, and/or actual physical confinement affect the ability of the elderly to shop and select food, to prepare foods, and sometimes, even to eat foods.

There is an age-related impairment in the general digestive and absorptive capacity without manifest GI disorder. However, this impairment may not be clinically significant, and the GI tract generally maintains an adequate level of functioning throughout life. Alteration or decrease in functions and change in anatomical structure may cause some digestive and absorptive problems, but major problems arise from diseases, medication, and/or alcoholism.

Aging-related changes in the oral cavity and GI system include declines in salivary secretion, salivary amylase concentration, and gastric hydrochloric acid secretion in the parietal cells. In addition, gastric secretion of pepsin and intrinsic factor, as well as the secretion of pancreatic amylase and trypsin, also decrease with aging. The mucosal absorptive surface area may be reduced with aging and cause absorptive insufficiency. The decrease in proliferation rate in mucosal epithelial cells may also contribute to absorptive insufficiency in old age.

Body composition changes with aging. Adipose tissue increases, whereas lean body mass and water content both decrease. As a result, the basal metabolic rate declines. Increased adiposity with aging results in a decreased ability to respond to insulin challenge and in altered lipid metabolism which leads to a higher incidence of cardiovascular disease and diabetes mellitus. A decrease in bone mineral results in a high incidence of osteoporosis among the elderly.

Aging results in changes in the physicochemical properties of protein, in some enzyme levels, in concentrations of some hormones, and in the ability of some endocrine control systems to respond to stress or stimulation. These changes influence intermediary metabolism. In addition, alcoholism and some drugs may exert a significant effect on intermediary metabolism.

REFERENCES

1. **Dunnill, M. S. and Halley, W.,** Some observations of the quantitative anatomy of the kidney, *J. Pathol.*, 110, 113, 1973.

2. **Darmady, E. M., Offer, J., and Woolhouse, M. A.,** The parameters of the aging kidney, *J. Pathol.,* 109, 195, 1973.

3. **Shock, N. W.,** Physiologic aspects of aging, *J. Am. Diet. Assoc.,* 56, 491, 1970.

4. **Wesson, L. G., Jr.,** Renal hemodynamics in physiological states, in *Physiology of the Human Kidney,* Wesson, L. G., Jr., Ed., Grune & Stratton, New York, 1969, 98.

5. **Lewis, W. H., Jr. and Alving, A. S.,** Changes with age in the renal function in adult men, *Am. J. Physiol.,* 123, 500, 1938.

6. **Linderman, R. D., Lee, T. D., Jr., Yiengst, M. J., and Shock, N. W.,** Influence of age, renal disease, hypertension, diuretics, and calcium on the antidiuretic responses to suboptimal infusions of vasopressin, *J. Lab. Clin. Med.,* 68, 206, 1966.

7. **Brandfonbrener, M., Landowne, M., and Shock, N. W.,** Changes in cardiac output with age, *Circulation,* 12, 557, 1955.

8. **Julius, S., Amery, A., Whitlock, L. S., and Conway, J.,** Influence of age on the hemodynamic response to exercise, *Circulation,* 36, 222, 1967.

9. **Bader, H.,** Dependence of wall stress in the human thoracic aorta on age and pressure, *Circ. Res.,* 20, 354, 1967.

10. **Gozna, E. R., Marble, A. E., Shaw, A., and Holland, J. G.,** Age related changes in the mechanics of the aorta and pulmonary artery of man, *J. Appl. Physiol.,* 36, 407, 1974.

11. **Muiesan, G., Sorbini, C. A., and Grassi, V.,** Respiratory function in the aged, *Bull. Physio-Pathol. Respir.,* 7, 973, 1971.

12. **Drinkwater, B. L., Horvath, S. M., and Wells, C. L.,** Aerobic powers of females, ages 10 to 68, *J. Gerontol.,* 30, 385, 1975.

13. **Chebotarev, D. F., Korkushko, O. V., and Ivanov, L. A.,** Mechanism of hypoxia in the elderly, *J. Gerontol.,* 29, 393, 1974.

14. **Brody, H.,** Aging of the vertebrate brain, in *Development and Aging in the Nervous System,* Academic Press, New York, 1973, 121.

15. **Tenney, S. M. and Miller, R. M.,** Deadspace ventilation in old age, *J. Appl. Physiol.,* 9, 321, 1956.

16. **Cohn, J. E. and Donoso, H. D.,** Mechanical properties of lung in normal men over 60 years old, *J. Clin. Invest.,* 42, 1406, 1963.

17. **Birren, E. J.,** in *The Physiology of Aging,* Prentice-Hall, Englewood Cliffs, N. J., 1964.

18. **Gutmann, E. and Hanzlikova, V.,** Basic mechanisms of aging in the neuromuscular system, *Mech. Ageing Dev.,* 1, 327, 1972/1973.

19. **Campbell, N. J., McComas, A. J., and Petito, F.,** Physiological changes in aging muscles, *J. Neurol. Neurosurg. Psychiatry,* 36, 174, 1973.

20. **Inokuchi, S., Ishikawa, H., Iwamoto, S., and Kimura, T.,** Age-related changes in the histological composition of the rectus abdominus muscle of the adult human, *Hum. Biol.,* 47, 231, 1975.

21. **Curtis, H.,** *Biological Mechanisms of Aging,* Charles C Thomas, Springfield, Ill., 1966.

22. **Lawton, A. H.,** Introductory remarks, in *Trace Elements in Aging,* Hsu, J. M., Davis, R. L., and Neithamer, R. W., Eds., Eckerd College, St. Petersburg, Fla., 1976, 3.

23. **Mochizuki, Y.,** Papilla foliata of Japanese, *Folia Anal. Jpn.,* 18, 337, 1939.

24. **Allara, E.,** Investigations on the human taste organ. I. The structure of taste papillae at various ages, *Arch. Ital. Anat. Embriol.,* 42, 506, 1939.

25. **Arey, L. B., Tremaine, M. J., and Monzingo, F. L.,** The numerical and topographical relations of taste buds to human circumvallate papillae throughout the life span, *Anat. Rec.,* 64(1), 1936.

26. **Kamath, S. K.,** Taste acuity and aging, *Am. J. Clin. Nutr.,* 36, 766, 1982.

27. **Cooper, R. M., Bilash, I., and Zubek, J. P.,** The effect of age on taste sensitivity, *J. Gerontol.,* 14, 56, 1959.

28. **Richter, C. P. and Campbell, K. H.,** Sucrose taste thresholds of rats and humans, *Am. J. Physiol.,* 128, 291, 1940.

29. **Grzegorczyk, P. B., Jones, S. W., and Mistretta, C. M.,** Age-related differences in salt taste acuity, *J. Gerontol.,* 34, 834, 1979.

30. **Ramsey, W. O.,** Nutritional problems of the aged, *J. Prosthet. Dent.,* 49, 16, 1983.

31. **Liss, L. and Gomez, F.,** The nature of senile changes of the human olfactory bulb and tract, *Arch. Otolaryngol.,* 67, 167, 1958.

32. **Smith, C. G.,** Age incidence of atrophy of olfactory nerves in man, *J. Comp. Neurol.,* 77, 589, 1942.

33. **Kimbrell, G. McA. and Furchtgott, E.,** Effect of aging on olfactory threshold, *J. Gerontol.,* 18, 364, 1963.

34. **Schiffman, S. S., Moss, J., and Erickson, R. P.,** Thresholds of food odors in the elderly, *Exp. Aging Res.,* 2, 389, 1976.

35. **Massler, M.,** Geriatric nutrition: the role of taste and smell in appetite, *J. Prosthet. Dent.,* 43, 247, 1980.

36. **Shank, R. E.,** Nutritional characteristics of the elderly — an overview, in *Nutrition, Longevity and Aging,* Rockstein, M. and Sussman, M. L., Eds., Academic Press, New York, 1976, 23.

37. **Sklar, M.,** Functional bowel distress and constipation in the aged, *Geriatrics,* 27(9), 79, 1972.
38. **Gershaff, S. N.,** The role of vitamins and minerals in taste in the chemical senses and nutrition, in *The Chemical Senses and Nutrition,* Kare, M. R. and Mallen, O., Eds., Academic Press, New York, 1977, 201.
39. **Hambidge, K. M., Hambidge, C., Jacobs, M., and Baum, J. D.,** Low levels of zinc in hair: anorexia, poor growth and hypogeusia in children, *Pediatr. Res.,* 6, 868, 1972.
40. **Roe, D. A.,** *Geriatric Nutrition,* Prentice-Hall, Englewood Cliffs, N.J., 1983, 84.
41. **Cooper, J. W.,** Food-drug interactions, *U.S. Pharm.,* December/November, 16, 1976.
42. **Roberts, N. J.,** Dietary management of the cancer patient, *Nutr. Cancer,* 1, 107, 1979.
43. **Roe, D. A.,** *Drug Induced Nutritional Deficiencies,* AVI, Westport, Conn., 1978, 149.
44. **Roe, D. A.,** *Geriatric Nutrition,* Prentice-Hall, Englewood Cliffs, N.J., 1983, 79.
45. **Friedman, J. W.,** Dentistry in the geriatric patient: mutilation by consensus, *Geriatrics,* 23(8), 98, 1968.
46. **Kohrs, M. B., O'Neal, R., Preston, A., Eklund, D., and Abrahams, O.,** Nutritional status of elderly residents in Missouri, *Am. J. Clin. Nutr.,* 31, 2186, 1978.
47. **Kaplan, H.,** The oral cavity in geriatrics, *Geriatrics,* 26, 96, 1971.
48. **Wical, K.,** Common sense dietary recommendations for geriatric patients, *J. Prosthet. Dent.,* 49, 162, 1983.
49. **Floch, M. H.,** Nutritional problems following gastric and small bowel surgery, in *Nutrition and Diet Therapy in Gastrointestinal Disease,* Plenum Press, New York, 1981, 151.
50. **Roe, D. A.,** *Geriatric Nutrition,* Prentice-Hall, Englewood Cliffs, N.J., 1983, 82.
51. **Shank, R. E.,** Nutritional characteristics of the elderly-an overview, in *Nutrition, Longevity, and Aging,* Rockstein, M. and Sussman, M. L., Eds., Academic Press, New York, 1976, 15.
52. **Rivlin, R. S.,** Nutrition and aging: some unanswered questions, *Am. J. Med.,* 71, 337, 1981.
53. **Shear, J.,** A geriatrician looks at the dental problems of the aged, *J. Md. Dent. Assoc.,* 10, 19, 1967.
54. **Meyer, J., Spier, E., and Neuwelt, F.,** Basal secretion of digestive enzymes in old age, *Arch. Intern. Med.,* 65, 171, 1940.
55. **Becks, H. and Wainwright, W. W.,** Human saliva. VIII. Rate of flow of resting saliva of healthy individuals, *J. Dent. Res.,* 22, 391, 1943.
56. **Soergel, K. H., Zboralske, F., and Amberg, J. R.,** Presbyesophagus: esophageal motility in nonagenarians, *J. Clin. Invest.,* 43, 1472, 1964.
57. **Bhanthumnavin, K. and Schuster, M. M.,** Aging in gastrointestinal function, in *Handbook of the Biology of Aging,* Finch, C. E. and Hayflick, L., Eds., Van Nostrand Reinhold, New York, 1977, 709.
58. **Mandelstam, P. and Lieber, A.,** Cineradiographic evaluation of the esophagus in normal adults: a study of 146 subjects ranging in age from 21 to 90 years, *Gastroenterology,* 58, 32, 1970.
59. **Hollis, J. B. and Castell, D. O.,** Esophageal function in elderly man. A new look at presbyesophagus, *Ann. Intern. Med.,* 80, 371, 1974.
60. **Farrell, R., Castell, D., and McGugan, J.,** Measurements and comparisons of lower esophageal sphincter pressures and serum gastrin levels in patients with gastro-esophageal reflux, *Gastroenterology,* 67, 415, 1974.
61. **Levin, E., Kirsner, J. B., and Palmer, W. L.,** A simple measure of gastric secretion in man: comparison of one hour basal secretion, histamine secretion and 12 hour nocturnal gastric secretion, *Gastroenterology,* 19, 88, 1951.
62. **Davies, D. and James, T. J. I.,** Investigation into gastric secretion of 100 normal persons over age of 60, *Q. J. Med.,* 24, 1, 1930.
63. **Vanzant, F. R., Alvarez, W. C., Eusterman, G. B., Dunn, H. L., and Berkson, J.,** The normal range of gastric acidity from youth to old age, *Arch. Intern. Med.,* 49, 345, 1932.
64. **Bloomfield, A. L. and Keeper, C. S.,** Gastric motility and the volume of gastric secretion in man, *J. Clin. Invest.,* 5, 295, 1928.
65. **Baron, J. H.,** Studies of basal and peak acid output with an augmented histamine test, *Gut,* 4, 136, 1963.
66. **Meyer, J. and Necheles, H.,** Studies in old age. IV. The clinical significance of salivary, gastric, and pancreatic secretion in the aged, *JAMA,* 115, 2050, 1940.
67. **Andrews, G. R., Haneman, B., Arnold, B. J., Booth, J. C., and Taylor, K.,** Atrophic gastritis in the aged, *Australas. Ann. Med.,* 16, 230, 1967.
68. **Coghill, N. F., Doniach, D., Roitt, I. M., Mollin, D. L., and Williams, A. W.,** Autoantibodies in simple atrophic gastritis, *Gut,* 6, 48, 1965.
69. **Rosenberg, I. R., Frieland, N., Janowitz, H. D., and Dreiling, D. A.,** The effect of age and sex upon human pancreatic secretion of fluid and bicarbonate, *Gastroenterology,* 50, 191, 1966.
70. **Lesher, S.,** Chronic irradiation and aging in mice and rats, in *Radiation and Aging,* Lindopp, P. and Sacher, G., Eds., Taylor & Francis, London, 1966, 183.
71. **Lesher, S., Fry, R. J. M., and Kohn, H. I.,** Age and the generation time of the mouse duodenal epithelial cell, *Exp. Cell Res.,* 24, 334, 1961.

72. **Lesher, S., Fry, R. J. M., and Kohn, H. I.,** Aging and the generation cycle of intestinal epithelial cells in the mouse, *Gerontologia,* 5, 176, 1961.

73. **Fry, R. J. M., Lesher, S., and Kohn, H. I.,** Age effect on cell-transit time in mouse jejunal epithelium, *Am. J. Physiol.,* 201, 213, 1961.

74. **Cameron, I. L.,** Cell proliferation and renewal in aging mice, *J. Gerontol.,* 27, 162, 1972.

75. **Suntzeff, V. and Angeletti, P.,** Histological and histochemical changes in intestines of mice with aging, *J. Gerontol.,* 16, 226, 1961.

76. **Lascalea, M. C.,** The digestive system in old age, *Excerpta Med.,* 2, 419, 1959.

77. **Moog, F.,** The small intestine in old mice: growth, alkaline phosphatase and disaccharidase activities, and deposition of amyloid, *Exp. Gerontol.,* 12, 223, 1977.

78. **Hohn, P., Gabbert, H., and Wagner, R.,** Differentiation and aging of the rat intestinal mucosa, *Mech. Ageing Dev.,* 7, 217, 1978.

79. **Webster, S. G. P. and Leeming, J. T.,** The appearance of the small bowel mucosa in old age, *Age Ageing,* 4, 168, 1975.

80. **Warren, P. M., Pepperman, M. A., and Montogomery, R. D.,** Age changes in small-intestinal musoca, *Lancet,* 2, 849, 1978.

81. **Berman, P. M. and Kirsner, J. B.,** The aging gut. I. Diseases of the esophagus, small intestine, and appendix, *Geriatrics,* 27(3), 84, 1972.

82. **Roe, D. A.,** *Geriatric Nutrition,* Prentice-Hall, Englewood Cliffs, N.J., 1983, 18.

83. **Farrell, R. L., Castell, D. O., and McGuigan, J. E.,** Measurements and comparisons of lower esophageal sphincter pressures and serum gastrin levels in patients with gastro-esophageal reflux, *Gastroenterology,* 67, 415, 1974.

84. **Hebbel, R.,** The topography of chronic gastritis in otherwise normal stomachs, *Am. J. Pathol.,* 25, 125, 1949.

85. **Andrews, G., Haneman, B., Arnold, B., Booth, J., and Taylor, K.,** Atrophic gastritis in the aged, *Australas. Ann. Med.,* 16, 230, 1967.

86. **King, E. S. J.,** Diverticula of the small intestine, *Aust. N.Z. J. Surg.,* 19, 301, 1950.

87. **Grant, J. C. B.,** On the frequency and age in incidence of duodenal diverticula, *Can. Med. Assoc. J.,* 33, 258, 1935.

88. **Holt, P. R.,** Effects of aging upon intestinal absorption, in *Nutritional Approaches to Aging Research,* Moment, G. B., Ed., CRC Press, Boca Raton, Fla., 1982, 170.

89. **Holt, P. R.,** *Malabsorption,* Ross Laboroaties, Columbus, Oh., 1977.

90. **Faloon, W. W., Fisher, C. J., and Duggan, K. C.,** Occurrence of a sprue-like syndrome during neomycin therapy, *J. Clin. Invest.,* 37, 893, 1958.

91. **Jacobson, E. D. and Faloon, W. W.,** Malabsorptive effects of neomycin in commonly used doses, *JAMA,* 175, 187, 1961.

92. **Faloon, W. W., Paes, I. C., Woolfolk, D., Nankin, H., Wallace, K., and Haro, E. N.,** Effect of neomycin and kanamycin upon intestinal absorption, *Ann. N.Y. Acad. Sci.,* 132, 879, 1966.

93. **Mehta, S. K., Weser, E., and Sleisenger, M. H.,** The in vitro effect of bacterial metabolites and antibiotics on pancreatic lipase activity, *J. Clin. Invest.,* 43, 1252, 1964.

94. **Schuckit, M. A. and Pastor, P. A.,** The elderly as an unique population: alcoholism, *Clin. Exp. Res.,* 2, 31, 1978.

95. **Myerson, R. M.,** Metabolic aspects of alcohol and their biological significance, *Med. Clin. N. Am.,* 57, 925, 1973.

96. **Klipstein, F. A., Berlinger, F. G., and Juden Reed, L.,** Folate deficiency associated with drug therapy for tuberculosis, *Blood,* 29, 697, 1967.

97. **Fryer, J. H.,** Studies of body composition in men aged 60 and over, in *Biological Aspects of Aging,* Shock, N. W., Ed., Columbia University Press, New York, 1962, 59.

98. **Parizkova, J.,** Body composition and lipid metabolism, *Proc. Nutr. Soc.,* 32, 181, 1973.

99. **Brozek, J.,** Changes of body composition in man during maturity and their nutritional implications, *Fed. Proc. Fed. Am. Soc. Exp. Biol.,* 11, 784, 1952.

100. **Myhre, L. G. and Kessler, W. V.,** Body density and potassium 40: measurements of body composition as related to age, *J. Appl. Physiol.,* 21, 1251, 1966.

101. **Forbes, G. B. and Reina, J. C.,** Adult lean body mass declines with age; some longitudinal observations, *Metabolism,* 19, 653, 1970.

102. **Novak, L.,** Aging, total body potassium, fat-free mass, and cell mass in males and females between ages 18 and 85 years, *J. Gerontol.,* 27, 438, 1972.

103. **Pierson, R. M., Lin, D. H. Y., and Phillips, R. A.,** Total body potassium in health; effects of age, sex, height, and fat, *Am. J. Physiol.,* 226, 206, 1974.

104. **Simms, H. S. and Stolman, A.,** Changes in human tissue electrolyte in senescence, *Science,* 86, 269, 1937.

105. **Keys, A., Taylor, H. L., and Grande, F.,** Basal metabolism and age of adult man, *Metabolism,* 22, 579, 1973.
106. **Tzankoff, S. P. and Norris, A. H.,** Effect of muscle mass decrease on age-related BMR changes, *J. Appl. Physiol.,* 43, 1001, 1977.
107. **Reigle, G. D. and Miller, A. E.,** Aging effects on hormone concentrations and actions, in *CRC Handbook of Biochemistry in Aging,* Florini, J. R., Ed., CRC Press, Boca Raton, Fla., 1981, 247.
108. **Feldman, S. A. and Glagov, S.,** Transmedial collagen and elastin gradients in human aortas: reversal with age, *Atherosclerosis,* 13, 385, 1971.
109. **Schaub, M. C.,** Qualitative and quantitative changes of collagen in parenchymatous organs of the rat during aging, *Gerontologia,* 8, 114, 1963.
110. **Briscoe, A. M. and Loring, W. E.,** Elastin content of the human lung, *Proc. Soc. Exp. Biol. Med.,* 99, 162, 1958.
111. **Medvedev, Z. A.,** Changes in proteins and nucleic acids occurring with aging and the problems of aging at the molecular level, in *Protein Biosynthesis,* Plenum Press, New York, 1966, 469.
112. **Sasaki, R., Ichikawa, S., Yamagiwa, H., Ito, A., and Yanagata, S.,** Aging and hydroxyproline content in human heart muscle, *Tohoku J. Exp. Med.,* 118, 11, 1976.
113. **Fujii, K., Yoshinori, K., and Sasaki, S.,** Aging of human bone and articular cartilage collagen: changes in the reducible cross-links and their precursors, *Gerontology,* 22, 363, 1976.
114. **Kohn, R. A. and Rollerson, E. J.,** Studies on the effect of heat and age in decreasing ability of human collagen to swell in acid, *J. Gerontol.,* 14, 11, 1959.
115. **Hamlin, C. R. and Kohn, R. R.,** Evidence for progressive age-related structural changes in post mature human collagen, *Biochem. Biophys. Acta,* 236, 458, 1971.
116. **Zwolinski, R. J., Hamlin, C. R., and Kohn, R. R.,** Age-related alteration in human heart collagen, *Proc. Soc. Exp. Biol. Med.,* 152, 362, 1976.
117. **LaBella, F. S. and Paul, G.,** Structure of collagen from human tendon as influenced by age and sex, *J. Gerontol.,* 20, 54, 1965.
118. **John, R. and Thomas, J.,** Chemical composition of elastins isolated from aortas and pulmonary tissues of different ages, *Biochem. J.,* 127, 261, 1972.
119. **Stidworthy, G., Masters, Y. F., and Shetlar, M. R.,** The effect of aging on mucopolysaccharide composition of human costal cartilage as measured by hexosamine and uronic acid content, *J. Gerontol.,* 13, 10, 1958.
120. **Hjertquist, S. O. and Lemperg, R.,** Identification and concentration of the glycosaminoglycans of human articular cartilage in relation to age and osteoarthritis, *Calcif. Tissue Res.,* 10, 223, 1972.
121. **Buddecke, E. and Ruff-Lichtenstein, E.,** Anionic polysaccharides and glycoproteins of human lung connective tissue in relation to age and silicosis, *Beitr. Silikose-Forsch.,* Suppl. 5, 173, 1963.
122. **Clausen, B.,** Influence of age on connective tissue: hexosamine and hydroxyproline in human aorta, myocardium, and skin, *Lab. Invest.,* 11, 229, 1962.
123. **Saltzman, H. A., Sieker, H. O., and Green, J.,** Hexosamine and hydroxyproline content in human bronchial cartilage from aged and diseased lungs, *J. Lab. Clin. Med.,* 62, 78, 1963.
124. **Bourne, G. H.,** Structure changes in aging, in *Aging: Some Social and Biological Aspects,* Shock, N. W., Ed., American Association for the Advancement of Science, Washington, D.C., 1960, 123.
125. **White, A.,** Some biochemical aspects of aging, in *Aging: Some Social and Biological Aspects,* Shock, N. W., Ed., American Association for the Advancement of Science, Washington, D.C., 1960, 137.
126. **Robinson, D. S.,** Changes in monoamine oxidase and monamines with human development and aging, *Fed. Proc. Fed. Am. Soc. Exp. Biol.,* 34, 105, 1975.
127. **Fitzpatrick, L. T., Brunet, P., and Kukita, A.,** The nature of hair pigment, in *Biology of Hair Growth,* Montague, W. and Ellis, R. A., Eds., Academic Press, New York, 1958, 255.
128. **Ross, M. H.,** Aging, nutrition and hepatic enzyme activity patterns in the rats, *J. Nutr.,* 97, 565, 1969.
129. **Zorzoli, A.,** The influence of age on phosphatase activity in the liver of the mouse, *J. Gerontol.,* 10, 156, 1955.
130. **Ellens, A. and Wattiaux, R.,** Age-correlated changes in lysosomal activities: an index of aging, *Exp. Gerontol.,* 4, 131, 1969.
131. **Wilson, P. D.,** Enzyme patterns in young and old mouse livers and lungs, *Gerontologia,* 18, 36, 1972.
132. **Barrows, C. H., Roeder, L. M., and Falzone, J. A.,** Effect of age on the activities of enzymes and the concentrations of nucleic acids in the tissues of female wild rats, *J. Gerontol.,* 17, 144, 1962.
133. **Britton, V. J., Sherman, F. G., and Florini, J. R.,** Effect of age on RNA synthesis by nuclei and soluble RNA polymerases from liver and muscle of C57BL/6J mice, *J. Gerontol.,* 27, 188, 1972.
134. **Kurnick, N. B. and Kernen, R. L.,** The effect of aging on the deoxyribose nuclease system, body and organ weight and cellular content, *J. Gerontol.,* 17, 245, 1962.
135. **Knock, D. L. and Sleyster, E. C.,** Lysosomal enzyme activities in parenchymal and nonparenchymal liver cells isolated from young, adult and old rats, *Mech. Ageing Dev.,* 5, 389, 1976.

136. **Wilson, P. D., Hill, B. T., and Franks, L. M.,** The effect of age on mitochondrial enzymes and respiration, *Gerontologia,* 21, 95, 1975.

137. **Brouwer, A., van Bezooijen, C. F. A., and Knook, D. L.,** Respiratory activities of hepatocytes isolated from rats of various ages. A brief note, *Mech. Ageing Dev.,* 6, 265, 1977.

138. **Gold, P. H., Gee, M. V., and Strehler, B. S.,** Effect of age on oxidative phosphorylation in the rat, *J. Gerontol.,* 23, 509, 1968.

139. **Wang, R. K. J. and Mays, L. L.,** Opposite changes in rat liver glucose-6-phosphate dehydrogenase during aging in Sprague-Dawley and Fisher 344 male rats, *Exp. Gerontol.,* 12, 117, 1977.

140. **Finch, C. E.,** Enzyme activities, gene function and aging in mammals (review), *Exp. Gerontol.,* 7, 53, 1972.

141. **Wilson, P. D.,** Enzyme changes in aging mammals, *Gerontologia,* 19, 79, 1973.

142. **Adelman, R. C.,** An age-dependent modification of enzyme regulation, *J. Biol. Chem.,* 245, 1032, 1970.

143. **Singhal, R. L.,** Effect of age on the induction of glucose-6-diphosphatase in rat liver, *J. Gerontol.,* 22, 77, 1967.

144. **Haining, J. L. and Correll, W. W.,** Turnover of tryptophan induced tryptophan pyrrolase in rat liver as a function of age, *J. Gerontol.,* 24, 143, 1969.

145. **Albeaux-Fernet, M., Bohler, C. C. S., and Karpas, A. E.,** Testicular function in the aging male, in *Geriatric Endocrinology,* Greenblatt, R. B., Ed., Raven Press, New York, 1978, 201.

146. **Vermeulen, A.,** The hormonal activity of the postmenopausal ovary, *J. Clin. Endocrinol. Metab.,* 42, 247, 1976.

147. **Blichert-Toft, M.,** The adrenal glands in old age, in *Geriatric Endocrinology,* Greenblatt, R. B., Ed., Raven Press, New York, 1978, 81.

148. **Reichel, W.,** Lipofuscin pigment accumulation and distribution in five rat organs as a function of age, *J. Gerontol.,* 23, 145, 1968.

149. **Dobbie, J. W.,** Adrenocortical nodular hyperplasia: the aging adrenal, *J. Pathol.,* 99, 1, 1969.

150. **Romanoff, L. P., Morris, C. W., Welch, P., Rodriguez, R. M., and Pincus, G.,** The metabolism of cortisol-4-C^{14} in young and elderly men. I. Secretion rate of cortisol and daily excretion of tetrahydrocortisol, allotetrahydrocortisol, tetrahydrocortisone and cortolone (20α and 20β), *J. Clin. Endocrinol. Metab.,* 21, 1413, 1961.

151. **Friedman, M., Green, M. F., and Sharland, D. E.,** Assessment of hypothalmic-pituitary-adrenal function in the geriatric age group, *J. Gerontol.,* 24, 292, 1969.

152. **Hess, G. D. and Riegle, G. D.,** Effect of chronic ACTH stimulation on adrenocortical function in young and aged rats, *Am. J. Physiol.,* 222, 1458, 1972.

153. **Riggle, G. D.,** Chronic stress effects on adrenocortical responsiveness in young and aged rats, *Neuroendocrinology,* 11, 1, 1973.

154. **Flood, C., Gherondache, C., Pincus, G., Tait, J. F., Tait, S. A. S., and Willoughby, S.,** The mechanism and secretion of aldosterone in elderly subjects, *J. Clin. Invest.,* 46, 960, 1967.

155. **Crane, M. G. and Harris, J. J.,** Effect of aging on renin activity and urinary aldosterone levels, Program 55th Annu. Meet. Endocr. Soc., A-74(Abstract 51), 1973.

156. **Weidmann, P., Myttenaere-Bursztein, S., Maxwell, M. H., and Lima, J.,** Effect of aging on plasma renin and aldosterone in normal man, *Kidney Int.,* 8, 325, 1975.

157. **Andres, R. and Tobin, J. D.,** Aging and disposition of glucose: explorations in aging, *Adv. Exp. Med. Biol.,* 61, 239, 1974.

158. **Dudl, R. J. and Ensinck, J. W.,** The role of insulin, glucagon and growth hormone in carbohydrate homeostasis during aging, *Diabetes,* 21(Suppl. 1), 357, 1972.

159. **Burgess, J. A.,** Diabetes mellitus in aging, in *Hypothalamus, Pituitary and Aging,* Everitt, A. V. and Burgess, J. A., Eds., Charles C Thomas, Springfield, Ill., 1976, 497.

160. **Chlouverakis, C., Jarrett, R. J., and Keen, H.,** Glucose tolerance, age, and circulating insulin, *Lancet,* 1, 806, 1967.

161. **Kilo, C., Vogler, N., and Williamson, J. R.,** Muscle capillary basement membrane changes related to aging and to diabetes mellitus, *Diabetes,* 21, 881, 1972.

162. **Hermann, J., Rusche, H. J., Kroll, H. J., Hilger, P., and Krushkemper, H. L.,** Free triiodothyronine (T3) and thyroxine (T4) serum levels in old age, *Horm. Metab. Res.,* 6, 239, 1974.

163. **Bermudex, F., Surks, M. I., and Oppenheimer, J. H.,** High incidence of decreased serum triiodothyronine concentration in patients with nonthyroidal disease, *J. Clin. Endocrinol. Metab.,* 41, 27, 1975.

164. **Ohara, H., Kobayaski, T., Shiraishi, M., and Wada, T.,** Thyroid function of the aged as viewed from the pituitary-thyroid system, *Endocrinol. Jpn.,* 21, 377, 1974.

165. **Gaffney, G. W., Gregerman, R. I., and Shock, N. W.,** Relationship of age to the thyroidal accumulation, renal excretion, and distribution of radioiodide in euthyroid man, *J. Clin. Endocrinol. Metab.,* 22, 784, 1962.

166. **Gregerman, R. I., Gaffney, G. W., and Shock, N. W.,** Thyroxine turnover in euthyroid man with special reference to changes with age, *J. Clin. Invest.,* 41, 2065, 1962.

167. **Hesch, R. D., Gatz, J., Pape, J., Schmidt, E., and von zur Muhlen, A.,** Total and free triiodothyronine and thyroid-binding globulin concentrations in elderly human persons, *Eur. J. Clin. Invest.,* 6, 139, 1976.
168. **Braverman, L. E., Dawber, N. A., and Ingbar, S. H.,** Observations concerning the binding of thyroid hormones in sera of normal subjects of varying ages, *J. Clin. Invest.,* 45, 1273, 1966.
169. **Snyder, P. J. and Utiger, R. D.,** Response to thyrotropin releasing hormone (TRH) in normal man, *J. Clin. Endocrinol. Metab.,* 34, 380, 1972.
170. **Snyder, P. J. and Utiger, R. D.,** Thyrotropin response to thyrotropin releasing hormone in normal females over forty, *J. Clin. Endocrinol. Metab.,* 34, 1096, 1972.
171. **Fujita, T., Orimo, H., Okano, K., Yoshikawa, M., Shimo, R., Inove, T., and Itami, Y.,** Radioimmunoassay of serum parathyroid hormone in post-menopausal osteoporosis, *Endocrinol. Jpn.,* 19, 571, 1972.
172. **Roof, B. S. and Gorden, G. S.,** Hyperparathyroid disease in the aged, in *Geriatric Endocrinology,* Greenblatt, R. B., Ed., Raven Press, New York, 1978, 33.
173. **Orimo, H., Fujita, T., and Yoshikawa, M.,** Increased sensitivity of bone to parathyroid hormone in ovariectomized rats, *Endocrinology,* 90, 760, 1972.
174. **Werkheiser, W. C.,** The biochemical, cellular and pharmacological action and the effects of the folic acid antagonists, *Cancer Res.,* 23, 1277, 1963.

Chapter 4

SOCIAL-PSYCHOLOGICAL ASPECTS OF NUTRITION AMONG THE ELDERLY

Part I

SOCIAL DIMENSIONS OF NUTRITION

Jon Hendricks and Toni M. Calasanti

TABLE OF CONTENTS

Part II

PSYCHOLOGICAL DIMENSIONS OF NUTRITION

Sandra E. Gibbs and Howard B. Turner

TABLE OF CONTENTS

I. INTRODUCTION

Food consumption is never merely a means of obtaining nutrition. Eating is a complex phenomenon which reflects myriad factors ranging across physical health, dentition, mood, sex, age, socioeconomic and family status, cultural background, or experience to the variability in time and circumstances of nutritional intake. Obviously, each of these factors, and many other social-psychological variables as well, undergo change as individuals move through the life cycle. To appreciate the current nutritional status of older persons it is essential to understand how their present situation and the entire package of their life experiences are reflected in their eating habits. Like so many other facets of our lives, eating behavior is a combination of expediencies, learned preferences, and opportunity. In turn, nutritional status will have a reciprocal effect on health and social and psychological functioning during all phases of life, particularly in the later years. A behavioral perspective on food and eating is an attempt to explicate just how this reciprocal relationship works and, in the case of gerontology, how it affects the quality of life among the elderly.[1]

The intent of Part I of this chapter is to review a number of the findings regarding the nutritional status of older persons available in the social gerontology literature. We begin with a brief overview of the demographic profile of the elderly in America. To fully appreciate the demands on nutritional specialists it is necessary to have an appreciation of the changing parameters of the target population. Included in this review is a brief summary of the economic and health factors relevant to eating behavior and any interventions which may be undertaken. Next, we look at some of the possible social factors influencing the nutritional status of older persons. From there the discussion will focus more particularly on lifestyle as a mediator of nutritional intake. Finally, we will touch on steps which have been taken to ensure or maintain nutritional adequacy.

II. DEMOGRAPHIC PROFILE OF THE OLDER POPULATION

The importance of demographic characteristics lies in part in their implications for the mutability of the aging process itself. As long as variability exists, the implication is that improvements in general well being as well as nutrition are possible for those who lie in the lower range of the spread. At the same time, the variance itself makes facile generalizations problematic. Without taking sociocontextual factors into account, including dietary habits, adequate explanations will not be forthcoming. There is every reason to assume that like other sociocultural aspects of the aging process, the role of nutrient intake in morbidity, mortality, and other components of well being is affected by relative position in the social structure. As has been pointed out repeatedly, the consequences of demographic distribution and its concomitants extend into every area of life. The resources, health, and social service programs, as well as the priorities accorded the elderly in society reflect shifting population structures and the responsibilities assumed by a society for the care of particular age groups.

For purposes of the discussion here, the demographics of the older segments of the population suggest three interrelated issues. First, what are the characteristics of the older population? Have increases in life expectancy had a differential impact on subgroups of people in the further reaches of the life cycle? Second, what is the quality of life like for those living into their 80s and beyond? What can be anticipated in terms of vigor and morbidity? Third, how is nutrition implicated either in the gains in life expectancy or in the health profile of older people?

Now, and for at least the next 50 years, the number of people over the age of 65 will continue to be the most rapidly growing segment of the population. By the year 2025 up to approximately 19% of the American population will be 65 or older, a significant jump from the current 11 to 12%. The number of those over age 85 is growing faster still, at nearly twice the rate of the rest of the population. The reasons for this are obvious. Life expectancy, or the average length of time a person can expect to live, has been increasing throughout history. A male child born today can anticipate 70.7 years of life; a female infant can expect to live 78.3 years into the future. The gap between males and females has long been evident and most likely will widen to 10 or more years in the next 50 years. As interesting as global life expectancy is, age-specific life expectancy, the average number of years left at any given age, is even more noteworthy (Figure 1). In the 10-year period ending in 1978, mortality rates for those already age 65 declined approximately 1.5% for males but more than 2.3% for females. These trends translate into an average remaining life expectancy of some 18.4 years for white women at age 65 compared to 14 years for white men.[2,3] While there may well be a genetic basis for the male-female differential, a variety of sociocultural factors have also been found to be involved. As an example, males drive more than females; they also drink, smoke, and are more likely to try illicit drugs than are women. In spite of recent dramatic declines in mortality due to cardiovascular diseases, we know men do not seek medical attention as frequently as women nor have they benefitted from the medical breakthroughs that have ameliorated early-life mortality among females. Also, despite the heartfelt trauma which may accompany the death of a spouse, women do not experience bereavement mortality at a rate as high as that found among men.

Whatever the reasons, mortality differentials between men and women in the later years have grown to the point where what was a 22% female advantage in age-adjusted death rates in 1940 had reached 78% by 1978. As a consequence, women over age 65 outnumber their male age peers 3:2. The further into old age one moves, the greater the differentials become: between 65 and 74 there are 31% more women; in the next 10-year interval this number rises to 66%; after age 85 there are 224 women for every 100 men still alive. As a result, the nutritional, health, economic, and social issues of old age are primarily problems of women.[2,4]

There is also a differential pattern of life expectancy between whites, blacks (Figure 2), and other racial categories, although in the last decade the gap has been narrowing. At the beginning of the century, average life expectancy at birth was some 16 years greater for whites than for blacks, but by 1978 the difference had shrunk to about 5 years. In fact, since 1970, nonwhite females have surpassed white male life expectancy. In age-specific terms, a 65-year-old nonwhite woman has an average 18 years left to live while her nonwhite male counterpart has 14.1 years. As a matter of fact, a "cross-over" occurs after age 75, i.e., a higher death rate is observed among whites. It has been hypothesized that this is due in part to a "survival of the fittest" phenomenon. While the overall changes are difficult to interpret, most experts agree higher nonwhite fertility rates in the past, greater access to health care, and improving socioeconomic status, including nutrition, relative to whites are indicated as the most likely causes.[2,5]

Among the 23 million older persons in the U.S. in 1980, roughly 90% were white, another 8.2% were black, and the remainder were comprised of native Americans (including Eskimo and Aleut), Asian Americans, and those of Hispanic descent. The latter may be of any racial category and include people of Cuban, Mexican, South American, or Spanish origin. With a few notable exceptions, primarily Cubans, these nonwhite elderly constitute some of the most disadvantaged members of society. Various categories of ethnic elderly have experienced lifetimes of lower incomes, depressed educational backgrounds, and a higher prevalence of serious illnesses. In general, the hard-pressed nature of their living conditions suggests that the quality of life they encounter in old age is going to be lower on nearly all counts than that of white, mainstream elderly.[6]

A. Economic Circumstances of the Elderly

What do these demographic patterns portend for the economic position of the elderly in

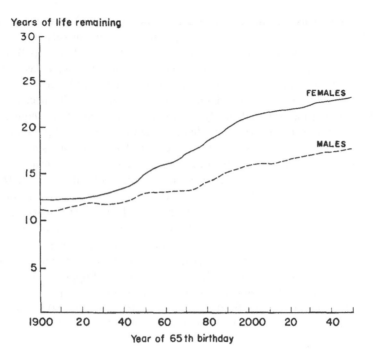

FIGURE 1. Life expectancy at age 65 for the years 1900 to 2050.[2]

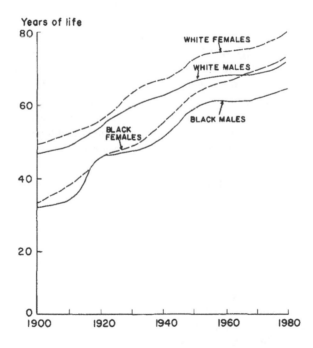

FIGURE 2. Expectation of life at birth by race and sex for the years 1900 to 1980.[2]

general? To begin with, it does not automatically mean greater demands will be placed on the working age population. While the proportion of the elderly will indeed expand, the aggregate dependent population being supported by those who are assumed to be "productive", and are between the ages of 18 and 64, is actually decreasing due to a declining birth rate. Current projections are that the total dependency ratio will continue to decline until the year 2020, when it will once again turn upward, but it is unlikely it will return to the peak of the 1960s and 1970s.[3] Still, the extensions in life expectancy spell longer periods of financial risk in the later years for individuals even though society as a whole is in a better position now to lend support to its elderly than it has ever been.[7]

The bulk of the elderly have limited sources of economic security because of inflation. They are thus among the most adversely affected, as the things they require for survival have been priced out of reach.[8] To claim that all citizens have an obligation to save for the "rainy days" of their old age ignores the structural changes in our society which affect financial well being. Societal pressures, technological changes, and educational expectations of the occupational realm have all contributed to the exodus of older workers from the labor force. So too, of course, have voluntary early retirement and self-employment opportunities. The net result is that among older men fewer than one fifth are still working after age 65, a dramatic decline from the one half of only 30 years ago. Even more surprising are the declines in the 55 to 64-year-old interval. According to recent statistics, unemployment in middle age and beyond is becoming an ever more likely prospect. Labor force participation of men between 55 and 64 years of age has been declining, dropping a total of 10% in the last decade. In fact, unemployment for those over the age of 55 is increasing more rapidly than for any other age group. By 1980 only 6 out of 10 men aged 60 to 64 were still part of the labor force. So precipitous has the shift become that the Commissioner of Labor Statistics for the New York federal region has suggested that the technological displacement accompanying structural changes in the nature of work is being settled disproportionately on male workers age 45 and over.[9] In fact, by the end of the 1970s almost two thirds of all new Social Security payments to men were awarded to those in the 62- to 65-year-old category.

The situation for women workers is somewhat better in terms of labor force participation rates. Among those over the age of 65 there has been a modest decline of 2% in the last 3 decades; now approximately 8% remain active. In the age interval between 55 and 64 those participating in the labor force have increased from 27% in 1950 to 42% in 1982. While the number of older woman workers has grown dramatically, they are employed primarily in the lower-paying rungs of the tertiary and service sectors, where they earn only marginally above the minimum wage. Like men, women are increasingly making early claims on Social Security; today roughly four fifths of all older women are requesting payments before the age of 65.[2,10]

Participation in the labor force represents both workers and those looking for employment. It does not, however, include older workers who hve been discouraged and are no longer actively seeking jobs. If the latter were tallied as well, unemployment among Americans over the age of 55 would reach 1.1 million, or roughly twice the current official figure. In addition, among the over 20 million retired workers drawing Social Security, over 60% receive reduced benefits because they "elected" to begin drawing prior to age 65.[2]

In terms of financial security in old age, both unemployment and underemployment bode ill. Since Social Security is based on earnings, those who experience occupation disruption, for whatever reason, are going to receive lower payments. Overall, older people have lower incomes and are much less economically secure than younger people. At best, only a scant minority maintain relatively high or even stable incomes throughout their later years. When all sources of income are taken into consideration, the median income for families headed by those over age 65 is approximately 90% of that of all families. Further, the income of black elderly families is only 44% that of their white counterparts. Among the so-called unrelated individuals over the age of 65, or those not living in a family context, income is about two thirds of younger unrelated persons and slightly more than 40% as much as older family income. Not surprisingly, most of those who live alone and are poor are women;

minorities too are disproportionately represented among the poor and near poor elderly. On the average, about one sixth of the elderly have incomes below federally established poverty levels and another one fourth are in the next higher category. Among black elderly roughly one third subsist on less than poverty-level incomes; among Hispanics the rate is about one fourth. When the "near poverty" level is taken into consideration the percentages rise dramatically. Needless to say, poverty rates increase the further into old age one moves; after age 72 the rate turns upward to nearly double the earlier level.[10]

Social Security was never intended as more than an income base, yet it now constitutes over one half of the monies received by over 50% of the elderly. In fact, for one fifth of the white elderly and two fifths of blacks, it accounts for over 90% of all income received; hence, much of the improvement in poverty levels in recent years is due to the indexing of Social Security payments. Nonetheless, about 20% of all couples and 46% of all single persons who rely solely on Social Security benefits live below the poverty level. Of course, poverty rates vary among subgroups of older people, climbing upward among white males, females, to black males, and then females, respectively. Fortunately, income may also derive from earnings, assets, private pensions, public assistance, and contributions. If one or some combination of these alternatives also provides income, poverty rates drop to a relatively low level. Yet when all sources are added together the average income of older people amounts to one half or less that of their younger counterparts. This same rate holds true among minority elderly, though it should be remembered that the relative earnings of blacks is less than 80% that of whites in all age categories. Across the board, two trends are apparent regardless of racial or other social characteristics. In the last decade there is less reliance on earnings, with a greater share of retirement income coming from public and private pension programs or the conversion of assets. The second pattern is that Social Security plays an increasingly important role as people move further into their old age. For many, the poverty that accrues with advancing age is a new experience. Among the old-old, the curtailed resources and the likelihood of major expenditures for health care impose real financial hardships.[2,11]

B. Health Care Expenditures in Later Life

While it is certainly the case that noncash benefits such as Medicare, Medicaid, housing assistance, etc. must also be taken into consideration when evaluating the quality of life among the elderly, the intent and the reality of these assistance programs is quite disparate. With the exception of Medicare, other noncash transfer programs designed to aid the elderly are means-tested; consequently, benefits are indexed to financial need. For example, in 1981, 86% of elderly food stamp recipients had incomes no higher than the near-poverty classification; the food stamps they received had a mean face value of $500 a year. Furthermore, of the nearly $200 billion earmarked in the federal budget for the elderly, 92% was set aside to fund entitlement programs underwritten by lifetime earnings contributions. Only 2.1% of all federal outlays actually constituted welfare payments to old people.[2,3,10]

Health care expenditures increase steadily across the life cycle, and as residents of the only remaining industrialized country without national health insurance, Americans face major out-of-pocket expenditures. Despite the fact that the elderly are healthier than most stereotypes suggest, health care expenses for older people represent just under one third of all health care costs in the U.S. With inflation for health care running well ahead of all other costs, figures for per capita medical care are obsolete before they are published. Still, an idea can be gained by examining the rate of expenditure.

Aggregated payments for medical expenses from all sources run twice as high among the elderly as among the rest of the adult population and some 7 times higher than for those 18 or younger. The reasons for this lie in the types of medical attention administered. While the exact figures vary by the agency doing the reporting, we know persons between 65 and

74 make physician office visits 4.8 times a year, climbing to 5.1 visits for those over 75; the comparable number of visits for younger people is 3.2. Once a person reaches age 65 they are over twice as likely to be hospitalized, have to stay twice as long, account for 20% of all surgeries, have twice as many prescriptions, and pay just about double out of their own pockets. Paradoxically, they see a dentist less frequently despite the fact that 60% or more have untreated dental problems. Interestingly, dental coverage is not included under Medicare benefits. Without providing a detailed breakdown of changes after age 65, suffice it to say that again, the further into old age one looks, the more the elderly are disproportionately represented. Turning to nursing home utilization, the most rapidly growing health care expenditure for older Americans, the number of older people residing in such facilities reflects longer life expectancy and is expected to increase by roughly 50% before the end of the century.[2,3,10]

C. Other Health Considerations Affecting the Aged

Detailed reviews of biomedical conditions among the elderly appear elsewhere in this volume; therefore, only a brief comment on health conditions is included here in the context of dependence and limitations brought on by the extended period of morbidity and diminished vigor accompanying extensions in life expectancy. Just what does the prevalence of chronic conditions mean for the elderly? We know that 80% of those over age 65 must contend with at least one chronic ailment, and over one third experience some limitation of their daily activities. Nonetheless, over two thirds of all old people still rate their health as good or excellent. We also know that whites are healthier than nonwhites, the wealthier are healthier than those with less income, and the diseases which affect aged men are most often those also seen as causes of death while those of aged women predominate as causes of illness. In terms of the latter, the most frequently reported chronic conditions after age 65 are arthritis and rheumatism, heart conditions, hypertension, impairments of the lower extremities, back, or hip, visual impairments, and diabetes. Each of these will affect nutrition and each may in turn be caused by nutrition-related diseases.[4,12,13]

Whatever the cause of a limitation of activities of daily living, most older people are not particularly constrained. As long as people are able to dress, bathe, use the toilet, and eat on their own, nutritional deficits are likely to be only modest (Figure 3). While the need for assistance in one or another of these areas undoubtedly increases with advancing age, the vast majority of even those over the age of 85 are able to perform these functions for themselves.

For years gerontologists have discussed the "rectangularization" of human survival curves with the implication being that the period of diminished vigor would decrease accordingly.[14,15] As Schneider and Brody[16] point out, however, many of the gains that have resulted in low death rates until quite late in life can be traced to reductions in infant mortality, infectious diseases, and to improvements in public health measures, sanitation, diet, as well as quality of life. With the number of those over age 85 growing more rapidly than any other segment of the population, the pattern of rectangularization may be changing due to a downward shift in exponential mortality rates in the later decades of life. Since this is also the age bracket most prone to chronic diseases, and no downward trend in morbidity has yet been noted, all evidence would suggest a lengthening of the number of years in which illness and disability are one of the companions of life. One implication of such a trend is the growth of a potentially dependent segment of the population which lacks the resources necessary to attend to its needs. As with so many other aspects of the aging process the ability to care for one's self and ensure proper nutrition is related to socioeconomic status and related sociological variables.[12,17-19]

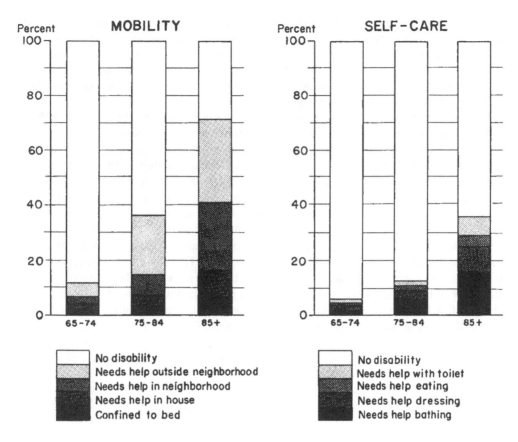

FIGURE 3. Impact of chronic conditions on daily living for the older population by age group, 1977.[3]

III. SOCIAL VARIABLES AS FACTORS IN DIETARY INTAKE

In examining the nutritional status of older persons it is important to consider the social context in which eating is embedded. Changes in living arrangements, shifts in social roles and support networks, alterations in mobility and psychomotor processes, plus many other dimensions of social interaction are implicated in nutrition practice.[20] In other words, eating is oftentimes more than a subsistence activity; it is somehow transformed into dining. If nutritionists did not have to deal with the symbolic significance of food, their goal of promoting sound diets would be considerably less complex. The social and the social-psychological variables intervening between merely opening one's mouth and adequate nutrition constitute what Margaret Mead[21] referred to as the "applied science of changing food habits"

A. Cohort Socialization

It would be a mistake to ignore the fact that we are all historically grounded beings. Today's elderly have brought with them into old age norms and values of a bygone era; the same will hold for those who become old tomorrow or the day after. Preference in any realm is an outgrowth of experience and opportunity. Historical changes and contemporary differences will each be incorporated into the dietary habits of the elderly; what they eat will reflect the combination of circumstances accreted over a lifetime. The elderly of today were reared at a time when idealized values were noticeably different than they are at present, and while they may have been learned 6 or more decades ago, they continue to exert powerful influences over the way the elderly view the world. Of course, part of that world view includes beliefs, customs, and preferences for particular dietary patterns. As Rux[22] so aptly notes, food is oftentimes a highly salient symbolic component of group membership. Its relevance derives in part from its representation of cultural identity; ethnic foods are ready examples but so too are family resemblances[23] and generational commonalities.[1,24-26] Reasonably distinct patterns emerge not only in terms of preferred foods, but in relation to those which are subject to normative proscriptions as well.[27]

While concise definitions of cohort or cultural food preferences may be difficult to obtain in a pluralistic society, there is consensus that attitudinal constellations do influence food choices.[27-31] Therefore, understanding the dietary patterns of any subgroup of elderly persons necessarily involves the consideration of early life as well as traditions built up over the life cycle. For the present generation of Americans over age 65, the symbolic meaning of food was initially acquired when the country was undergoing the rapid transitions associated with the early industrial era of the turn of the century. Breakfast and large midday meals were the norm, processed foods were unknown, optimal cooking times were thought to be considerably longer than at present, and the perceived therapeutic value of food groups are almost humorous when compared to the standards today.[32] This is not to say alterations have not been incorporated into food preferences over the years, but merely to assert that care must be taken before attempting to interpret eating patterns. Without ascertaining the nature and type of early life experiences it is difficult to know how to intervene in dietary regimes.

As an illustration, those elderly who lived through the Depression during their formative years carry with them an indelible mark that continues to color their orientation toward life. The deprivation experience of the pre-World War II mobilization years promoted adaptive responses appropriate under the circumstances of economic scarcity but slightly out of place in "a culture shaped by abundance".[33] With the addition of the ethnic, regional, and urban-rural differences, which were at that time more salient, it is impossible to discount early life experience.

B. Ideology of Independence

Age cohort differences in attitudes also become apparent in what can best be described as an ideology of independence. While generalizations are risky, among those who interact daily with old people stories of their autonomy quickly become legend. Many aged find it distasteful to even seek the services to which they are entitled by law as it represents an admission of dependency. Unclaimed Social Security payments, the unwillingness to file for medical assistance benefits, and self-imposed hardships from nonutilization of local or federal public services are just one side of the rugged individualism inherited from the social Darwinist ethic so prevalent during their youth.

The results of both the HANES I and the Ten-State Nutrition surveys have identified age-related nutrient and vitamin intake patterns which may stem in part from eating practices linked to beliefs and values of independence.[34-35] Another facet of this same philosophy is apparent in a widespread tendency of self-medication. In terms of their day-to-day health and physical well being, there is an entire realm of practices which remain largely out of sight from nutritionists and health care professionals alike. While the dimensions of self-prescription practices in either nutrition or medication are difficult to judge, one study in Philadelphia suggests only about 1% of all physical symptoms were reported to health professionals and that as much as 60% of all therapeutic remedies are self-selected.[36] Whether similar tendencies exist in regard to perceived avenues for adequate nutrition is an issue deserving of far greater attention by nutrition specialists since a pervasive pattern of self-care and autonomy has been recognized in many other aspects of the lives of the elderly.

C. Sociability and Eating Behavior

Regardless of the prominence of early learned behaviors, they are, of course, mediated by the social processes that accompany aging. As a person moves through various stages of the family life cycle and through friendships or associations that are affected by relocation, illness, and death, the sources of consensual validation undergo alterations as well. Without a spouse, family, or friends to share meals, not only does their meaning change, but the very basis of age identification may be distorted so that a person may come to feel older than his or her peers.[37,38] Intimacy and morale, built around confidants, are among the more crucial symbolic concomitants of eating; as the former change, it is only reasonable to expect changes in the latter. The social losses of later life have repercussions which eddy across all areas of life. Appetites suffer, interest flags, erraticism becomes common, and knowledge may prove to be out of date; each can be set in motion by changes in interaction patterns. While it is erroneous to believe all old people are bereft of close friends, family ties, and daily contacts, when breaches do occur in expressive social networks they have a profound effect on eating behavior.[39-42] While in many cases these effects may turn out to be only temporary, they nonetheless have an impact on nutrition.

Among the numerous components of expressive interaction that affect dietary behaviors, family life stage, marital status, and social support networks have received the greatest attention. The family unit is the principal consumer of goods and services in American society. As families move through the stage-like trajectory of their existence, their position along the continuum of constitution to dissolution affects food consumption patterns. Time spent on food acquisition and preparation, dietary preferences and needs, patterns of eating out — the entire locus of food decision making reflects stage and type of family life.[43-45] At each juncture, the relative priority given to health, economy, and convenience of food is rearranged and correlates with other sociological variables characteristic of particular stages. During the later phases budgetary concerns become increasingly common, while issues of health, such as fats and cholesterol, are paradoxically less frequent than they are for the middle aged.[46]

In the last stages of the family, when one or the other spouse is deceased, dietary patterns are different still. As might be expected, male-female differences are likely to be marked with males relying more on convenience foods and eating away from home. Women, on the other hand, appear to eat more frequent meals in the home and consume more snacks than their male counterparts.[47-49] In addition, there may well be gender-linked tendencies in the utilization of congregate eating services or nutritional counseling. Certainly there are major research questions about late-life families yet to be addressed. While we know that the elderly who live alone follow distinct regimes, and in many instances have less adequate diets, large-scale investigations of living arrangements, marital status, and other secondary factors influencing food intake are still needed.[20]

Maintenance of friendship ties becomes increasingly problematic among the very old; consequently, opportunities for social contact also decline. Immediately following retirement social involvements increase but decline gradually thereafter.[50,51] Since interaction with friends is a key factor in morale and life satisfaction, and these in turn have been linked to nutritional adequacy, it is often assumed that those who are alone are nutritionally at risk. Not only is interaction linked to social-psychological well being, it provides access to a wider range of eating opportunities as well.[20] In fact, current federal legislation (Titles III and VII) incorporates the principle of eating in the company of others as a justification for congregate service programs aimed at ensuring appropriate sustenance levels.[52,53] There is a difference, however, between being alone and being desolate; closer inspection of the dynamics of this issue is necessary before any definitive statements can be made.

What social scientists refer to as expressive and instrumental support constitute distinct components of an older person's social network. Clearly, the qualitative dimensions of social interaction may exert more influence on eating behavior than the mere number of contacts per se.[20,40,52] As will be seen in the following section, integration into various types of groups has an effect on nutrition. Further, making sense of the patterns observed in any particular group of older people necessarily involves an examination of key lifestyle factors. Daily life is a multidimensional reality which does not lend itself to quick or facile treatment, yet is is essential to investigate the components of lifestyle if we are to understand the social context of nutrition among the elderly.

IV. LIFESTYLE FACTORS AFFECTING NUTRITION

While one third of the elderly surveyed in 1976 attributed changes in their diets to alterations in their lifestyle,[54] this concept is deceptively complex. What does the notion of "lifestyle" entail? Does it reflect changes in work schedule? In income? Housing? Differences in culture? Lifestyle implies these as well as many other factors, plus relationships among them. While the following discussion will attempt to discuss each of these influences in turn, the dynamics between them must be kept in mind. It should also be remembered that studies of the effects of lifestyle on the diet of the elderly are not only relatively new, but generally involve mere description of demographic factors without integrating them into an overarching conceptual model.[31] Furthermore, theoretical, methodological, and research disparities make comparisons and generalizations across studies difficult.[55]

A. The Structure of Lifestyle

Socioeconomic status (SES) is potentially the greatest component part of lifestyle. Based upon the interrelated variables of income, occupation, and education, SES affects not only where one lives and what one is able to buy, but also such food-related factors as food preferences, food habits, adequate storage, cooking facilities, etc. While it is apparent that the literature focusing on nutrition among the elderly recognizes the importance of SES for dietary quality, most studies focus on participants in meal programs or low-income areas, thus limiting the variability of social class categories under investigation. This omission is particularly significant. If, as was demonstrated in a survey of New Mexican elderly who were financially solvent, the dietary intake of higher-income aged is similar to the rest of the population, it may be that an exclusive focus on the elderly as a group nutritionally at risk would be misleading and, in essence, culpable for viewing the aged as a homogenous group.[55,56] It would be a mistake to assume socioeconomic components of lifestyle are randomly distributed across the population. Rather, there are structural factors shaping individual life scripts which exist independently of an individual's personal knowledge or characteristics.

Research on the present economic system in the U.S. suggests wide ranging changes in the workplace,[57,58] which have important effects on the social class structure. Characterized as a dual economy,[59,60] modern industrial production has been effectively divided into core and peripheral sectors. A worker's location in one or the other of these structurally determined spheres exerts a major influence on level of remuneration, lifestyle, attitudes toward health, access to health care, and world view, above and beyond one's occupational title.[61] Further, while workers in the core sector can generally expect to receive pensions in addition to Social Security, former employees of the peripheral sector have no such guarantee.[62] If, as Myles[63] contends, social inequality in old age is reflective of a life-long pattern, low-income elderly are most likely from this peripheral work force, and have been members of relatively lower social classes throughout their lives. Thus, other factors which will influence nutrition, such as cohort socialization, culture, and religion, must be viewed with an eye to these structural considerations. The socioeconomic realities of one's life will be weighed against each of these factors. For example, an Orthodox Jew may desire to keep a kosher kitchen and menu which may be perfectly in agreement with good nutrition. If one is a peripheral worker, however, there will be a compromise between cultural heritage and how much of this can be obtained given a limited income, neighborhood, etc. These problems will not only be exacerbated in old age, but the results may have a strong impact on nutrition.

B. The Effects of Socioeconomic Status on Late-Life Nutrition

While structural components are generally lacking in many published studies addressing SES and nutrition among the aged, SES in a more traditional sense has been incorporated. Roe[64] examined the importance of social class in the formation of food preferences, noting people are more likely to select foods connoting higher perceived levels of prestige. For example, choosing butter over margarine, or steak over hamburger may symbolize upward social mobility or higher prestige. In a study of low-income young women in Canada, Reaburn et al.[65] found a relationship between perceived prestige of some foods and subsequent selection; however, this only held true for a minority of the foods under study. The role such perceptions play among the elderly remains unclear and the almost exclusive focus on low-income aged precludes broader comparisons necessary to assert that a class preference exists for the entire elderly population.

While income and educational level are often included in analyses, occupation, past or present, is noticeably absent. While Roe[64] suggests that unemployment or underemployment can lead to overconsumption of junk foods "as a means of breaking the daily monotony", type of occupation has not been given serious consideration in analyses of the elderly. Indeed, the only nutritional study which even incorporates the SES index appears to be that of Grotowski and Sims.[66] In their survey of three groups of white elderly from various social class backgrounds, they confirmed that SES was a key variable in explaining dietary quality. The exclusion of minorities, who are disproportionately represented in the periphery,[61] keeps these results from being generalizable to the elderly as a whole, however.

The relationship between income and the dietary quality of the elderly is still hazy, even among the poorer segments of the elderly population. Although a lack of monies may limit dietary choice,[64,65] and many elderly would buy more food if funds allowed,[67] the practical consequences of financial constraints are not totally clear. In some studies, income has been shown to be positively related to dietary intake,[68-71] as well as the overall number of eating episodes per day.[72] While this relationship may at first appear to be common sense, it is actually quite complex. At least one study of participants and nonparticipants of a meal program in central Maine found income level is not associated with nutrition;[73] still another found age and education were more vital than income.[74] Incorporating structural variables into such research may help clear up many of the inconsistencies. For example, a lack of association between income and nutrition could be traced to a lifetime of mediation between low income and food choices if the respondents had been (or were married to) peripheral employees. In addition, income levels cannot be isolated from the economy as a whole; historical fluctuations in the overall economic picture will effect the purchasing power of limited monies at any particular point in time.[31] Finally, the association between income and dietary quality is also mediated by nutritional knowledge, a subject which will be discussed below.

The mixed findings on the relationship between income and nutrition are mirrored when the investigation turns to educational background. In two separate studies, the elderly showed no association between education and dietary quality,[54,73] while other research has documented dietary quality improving with more education.[67,69,75] Since education is, in fact, only a part of the SES index, the contradiction suggests that this factor cannot serve as a proxy for the structural aspects of social class or how this in turn affects lifestyle. It has been argued that "there are men and women of advanced educational background who select foods inadequate in key nutrients because they adhere to an exotic food philosophy".[64] Members of higher social classes, generally those with more education, have been found to be more likely to purchase "health foods". However, this may be a function of having more money to spend for such items.[66] Hence, when education is incorporated into research merely as an attribute of an individual, apart from its social context, its explanatory value is diluted.

C. Ethnic Identity and Food Preference

One of the strongest influences on the way in which SES conditions daily existence is one's cultural background. This is a reciprocal relationship: as was demonstrated earlier, the likelihood of poverty is increased given certain ethnic affiliations. In turn, cultural heritage will influence the patterning of daily lives within class categories. For example, in deciding which foods are "prestigious", many Americans of Anglo-Saxon descent would choose high-quality steaks and not liver. By the same token, liver is considered to be a "delicacy" by some of European descent.[65] These selections represent cultural preferences, and are not merely a function of the quantity of disposable income, education level, or other such factors.

Concern with culturally based consumption patterns was evidenced as far back as the 19th century work of LePlay in Europe, who concluded that the structure and values of a particular society had an important effect in this regard. Subsequent studies in the U.S. have concurred; for example, the cultural eating patterns of blacks have been maintained across generations. In a recent comparison of major ethnic groups among the population as a whole, cultural differences were found to exist, regardless of income, with the most striking disparities existing between blacks and whites.[76]

The retention of ethnic eating patterns will have a special effect among some groups of elderly especially as they are faced with such age-related changes as institutionalization and alterations in the "old neighborhood". Culture not only involves a set of values, beliefs, and customs, it also encompasses a sense of *identity*. As people age, experience role changes, and lose such things as spouse, income, and physical mobility, the need to retain some stability in one's sense of self becomes increasingly important. For those with a strong sense of ethnic identification, being surrounded by familiar foods and cultural eating practices renders a sense of security and comfort.[77] This is especially true on days or occasions that the culture defines as important such as holidays, weddings, deaths, or whenever a ritual is prescribed.[22]

The effects of culture on the diets of the elderly can be investigated in a variety of ways: whether the diet itself is nutritionally sound, which ethnic groups have, in general, retained their eating patterns, and whether or not the elderly are able to continue such practices and maintain their sense of self. Though research in this area is sparse, present indications are that ethnicity does indeed play an important role in the diets of the elderly. This seems to be especially true among Chicanos and Chinese Americans who, according to one study, are often first generation and have almost totally retained the eating patterns of their original culture.[78,79] Findings among blacks, on the other hand, are not as definitive. While the aforementioned research indicated that this group "was well assimilated in regard to American food habits",[78] a national survey of blacks and whites of all ages found substantial differences in terms of food choices.[76] Furthermore, overall nutritional differences between blacks and whites remain. Utilizing the Ten State Survey conducted by the Department of Health, Education, and Welfare from 1968 to 1970, Munro[79] found white men and women, regardless of income, demonstrated better dietary quality than either blacks or Spanish-Americans. Indeed, while gender differences exist, and income level had some effect, the most important differences in selected nutrients appear to be based upon ethnicity. From Munro's comparisons it cannot be concluded that either blacks or Spanish-Americans are "worse off" than the other. For example, while Spanish-American men reveal lower levels of vitamin A intake than blacks, higher-income members of the latter group consume more calcium. Similarly mixed findings hold true for women.

Among those elderly for whom adherence to cultural food choices is important, significant difficulties can arise. For example, Natow and Heslin[77] have pointed to the need for recognizing the importance of food for the mental well being of observant Jews; institutionalization frequently precludes eating kosher meals. The loss of self-esteem and identity that may come with institutionalization is thereby exacerbated. On the other hand, furnishing the appropriate foods can allow these individuals to retain more sense of self. Rux[22] has indicated that not only do nursing home residents "thrive in situations in which their culture and heritage are honored" but also that senior centers which are staffed by members of ethnic groups and recognize such cultural traditions are far more successful in soliciting attendance from the older portion of this population than those which ignore such diversity.

Independently living elderly face an additional problem in trying to retain their cultural heritage. Many aged have lived in the same location for many years and, while they have stayed, things around them may have changed quite drastically. The corner store which used to carry kosher food, or the Italian deli, or the Chinese grocery, have long since been replaced by the chain supermarket which may not provide the requisite foods.[80] Transportation to more distant markets is a problem with the result being that ethnic aged are isolated from their cultural ties. The inability to maintain customary eating patterns may consequently lead to loss of self-esteem and disinterest in food.

D. Geographic Locale, Lifestyle, and Nutrition

Region of the country, degree of urbanization, and even location of neighborhoods all influence and are influenced by lifestyle. In turn, each will affect eating patterns. To take region of country as an example, it seems reasonable to expect a Midwesterner to eat more beef than a Southerner, a Southerner to eat more country ham than a Westerner, etc. These differences in preferences and availability may be exacerbated by regional disparities in income level leading some researchers to regard geography itself as a "cultural generator" of consumption patterns.[76]

Past research has highlighted geography as an important contributor to eating patterns among the elderly.[80] While important regional variations were documented by the U.S. Department of Agriculture Food Consumption Surveys (USDA-FCS) through 1966, recent demographic changes may weaken the effect of geography on consumption. According to Guseman et al.,[76] regional income variation has decreased while a substantial migration between regions has occurred. Pinpointing potential changes in consumption patterns is complicated by the limited availability of foods in specific regions. In attempting to distinguish between regional differences resulting from food preferences and those from food availability, the findings of Guseman and co-workers prove inconclusive. While there appeared to be some disparities between regions when they compared broad food categories, these variations lessened when their broad classes, such as "meat", were broken down. Nonetheless, they conclude regional patterns are becoming increasingly similar.

E. Dietary Patterns of Rural and Urban Elderly

In their study of rural elderly, Bazarre et al.[81] maintain that while large segments of our older population live in rural areas, little is known about their nutrient intake. To date, sample sizes have been small and findings inconsistent. It is generally thought rural elderly are disadvantaged due to income and transportation problems that are made worse by lack of services and isolation.[26,81] In rural Pennsylvania, Guthrie et al.[82] found that income is indeed crucial for sound nutrition. However, both low-income elderly and those older persons who did not qualify for food assistance had less nutritious diets than low-income families of other ages. Rawson et al.[68] also found income as well as public or private transportation and information about food assistance programs and policies to be important for dietary intake. LeClerc and Thornbury,[73] on the other hand, found no relationship between income and overall nutrition. Further, the extent of nutrient inadequacy among rural elderly is still sketchy; in fact, Rawson et al.[68] were not able to differentiate their rural respondents from urban elderly. Vitamin A and calcium appear to be less than the RDA for a significant number of rural elderly, however.[81]

Aside from all other considerations of residence, urban living as a whole has an effect on nutrition. For example, Slesinger et al.[72] found among elderly living in Milwaukee, income level and living alone were significant only in relation to fruit and vegetable intake. Interestingly, some researchers have noted a "cosmopolitan" effect on nutrition. In one case, the size of the town in which respondents grew up was positively related to educational level, nutritional knowledge, fewer meals eaten with children, and a willingness to try new foods.[54] Similarly, a "cosmopolitan lifestyle" has been associated with a "rational orientation" toward food. Such a disposition is characterized by food choices based on outside information sources and a generally appropriate nutrient intake. The feeling is that this cosmopolitan lifestyle leads to a greater exposure to mass media, education, and food fashions.[83] On the negative side, urban living in low socioeconomic areas is also associated with such things as fear of crime which, as was shown in a Boston area study, may prevent elderly persons from going to the store as often as necessary for optimal nutrition.[84] If this fear is coupled with transportation problems, the result on diets can be devastating.

F. Neighborhood Location: Transportation and Mobility

As the preceding discussion makes clear, neighborhood location is a pertinent factor in diet and nutrition. Lack of transportation in both rural and urban areas can have obvious effects.[68] Location of a dwelling itself, aside from the fear of crime, may also prove to be an impediment to good nutrition. Loss of mobility is often a concomitant of aging; aside from possible physical decrements, automobile driving eventually becomes difficult due to physical impairment and/or cost, and public transportation is often nonexistent or provides logistic problems for the aged.[80] A further complication is that much of the urban landscape is changing and becoming unfamiliar thereby inhibiting travel to shop.[85] For example, Sherman and Brittan[71] have documented the effects of the disappearance of corner grocery stores on the elderly as supermarket chains become more popular. While many franchise stores may have lower prices and greater selection, they are generally located in suburbs inaccessible to low-income, inner-city elderly. Distance is a crucial variable in determining where older people shop; Singleton et al.[74] found about half their group of both participants and nonparticipants in a meal program lived less than 1 mi. from a good store. While low income did not explain shopping at neighborhood groceries vs. supermarkets in this particular study, others have found it to be quite important. Due to cost, most low-income elderly walk to stores more often than their higher-status counterparts.[86] In Denver, it was found that 78% of the low-income group walked compared to only 15% of upper- and middle-class elderly. The majority of walkers lived less than 4 blocks from where they shopped, and 75% were less than 8 blocks; since most supermarkets were in the suburbs, they were restricted to corner groceries. While walking may be healthful, it also presents hazards related to crossing intersections, carrying groceries, limiting the items that can be purchased, and weather.[71] These difficulties in obtaining high quality, cost-effective groceries obviously have an impact on nutrition.

G. The Impact of Living Arrangements on Nutritional Intake

1. Physical Aspects

Location, but also type of dwelling, either in urban or rural areas, will affect nutrition. To begin with, cooking, storage, and other facilities are important mediators of consumption.[64,87] Lack of workable cooking areas not only limits food selection,[88] but can also force the elderly to eat out; men especially eat at restaurants more often than women. While the nutritional benefit of eating out is dictated by menu and income, it may also limit food selection.[64] Lack of appropriate storage space and/or facilities can also have an effect. On the one hand, Roundtree and Tinklin[86] found inadequate storage forced low-income elderly to shop more frequently than their wealthier counterparts. On the other hand, infrequent shopping trips reduce consumption of perishables, such as milk and fresh produce, both important for a good diet.[80] Regardless of number of trips to the store, appropriate storage for perishables is mandatory.

2. Social Aspects

While the ability to maintain independent living is associated with better dietary quality,[89] there is a great deal of variation in living arrangements among noninstitutionalized older persons. As was noted above, diversity stems from marital status, income, and sex. While our understanding of all these variables is not complete, some relationships have become clear. Since nearly 30% of the elderly live alone,[90] much speculation has centered around the consequences of social isolation on nutrition, often concluding that elderly who live alone have poorer diets.[64,69,84] However, as has also been pointed out, social isolation is not the same as loneliness. Edwards[80] provides a useful dichotomy: "isolates" choose to live alone, while "desolates" are forced into this situation. Even though both groups do not live with others, the depression or loneliness which ostensibly leads to lack of motivation for food preparation and eating[54,91] is not necessarily characteristic of both groups. While some research has shown the elderly who live alone consume food which involves less preparation, have less varied diets, and lower nutrient intake,[31,66] other studies have found no significant dietary differences resulting from household composition.[69,92] This relationship becomes more complicated when marital status is introduced. Looking at single women vs. married couples, Schafer and Keith[75] found the former group had better diets. LeClerc and Thornbury[73] also found living alone or with others had no effect on diet except for divorcees, who consumed significantly greater amounts of carbohydrates, calcium, and riboflavin than either married or widowed groups.

Much of the concern with social isolation stems from the notion that since eating is a social activity, partaking of food by oneself becomes more of an obligation than an opportunity for interaction. One recent study of elderly who live alone found that the vast majority expressed nutritional attitudes which reflected the belief that food is a source of sociability, adventure, and enjoyment. Only 14% felt eating served only a "utilitarian" purpose.[90] Interestingly, these attitudes were not associated with income. While some researchers suggest diet does improve when meals are shared,[66] LeClerc and Thornbury[73] discovered no nutritional effect for the number of meals eaten alone. The sex of the elderly loner is also of interest; however, there is little research in this area. Elderly males are often unable to cook, and at least one study has found elderly single-living men to be at greater nutritional risk than their female counterparts. However, Schafer and Keith[75] found that among married couples, when men contributed in "diet decision making", overall nutrition improved.

H. Institutionalization

Institutionalization may affect nutrition in two ways: emotionally and practically. Loss of independence and status often accompanies placement in a care facility because the residents lack power. They must accede to the decisions and schedules of others, and often have little role in decision making themselves.[80] Subsequent loss of morale is then reflected in poor eating habits. Further, while research indicates the well being of elderly in institutions that honor their culture can be greatly enhanced,[22,77] generally this is not the case. The sense of aloneness and dependency felt by ethnic elderly is intensified when the foods are strange;[22] consequently, nutrition suffers.

On a more practical level, nutrient intake for elderly residents of nursing homes is limited by what the institution provides.[64] One survey of women in a nursing home in Colorado found them to be below the RDA in energy, thiamin, calcium, and iron intakes. In fact, they were no better off than community-based low-income white or ethnic women,[79] despite the fact their diet was monitored. Does this result from poor planning by dieticians? Since state health departments set nutritional guidelines for resident meals,[93] this cannot explain the dietary lack. Providing nutritious meals does not ensure that they will be eaten.[94] Research conducted in institutions in rural southern Illinois and in Wisconsin found that while some menus did reflect approximately 100% of the RDA for both sexes, residents consumed less than one third of the necessary energy, calcium, and thiamin at the main meal.[93,95] Thus, it appears crucial to not only increase nutrient density, but to undertake periodic monitoring of actual intakes and modifying of food service to be more appealing.[93,95] Since mealtime is often the social highlight of the day for nursing home residents,[80,84,96] community meals should be encouraged whenever possible with menus planned to enhance the conviviality of the situation. Food demonstrations, which allow residents to actually watch the preparation of dishes, has facilitated both greater interest in eating as well as sociability.[96] Furthermore, research in dining room management has found that mealtime is more enjoyable when individuals are given more attention. This feeling is enhanced when the dietary staff is actually present in the dining room and stop to converse with residents.

I. Role Changes

Loss of spouse or retirement can spell significant changes in lifestyle that may have an impact on nutrition. These role changes can be examined in two ways: situational variations, such as loss of income, living alone, etc., and, as will be seen in Part II, psychological effects and stresses. The effects of such changes can only be assessed longitudinally. To date, however, no such nutritional study has been undertaken. Nonetheless, we can speculate a bit based on the lifestyle changes known to be associated with loss of spouse or retirement.

As was noted earlier, living alone does not, by itself, have an adverse affect on nutrition. Nor does losing a spouse automatically place the survivor in the category of social isolate nutritionally at risk. Indeed, the study conducted by Schaefer and Keith[75] compared the diets of married couples and single women; 87% of the latter were widowed, yet this group had a significantly better diet. Further, widowhood does not mean one no longer partakes of meals with others. Since most elderly who live alone still maintain ties with adult children who live in close proximity, it is possible that much social activity is geared to take place around meals.[54,80] Of course, socioeconomic status plays a role here, as the better off are more able to afford to entertain as well as to dine out. Overall, widowhood probably does not have too negative an effect on nutrition, unless inadequate finances intervene.

Retirement involves a wide variety of changes which may affect nutrition. Not only is there a general loss of income, but the absence of a daily work structure may alter mealtimes. Meals can become the highlight of the day, be used to break monotony, be skipped because of later rising, or more time and energy can be expended in their preparation. If some meals were the basis of social activity between co-workers they become less frequent and poorer nutrition can result. Still, little research has delved into the impact of retirement on nutrition. While retirement does spell differences in income it does not necessarily mean a lower percentage of funds will be spent on food. Since overall expenditures for retirees were only one quarter lower than workers, it is likely that disposable income alone does not explain poor nutrition. Actually, age has been found to be more important than retirement in accounting for overall expenditures; therefore, the number of years one spends in retirement could have an important impact on nutrition.[97]

In terms of changing meal patterns with retirement, even less is known. On the average, the elderly tend to skip fewer meals than younger people; they are also more likely to eat breakfast.[72] On the face of it, it would appear that at least in the early years, retirement does not noticeably affect meal patterns, but it is possible that there is a cohort effect: i.e., that the present generation of elderly have been socialized into particular eating habits, such as having breakfast, that they maintain throughout life. Without longitudinal analysis, it is difficult to pinpoint the sources of these meal patterns.

V. MAINTAINING ADEQUATE NUTRITION

A. Federal Programs

1. Titles VII and III

In response to both the growing numbers of elderly and their special needs, the federal government has enacted programs designed to alleviate some of the problems of growing old. In recent decades much of this legislation has focused on nutritional needs. Although the Older Americans Act was passed in 1965, dietary issues only came to the fore in 1969. At that time, both the Senate Select Committee on Nutrition and Human Needs and the White House Conference on Food, Nutrition, and Health identified the aged, especially low-income elderly, as a group nutritionally at risk. With the amendment of the Older Americans Act in 1972 came Title VII (P.L. 92-258) and the creation of the National Program for Older Americans (NPOA). Through NPOA, states received up to 90% funding for programs serving one hot meal a day providing one third of the RDA, 5 days a week. To qualify for these meals, individuals had to be over 60 (spouse could be of any age) and suffering malnutrition due to financial or physical constraints, inadequate cooking facilities or skills, or suffering loneliness which leads to a disinterest in eating. By mid-1976, 845 such congregate projects existed. While these programs were not designed to deliver food, the Meals-on-Wheels Act was introduced in 1973.[98]

In 1978, further amendments to the Older Americans Act restructured the various senior citizens programs and subsumed social service projects previously under Titles III, V, and VII under a new Title III. The overall goals of Title III as it relates to nutrition "include providing low-cost, nutritious meals, social interaction, counseling, referrals to other supportive programs and agencies, and access to transportation",[73] in addition to nutrition education. By 1983 expenditures for the nutrition intervention program were projected to exceed $500 million. In 1980, 165 million meals were served; the majority of recipients were below the poverty line, and roughly one quarter were minority group members.[99] The effectiveness of these projects, as seen from such things as frequency of participation, subsequent dietary intake, and changes in dietary habits, are still being investigated and will be discussed below.

2. Food Stamps

An additional federal program with a clear impact on nutritional status of the elderly is the food stamp program. As originally established, it was designed to help low-income households achieve nutritional adequacy. In the early years, recipients were required to "purchase" the stamps by trading cash for an equal amount of food stamps plus a "bonus". Some researchers have suggested that this provision helps explain why the elderly participated less than other low-income groups.[99] Sherman and Brittan,[71] for example, found that while two thirds of their respondents were aware of the food stamp program, only 10% used them despite the fact that all were eligible. The purchase requirement for food stamps was eliminated in January 1979. By April 1979 elderly participation in the program had increased 32% from the previous year. Provisions for mailing certification, and larger allowable deductions for medical and shelter costs have also helped increase the numbers of elderly utilizing food stamps. It has been estimated that at the present time, one fourth of the households receiving food stamps contain at least one aged member.[99] The effectiveness of food stamps for nutrition is not conclusive, however. While Guthrie et al.[82] found that those elderly who received food stamps had a more nutritious diet, other studies report little or no effect.[99] Even if food stamps are found to increase nutritional adequacy, the problem of participation by older persons remains. While the provisions have increased the number of elderly in the program, not all of those who are eligible participate.[66] Physical and financial constraints have been alleviated to some degree but the feeling that food stamps ''smack of charity'' still act as a barrier.[71]

B. Effectiveness of Meal Programs: Frequency of Participation and Dietary Intake

Krout aptly described the relationship between the existence of meal programs and participation when he stated that the mere existence of services does not ensure that those in need will utilize them. An awareness of the program is, of course, crucial; in her study of participants, Caliendo[101] found that the majority heard of the meal program from friends. A substantial number, nearly one third, were personally contacted by administrators of the project. However, awareness of services does not always correlate with knowledge of what the program entails. To date, little is known about the demographic correlates of such knowledge except that educational level plays a significant role.[100] No consistent patterns of demographic characteristics have yet been found to be important in explaining participation;[101,102] about all we do know is that the provision of transportation promotes participation, that self-rated health is implicated, and that frequency of food shopping, food expenditures, and subjective assessments of sufficient funds for food were positively related to engaging in the meal program.[67,69,101]

Still, several explanations for lack of participation in meal programs have been offered. In discussing utilization of all services by the elderly, Krout[100] has indicated that the impersonal nature of the bureaucratic structure as well as the perceived demeaning attitudes of professionals may impose an obstacle. He also suggests that socializing may be an important factor — that those who value and are involved in other social networks are more likely to take advantage of services. Burkhardt et al.[102] investigated elderly attendance at congregate meal sites and found that the quality of services was the most important predictor even when level of need was considered. On-site cooking was preferred over prepackaged meals (similar to those served on airplanes); nutrition sites at senior centers were viewed more favorably than those located in low-income housing areas if the elderly did not happen to reside there. The number of other nutritional programs in the community also played a role. The notion of quality of service is an important one, but an equally salient issue is culture. Not only should the staff reflect the ethnic nature of elderly within a particular community, but the kind of food served may be important in soliciting the highest level of participation.[78] It must reflect their cultural preferences.

The effectiveness of meal programs in increasing nutrient intake is still a bit sketchy. LeClerc and Thornbury[73] found no significant differences in dietary quality between meal program recipients and nonrecipients, although the nutrient intake of the former group was somewhat higher. Similarly, Caliendo and Batcher[69] found no relationship between frequency of participation and dietary quality. On the other hand, most data do seem to indicate that the mean nutrient intake of participants at the time of attendance at a meal site is higher than for nonparticipants.[99] When nutrient intake at meal sites is compared to the total daily intake for participants, it is clear that the congregate meal has provided the majority of nutrients for most of these individuals.[101] Where meal programs tend to fall short is in the area of nutrition education; it appears that participants in these projects fail to carry new eating habits over to meals eaten away from the site.[99]

C. Nutritional Knowledge and Attitudes

Nutritional knowledge is a key in determining both attitudes about nutrition and overall dietary quality.[66] While the following discussion will focus on the sources of nutritional knowledge and beliefs, it is important to recall that not all people begin from the same point. Numerous studies have found socioeconomic status or educational level to be closely related to nutritional knowledge; however, sources of knowledge must be seen as intervening between demographic characteristics and dietary understandings.[54,64,66,73,94] In looking at the population as a whole, men appear to have less nutrition knowledge than women; men over the age of 50 and/or with a low SES background were especially likely to be lacking.[94] However, LeClerc and Thornbury[73] found no sex-based differences in their population of older respondents. Older people, in general, are likely to score low on nutritional knowledge. Since education is positively related to knowledge, and the elderly generally are less educated than younger cohorts, it is not surprising older age itself presents a situation that makes the elderly even more susceptible to nutritional misinformation and fad diets.[88]

While it is assumed that nutritional knowledge and attitudes are important in predicting dietary quality among the elderly,[73,84] such a view is not necessarily borne out in research. Caliendo and Batcher,[69] for example, found no relationship between nutritional knowledge and dietary quality among meal program participants. Similarly, educational level may be associated with nutritional knowledge, but it may not be related to diet.[54] Even among a sample of elderly who were fairly well informed nutritionally, such knowledge was not found to translate into good dietary practices.[101]

One possible reason for these confusing results is the role of advertising in the choice of food. Consumer attitudes have changed over the years and reflect more conservative beliefs about such things as government regulations. There seems to be a feeling that products need to be safer, for example, but not necessarily absolutely safe. Communication is taken to be a panacea; thus, labeling is preferred to banning products.[103] Reality belies this belief, however. Even among a relatively educated group of elderly — no fewer than 63% had at least some college — less than half even read labels.[54] Further, the merchandising, advertising, and display of foods utilized by the food industry has been successful in influencing the food choices made by the total population, not just the elderly.[64] Given the immense changes in this industry over the lifetimes of older persons, such as new food products and imitation foods, food selection is even more difficult if one is worried about nutrition.[88]

Sources of nutritional knowledge are also important. While many elderly report that their own past experience is a major source of nutrition knowledge and attitudes, many other means are also cited.[73,101] Mass media in general, and television in particular, are felt to be nutrition information sources.[66,73,75] While Grotkowski and Sims[66] found no relationship between television viewing and nutrition knowledge, cookbooks and magazine articles were significantly associated with both nutritional knowledge and attitudes. Caliendo and Batcher[69] found that both television viewing and newspaper reading are related to higher-quality diets; however, they did not test for nutritional knowledge. Whereas physicians were cited by many as a source of knowledge and were related to nutrition attitudes in one survey, only 14% identified doctors or nurses and 3% identified dieticians or nutritionists in another.[66,101] Finally, significant others, including spouse, friends, and neighbors, but not relatives, have been found to positively influence diet and provide sources of information.[73] Either limited or closed social networks have been found to constrict information flow, resulting in a lack of knowledge about nutrition.[68]

Finally, shopping habits may be a factor when trying to convert nutritional knowledge into actual diet. While we have already discussed the role of income, facilities, and transportation, advertising will again come to the fore, given the number of elderly who cite frequent television viewing or radio listening. For example, Caliendo and Smith[104] found 73% of their respondents said they sometimes purchased foods advertised by the media; only 27% said they never did. It appears advertising is not only potent, but even inhibits adequate nutrition among those with baseline knowledge.

D. Nutrition Education

Nutrition education should be a twofold endeavor: not only must consumers learn the nutritive value of foods, but this knowledge needs to be incorporated in daily practices. Given the vacillation of the economy, such education is especially crucial for those living on fixed incomes. Thus, one aspect of nutrition education is to equip consumers with the knowledge they need to make wise food choices given changing markets and prices. Furthermore, elderly consumers need to learn how to select foods that will be both nutritious and allow weight maintenance. At the same time, educators are faced with powerful competition from the media and the food industry, barriers which are often difficult to overcome.[64,88,94] Perhaps the major obstacle is the fact that in the past many health professionals believed the mere dissemination of nutritional knowledge would change eating habits regardless of the way the information was delivered.[64] Research among the elderly has not borne out this assumption. Caliendo[101] found her respondents demonstrated a fair degree of nutrition knowledge but their actual diets were not congruent with their informational level. In their investigation of Title III meal program participants, LeClerc and Thornbury[73] also concluded the nutritional education component had been ineffective. While participating elderly displayed a good deal of knowledge about nutrition, it had not been applied to their daily lives nor was it reflected in their attitudes. The goal must be not only to provide information but to change attitudes.[66]

In recent years, a renewed emphasis has been placed on evaluating nutrition education among the elderly in order to ascertain the most effective teaching methods. In a test program in Canada, 14 "marker" foods were used in consumer issues discussions, food demonstrations, and recipes. While the results were modest, consumer awareness was increased; not only did overall knowledge show some improvement, but there were also some small changes in perceptions of foods and overall practices.[105] The most beneficial education programs seem to be those which incorporate demonstrations: lectures by themselves often have little impact. In addition, education programs must have relevance to the target group; e.g., culture-specific foods need to be utilized and discussed when dealing with ethnic elderly. The most important factor seems to be involvement in the educational process; methods of incorporating the elderly into discussions, demonstrations, and decision making all seem to have the greatest impact both for learning and for subsequent change.[87] Planning of these educational activities must incorporate and reflect the lives of the clientele; physical, psychological, and sociological aspects must all be considered.[67]

Finally, nutrition education must reflect geographic locale. Rawson et al.[68] maintain that programs need to be designed to account for local food patterns and nutrient deficiencies of a region. Among their group of rural elderly, for example, it was discovered that they were lacking in calcium. Thus, not only were menus altered to include different dairy products for those who did not drink milk, but the nutritionist focused on dairy products in the presentation of nutritional material. Further, innovative means of education, such as Food Bingo, were effectively utilized. Interestingly, the fact that the nutritionist was perceived to be a younger woman with less experience combined with a feeling of independence to make these older people somewhat resistant to nutritional education. Consequently, teaching methods revolved around informal discussions of topics rather than traditional lectures on nutritional education.

VI. MAINTAINING ADEQUATE NUTRITION AMONG THE ELDERLY: SUGGESTIONS FOR THE FUTURE

Clearly, more research into the social factors affecting not only nutritional intake but effective participation in meal programs is necessary before any definitive statements can be made. It is obvious that simply identifying nutritional problems and offering ameliorative services does not guarantee they will be utilized.[100] At the same time, several new and innovative programs designed to educate the elderly about nutrition can suggest directions for the future.

As our society continues to age, a preventative approach to malnutrition through education will become increasingly important. While such programs do exist, they are limited in number and scope and need to be expanded. Such an increase is futile, however, if the education is ineffective. The design of nutrition education projects must include factors that will motivate the elderly to continue to participate. To this end, it must be remembered that research has indicated that incorporating the elderly into the planning and carrying out of educational activities has served to increase attendance and motivation. Based on this it has been suggested that groups of elderly be trained as peer educators. The key is to ensure they have positive attitudes toward nutrition, a sound knowledge base, leadership abilities, and common experiences with the clientele. Additionally, training should take place in an area which would not demand that they spend the night away from home. Funding to defray some of the "hidden costs" involved would also be helpful.[67,107] Particular and local needs and experiences of the target group must also be considered in formulating nutrition education. This involves not only cultural considerations, but also possible sex-based differences. One nutrition education program found women to be more concerned about basic nutrition education, while men care more about food preparation.[105] In their study of rural elderly, Rawson et al.[68] also point to a need to teach men more about cooking. Perhaps the best approach would be that of an "emergent curriculum" as has been suggested by Ludman.[108] Basically, this method attempts to assess what the elderly want to know about food and nutrition, then allows the curriculum to evolve over time through discussion. Such dialogue points to topic areas unique to the group and seen by them as important.

Rural areas present special difficulties because of transportation problems and/or isolation. To deal with this, one nutritional education team ran 1-day workshops (7 hr long) in five rural areas. While these sessions were evaluated highly by the elderly participants, many also said that 7 hr proved too taxing. Federal funding, however, mandated the duration. After 2 weeks, it was found many had improved something about their diet as a result of their participation. Follow-up workshops could help to reinforce these changes and provide additional support to the group.[88]

The amount of television viewing reported by many elderly has led to suggestions that television be used to promote better nutrition. Since news and information shows are preferred by the elderly, Natow and Heslin[109] advocate the use of public service announcements to advertise nutrition education programs, give nutritional information, and increase consumer awareness. In Canada, eight 10-min television shows were broadcast with each containing nutritious recipes incorporated around a discussion of a complete meal. While it is unclear how many elderly actually watched these shows, evaluation of the program seemed generally positive.[87]

One issue which was stressed at the 1981 White House Conference on Aging was the desired shift from federal to local and corporate responsibility for service delivery.[110] In recent years, various nutrition programs have sprung up across the country that attempt to go beyond the scope of traditional meal programs. One innovative effort geared at helping the elderly lower food costs and make shopping easier is the Alternative Food Delivery Systems Program in Michigan. Through this project, the elderly have been able to establish food cooperatives and gardening programs. Overall, efforts have aided area elderly in obtaining smaller food units for purchase as well as alleviating transportation, food quality, safety, and financial problems. At the same time, it has given the elderly the opportunity to be independent, use their skills, and socialize with others.[111] Similar food cooperatives and gardening projects have been established in other parts of the country.[109] In the Tenderloin district of San Francisco, the elderly in one SRO hotel have convinced a local farmers' market to sell low-cost produce to the residents every other week.[112] Sunkist Growers, Inc. publishes a newspaper entitled "Happiness Is" containing nutritional information for the elderly. Similar types of information could be disseminated in local newspapers as well as organizational newsletters. Finally, restaurant owners could be encouraged to include menu selections compatible with dietary restrictions,[109] while more grocery stores, especially in rural areas, could establish discounts for the elderly.[68]

In general, practitioners need to be both flexible and cognizant of the pride of the elderly when trying to improve nutrition. For example, ethnic background is not simply "something one must deal with", it may well be that further investigation into cultural eating patterns which emphasize many daily meals will render valuable knowledge about nutrition.[22] Perhaps of even greater value is the conclusion of Burkhardt et al.[102] in their study of the factors which determine attendance in meal programs. They warn that such social services must be viewed as consumer goods: "that program planners. . . (should) stop considering the elderly as so needy that they will attend any service provided for them and start considering the elderly even those who are severely disadvantaged, as rational consumers who are highly sensitive to the qualitative aspects of consumer programs. . . ."

PART II: PSYCHOLOGICAL DIMENSIONS OF NUTRITION

I. INTRODUCTION

As was shown in Part I of this chapter, there are several direct and indirect social-behavioral influences on nutrition. Part II of this chapter shall further explore these influences from a psychological viewpoint. Life course changes which influence health and hence nutrition are described, along with age-related changes in sensation that influence eating behavior. The psychological symbolism of food is discussed in relation to how this symbolism affects eating behavior. Social and emotional factors influencing nutritional intake are also described. Lastly, nutritional deficiencies which have been shown to manifest themselves as "psychological problems" are explored.

A more comprehensive picture of psychological factors which may influence nutrition is currently presented, in comparison to previous approaches taken by psychologists. Traditionally, psychologists have primarily concerned themselves with the identification of situational and emotional factors associated with eating behavior. Currently, much research has been focused upon eating disorders such as anorexia nervosa and obesity. Psychologists have not concerned themselves with developing a model which is focused on "everyday factors" influencing typical eating behavior. Even the study of eating disorders such as anorexia nervosa or obesity adds little to our knowledge of factors influencing nutrition in late adulthood, as anorexia nervosa is most frequently manifested in adolescence and young adulthood, and obesity is the most prevalent eating disorder for *all* ages of Americans.[113,114] Psychological factors associated with typical or "normal" eating behavior, specifically in late adulthood, have not been addressed as a uniquely salient topic of empirical study. Certain psychological factors which influence eating behavior and hence nutritional status are no doubt present for all ages. However, as development proceeds from young to late adulthood some variables influencing nutrition gain importance, such as health status, sensory abilities, subjective perception of self-aging, ability to adapt to loss, and opportunity for mediation of social isolation. The following sections will address such factors.

II. LIFE COURSE CHANGES INFLUENCING HEALTH AND NUTRITIONAL AWARENESS

A. Societal and Self-Perception of Aging

A previous review of research by Bennett and Eckman[115] reported that a negative stereotype of the aged is held by both young and old persons. This stereotype of old age, characterized by passivity and illness, affects not only the older person's self-perception, but also how others behave toward the older person.[116] With increasing age, the older adult's subjective perception of his/her body and overall health is affected by this negative stereotype of aging.[25] Age-appropriate behavior for the older person has been described by Howell and Loeb[25,117] as a preoccupation with health, digestion, constipation, and "iron-poor blood". As one ages, these "norms" for age-appropriate behavior are incorporated into everyday functioning. The older person may come to believe that he/she should be more concerned with health and that health should be declining, just because a certain chronological age has been reached. Stereotypic expectations of increasing poor health with aging may become a self-fulfilling prophecy if they are regarded as true by the older person. Concurrently, others may expect old age to be associated with poor health. Due to this stereotype of old age, sometimes poor health is reacted to as an "untreatable", inevitable consequence of aging instead of a possible pathology which possibly could be treated by health professionals.[118]

B. Health Changes and Health Monitoring

As was pointed out in Part I, with increased age there is an increased frequency of chronic illness as well as a growing occurrence of subjective complaints of multiple disability.[119] Shanas and Maddox[119] have reported that as much as 37% of individuals 65 years and older have some limiting chronic disability. However, Maddox[120] has also reported that older people tend to assess their personal state of health in terms of subjective, functional criteria rather than in terms of more objective criteria, such as a physician's examination. Additionally, older individuals' subjective health assessment has been found to be more predictive of their functioning and their feelings about themselves than are the objective medical ratings.[119] Thus, subjective assessments of health are determined not only by disease state, but also expectations of health. These expectations are influenced by one's reference group. That is, the older person compares his/her health with age peers who are stereotypically perceived as being in poorer health.[121] Therefore, the older individual in good health will see him/herself as an exception to the stereotypic view that "poor health is associated with aging", and hence subjectively assess their own health as "good", or in some cases better than it may have appeared from an objective medical examination. Conversely, an objective medical examination may determine that an individual is in good health, yet the stereotype of aging dictates "poor health", hence subjectively, health may be rated "poor".

In late adulthood there is an increased monitoring of health, due to age expectations to do so, and greater health concerns of self and others. Whanger[122] reported that vitamin supplementations increase with age. He presented data from a random sample of community-dwelling adults which showed that 14% of individuals below age 65 were using supplements, increasing to 25% at age 65 and 50% of those over age 90. This increase in vitamin supplementation may be related to increasing health concerns with age, although other causative factors may be implicated. In mid-life, awareness of self-mortality is triggered by several factors such as death of age peers, changes in physical appearance, and changes in health status of aged parents. Increased health monitoring of self appears to begin in mid-life and tends to be more characteristic of women than men.[123-125] Not only are middle-aged men described as more likely to deny illness symptoms, but also men in old age (60+).[124,125] This behavior has been linked to the stereotypic belief that illness is equated not only with passivity and old age, but also femininity.[124] Thus, health monitoring, admittance of health problems, and vitamin supplementation appears to be more frequent in women than men.[121,122] In fact, Neugarten[123] found that not only do middle-aged women monitor their own health, but also the health of their husbands. As it will be shown below, health monitoring may directly affect how well one cares for oneself. Nutritional status may be a reflection of this concern.

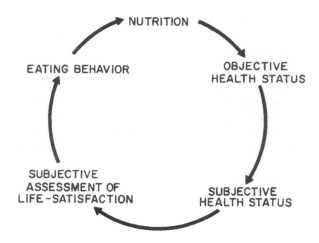

FIGURE 4.   Illustration of the relationship between nutrition, health, and life satisfaction.

C. Nutrition, Health Status, and Life Satisfaction

Age changes in health and health monitoring are important areas to examine as it has been estimated that between 30 to 50% of the health problems of the elderly adults in the U.S. stem directly from malnutrition.[126] Self-assessment of health in turn indirectly affects subsequent behavior associated with good nutrition. Subjective assessment of one's health as "good" has consistently been strongly related to personal life satisfaction.[127-130] Elderly individuals with more positive self-concepts and higher levels of satisfaction have been reported to have higher quality diets, and eating behavior associated with better nutrition.[131,132] This relationship is illustrated in Figure 4.

D. Behavioral Implications of Age Awareness and Health Changes

With increased age, changes in health and body appearance (e.g., the graying of hair, the gaining of wrinkles, and the increased ease of weight gain) act as salient cues for triggering how one feels about growing older.[133] Given that aging is usually associated with negative images, it is not uncommon for these age-related changes in body appearance to trigger behaviors which attempt to counteract these changes, such as dieting.[133] Elderly individuals who fear aging may be especially vulnerable to fads in dieting if such a diet is advertised to slow or prevent signs of aging or as a protection from acute or chronic disease.[53,134]

Food faddism has been described as serving certain psychological needs for an individual. Most often these fads are associated with a belief that certain foods provide special vigor or have special health benefits.[135,136] The belief that certain foods can eliminate disease may be more prevalent in the current generation of elderly individuals, given that nutrition education was not an integral part of their formal educational experience.[29,31] Also, the increase of chronic medical problems with increased age, which are not "curable" by traditional medical practices, may lead the elderly to try "nontraditional cures".[31]

III. AGE-RELATED CHANGES IN SENSATION INFLUENCING EATING BEHAVIOR AND NUTRITION

A. Taste and Smell

The gustatory and olfactory senses are very interdependent and therefore will be treated concurrently. Psychophysical investigations have traditionally been used to measure the acuity of the senses. Varying amounts of concentrations of substances are presented to individuals who then are required to discriminate differences in concentrations of a single substance or qualitatively different substances.[137] In summarizing previous research, Troll[138] has reported that taste sensitivity does not appear to change until approximately age 50, which has been associated with a progressive loss in the number of taste buds through the adult years. This loss first occurs in the anterior part of the tongue, according to Busse,[139] resulting in a decreased sensitivity to salty and sweet tastes prior to experiencing a decreased ability to detect bitter and sour sensations. The increased threshold, or increase in concentration of salt necessary before an individual indicates they have detected a difference in concentration, is not only associated with smoking and increased age, but also gender and illness.[140,141] Baker et al.[140] concluded that ability to detect a salty taste is decreased in smokers, in people with hypertension, and in men. The latter finding was clarified by Baker et al., who found that more elderly men than women smoke. Moore et al.[142] confirmed an age-related decline in sucrose sensitivity, but also noted a large variability within their sample of older persons. The decreased sucrose sensitivity was not found for all older people and therefore perhaps is not part of normal aging. Other research using both threshold and identification of pureed and blended foods via tasting, reported that older adults were at a disadvantage when compared to young adults, on both measures.[143,144]

The decreased sensitivity of smell is thought to be influenced by the decreased sensitivity to taste since the ability to differentiate between food odors and concentrations of chemicals has been reported to lessen with increased age.[143,145,146] Research by Schemper et al.[145] has indicated that this may not be solely a change in sensation, but also a cognitive limitation. Schemper and co-workers found that older persons displayed a poor ability not only to identify smells but also to generate labels for the smells during repeated presentations of the odors, when compared to younger adults. They also noted that older subjects failed to benefit in this task when they used self-generated labels, but they did benefit when they were given correct labels in their identification of odors. This suggests that the decreased ability for odor identification is not solely a sensory deficit, but also a cognitive deficit involving a memory and information processing component. Schiffman and Pasternak[143] also found that the ability to differentiate between food odors lessened with increased age, but the elderly exhibited differences in this ability depending upon which odor they were presented. They found the elderly were not only best at discriminating fruits from the rest of the stimuli, but also preferred the odor.

In general, food palatability has been found to be a very important factor influencing eating behavior.[147] Age-related losses in the senses of taste and smell consequently may significantly affect diet and nutrition.[148] Nutritious foods may very well taste bland to the elderly, encouraging a preference for strongly seasoned foods. Highly salted or sweetened foods, in turn, may work against the general welfare of this population by interacting with and accentuating other common health problems such as hypertension and obesity. Nutritional intervention by professionals must consider not only making the diets of older individuals healthy, but also "tasty" if one's appetite is to be maintained.

B. Vision

With increased age, gerontological researchers have noted an increased probability of visual impairment. Most notable are problems related to increased susceptibility to glare and reduced visual acuity. Fozard et al.[149] have pointed out that the lens of the eye becomes more opaque and less elastic as it thickens with increased age. Concurrently, the size of the pupil tends to decrease around age 50, which reduces the amount of light that reaches the retina. Therefore, brighter and more direct lighting is needed. Color vision appears to be unaffected through middle age. In late adulthood though, more light may be necessary to appreciate differences in colors. Also, colors will have to be made brighter and more intense so as not to appear dull. Gilbert[150] found a decreasing loss of sensitivity in individuals 60 years and over in the ability to discriminate between colors at the blue end of the color spectrum. Therefore, reds, oranges, and yellows are perceived the most distinctly by older individuals. The sensory ability of vision directly affects how attractive foods appear. Foods may appear similar if they are close together on the color spectrum and presented on the same dinner plate. Visual ability also directly affects the ability to prepare and eat nutritious meals. Kitchens demand good visual ability to avoid accidents and efficient functioning.

C. Hearing

Hearing ability begins to decline in the early 20s, with a gradual loss occurring over time. This loss appears to affect high-frequency tones more than low tones. Corso[151] has indicated that, as with vision, hearing deficits are not a universal phenomena of age-related change. Presbycusis (a decreased ability to detect and differentiate high-frequency sounds due to loss of neural function), has been reported to affect approximately 13% of persons 65+.[151] Overall, it has been estimated that approximately 30% of individuals in late adulthood exhibit some hearing loss.[151]

Hearing loss may significantly alter an individual's social and interpersonal environment. Presbycusis in particular has been shown to be associated with speech discrimination ability.[152] An older person with a hearing impairment may become self-conscious and even confused if conversation is misunderstood. An inappropriate response made by a hearing-impaired older person, because several words have been missed in a conversation, may be misconstrued by others as "senility", or even antisocial behavior instead of a hearing impairment. For some this may lead to an avoidance of social situations and social isolation. Given that meals are social situations, this may result in the hearing-impaired individual avoiding eating meals with others, or responding to meals negatively due to their association with frustration and embarrassment. A hearing impairment may also directly influence food preparation as many kitchen appliances make use of auditory bells and buzzers to signal food has reached some stage in its preparation. Auditory feedback is also used to judge whether an appliance is working properly or not (e.g., mixer, garbage disposal, electric coffee maker, refrigerator, etc.). Though the lack of hearing ability need not keep one from effectively and efficiently performing his/her culinary skills, it may be quite frustrating for an individual who has, for years, learned to rely on sound in the kitchen. Coupled with other often impending physical difficulties (e.g., increased difficulty in opening tightly closed lids due to arthritis, or reduced ability to read labels on food products), the loss of auditory sensitivity may unwittingly prove to be a significantly negative factor in the nutrition of the elderly.

IV. PSYCHOLOGICAL SYMBOLISM OF FOOD

A. The Association of Food and Early Attachment

The first interpersonal relationship experienced by an infant is placed in the context of caretaking and feeding. The caretaker's sensitivity to fulfillment of the infant's basic needs (e.g., feeding) have been described by Erikson[153] as an extremely salient factor in the formation of a secure, trusting attachment between the caretaker and infant. Within this first attachment relationship an infant experiences warmth, closeness, and security. Concurrently, food becomes associated with these emotions and lasting attitudes toward food are learned.[154-156] In fact, Freudian theorists have described later eating disorders as rooted in the early mother-infant attachment, whereby various impulses are triggered by the feeling of hunger.[157] Therefore, a caretaker's use of food as an inappropriate pacifier may result in obesity in the developing child (and later the adult) who has learned to decrease tension and anxiety by eating.

Adult food habits and preferences have been shown to be influenced by family feeding patterns experienced in childhood. The perpetuation of these food habits and preferences into adulthood, which have previously been associated with the family, may serve to enhance feelings of warmth and security.[24,158] Foods used as rewards or expressions of affection in childhood may have achieved special significance for a given person and be viewed by that individual as highly desirable. These "special foods" may continue to be used as a reward or pacifier in adulthood in times of stress, representing a positive link to an individual's past, and hence a source of comfort.[32,159,160] These foods that have some significance to an individual satisfy more than hunger; thus they can directly influence the food habits and nutrient intake of an individual. Berger[161] has cautioned professionals working with older adults that a major dietary modification for health reasons may potentially be very psychologically damaging to the elderly because of the symbolic association with certain foods.

B. Attitudes of Food in Adulthood

Pumpian-Mindlin[162] referred early on to the psychological symbolism of food. It has been suggested that food is a symbol for congeniality, sociability, friendship, security, praise, punishment, prestige, status, and power.[161-163] More recently, research by Axelson and Penfield[90] indicated that food and nutrition attitudes could be characterized by four independent factors: (1) social-adventuresome, (2) frugal-utilitarian, (3) qualitative-pleasurable, and (4) nutritious-healthful. These results clearly indicated that food is thought of and used for several purposes beyond the maintenance of good health or the satisfaction of hunger.

Food also appears to be saturated with age and sex symbolism in the American culture. Saxon and Etton[164] have noted that certain foods appear to be associated with masculinity, such as meat, while others, such as fruit and vegetables, appear to be associated with femininity. Foods associated with infancy, such as milk, cooked cereal, and pureed items, tend to be rejected as everyday food choices by adults.[164,165] Harrill et al.[165] found that female nursing home residents did prefer these latter items in times of stress, which was hypothesized to be due to some psychological significance of these items and their association with an earlier life stage.

The psychological symbolism inherent in food has several direct implications for the elderly. Chronic health problems may dictate that an older person eat foods "associated with infancy" or that they can no longer eat what is perceived as appropriate. This may reduce appetite and possibly threaten the older person's nutritional status.

V. SOCIAL SUPPORT AS A MODERATOR OF PHYSICAL AND PSYCHOLOGICAL HEALTH

It has been indicated that one attitude associated with food and eating is that of sociability, hospitality, and friendship.[90,163] Not only is food used as a medium for sociability, but perhaps more importantly, interaction with others has been shown to influence eating behavior, nutritional status, and health.[166] Cobb[167] has reported that individuals who are emotionally and instrumentally supported by others tend to be in better health when compared to individuals who lack a social support network. This finding is supported by mortality data which show the single, widowed, and divorced to have significantly higher death rates than married individuals.[168] Obviously, emotional adjustment influences differences found between those who are married vs. those are are not. However, even those individuals who live alone, and who are socially and emotionally adjusted, may experience poor health due to a lack of motivation to shop, cook, or eat by themselves.[54] Brockington and Lempert[169] concluded from their Stockport Survey that older individuals who ate at clubs or in the company of others ate more nutritiously. Eating alone is by definition incongruous with the attitude or belief that eating and food are mechanisms of sociability. Indeed, social isolation has been found to be correlated with poor nutrition.[84,170] Guthrie et al.[82] found that elderly individuals living in two-person households in rural Pennsylvania had a significantly more adequate intake of vitamins than individuals living in one-person households. Perhaps this finding was related to the earlier report by Neugarten[123] which indicated that women monitor the health of their husbands. Individuals who live with a spouse, or another person, may thus exhibit better health and nutrition for two reasons. First, the opportunity to eat with another person may make meals more pleasant as they offer opportunities for sociability. Second, health maintenance may be more active in a two-person household due to health monitoring by one's housemate.

Bradburn and Caplovits[171] have indicated that participation in social networks may lessen some of the negative events which are associated with the aging process. As it was described in Part I, the Title VII Nutrition Program for Older Americans (now Title III) that established congregate meal programs for individuals 60 years and older, recognized the saliency of social support as one of the four target goals of the program. These nutrition programs were implemented to (1) reduce social isolation, (2) promote better health, (3) promote the dignity of the individual, and (4) promote better nutrition. Evaluations of various congregate meal programs supported by Title III have reported, with some consistency, that participants exhibit improved morale, self-esteem, and life satisfaction, even more so than improved health or nutritional status.[172-174] Kim et al.[175] implemented a nutrition education program for nursing home residents. After 4 weeks, a significant improvement in eating habits was evidenced. Interestingly, Kim and associates concluded that their nutrition education program was probably not the sole source for the improved eating habits of the elderly nursing home residents, but also the social contact of the group itself, which no doubt increased feelings of sociability and morale.

VI. EMOTIONAL FACTORS INFLUENCING NUTRITION INTAKE IN THE ELDERLY

A. Role Loss, Depression, and Loneliness

The latter half of the adult life span has been characterized by many gerontologists as a time when loss is increasingly experienced. While role losses due to death of spouse, physical disability, and retirement are by no means negative events for all individuals, it is quite common for older individuals to report a period of unhappiness accompanying adjustment to such losses. The grief reaction that accompanies loss is frequently associated with depression, moodiness, and anxiety. It has been estimated that approximately 20% of individuals 65 years and older exhibit symptoms of depression at any one point in time.[176] Depression has consistently been associated with disturbances in eating behavior, resulting in a poor nutritional state.[177] Depression has been cited as one of the most common causes of nutritional inadequacy in older persons.[178] In an extreme grief reaction, individuals sometimes choose to isolate themselves from their friends and relatives, which may lead to inadequate dietary practices and changes in health status.[179,180] Epstein[181] has reported that many of the depressed aged also exhibit anorexia and weight loss, which in turn exaggerate depressive behavior. Under times of stress individuals may deny themselves enjoyable food if they are feeling worthless, guilty, or exhibit low self-esteem.[182] Harrill et al.[165] reported that female nursing home residents who were the most dissatisfied with their lives also had lower than normal calorie-nutrient intake for all nutrients except ascorbic acid.

Feelings of loneliness and depression have been associated with isolation.[180] Yet Ernst et al.[180] have distinguished between emotional, social, and physiological isolation. This distinction is important as all older individuals who live alone are not isolated. Likewise, elderly who live with others may feel isolated or be physically isolated due to illness or disability. It has been reported that the most severely malnourished elderly among the independent community dwellers are emotionally and socially isolated.[183] Exton-Smith,[184] in his classic longitudinal study of nutrition in Great Britain, found that malnutrition in the community-dwelling British elderly was very low, only exhibited by 3% of his sample. Similar findings were reported in a study conducted in Sweden by Anderson et al.[185] Their results indicated that subclinical malnutrition was more frequent in the elderly than actual malnutrition, especially in situations associated with stress, such as illness and widowhood. In cases such as these, feelings of loneliness and/or depression have been found to be associated with a disinterest in food and behavioral apathy with regard to eating.[186-188] Baum[189] and Hale[190] have also concluded from their independent research efforts that loneliness is also correlated with poor physical health.

Individuals with more positive self-concepts and an internal locus of control have been found to have better-quality diets.[75,132] Learner and Kivett,[42] in their study on perceived dietary adequacy among the rural elderly, noted a significant relation between low morale and frequency of diet problems.

B. Psychological Factors Associated with Eating Disorders

It has been indicated that obesity is a more pervasive problem in the elderly than is malnutrition caused by undernutrition.[1] Albanese[191] has indicated that an excess of calorie intake may be correlated with worry, boredom, or loneliness. Several researchers have empirically demonstrated that both humans and subhumans (e.g., rats) tend to increase food intake during emotional upset, boredom, and periods of stress in an attempt to attenuate the emotional responses to unpleasant stimuli.[192-194]

Early research by Schacter and Gross,[195] which was later supported by Rodin,[196] found that obese persons are more responsive to external stimuli associated with food than are normal weight persons. Obese individuals were found to eat more food than normal-weight individuals when they were led to believe more time had elapsed since their last meal than in fact had actually elapsed. This may partially explain why obesity is the pervasive problem in later adulthood when days become less time structured after a role change such as retirement.

C. Chronic Psychological Disorders Affecting Nutrition

It has been estimated that approximately 15% of older individuals (65+) exhibit some organic or functional mental disorder.[132] The chronically depressed or senile geriatric patient frequently exhibits not only disturbances in eating, such as decreased food intake, but also decreased physical activity.[132] Roth[197] has stated that individuals with organic brain syndromes always exhibit a diminished ability early on to obtain and prepare food; later, they exhibit a severely reduced ability to eat. Stahelin et al.[198] found two out of three meals to be refused by elderly patients with senile dementia (breakfast being the most accepted meal). Roth,[197] Stahelin et al.,[198] and Vincent and Gibson[199] all concluded that individuals in the later stages of senile dementia need nutritional supplementation (e.g., tube feeding or parenteral nutrition) to survive. Individuals with more severe mental disorders have been found to have reduced intakes of riboflavin, nicotinic acid, ascorbic acid, potassium, and iron.[200] Severe mental disorders have been associated with reduced nutrient intake due to an individual's inability to communicate needs, a loss of appetite, cognitive confusion, and in some cases as evidence of indirect self-destructive behavior.[199-201]

VII. NUTRITIONAL DEFICIENCIES AND ASSOCIATED PSYCHOLOGICAL EFFECTS

Emotional states not only influence eating behavior and nutritional status, but circularly diet and nutrition influence a person's psychological functioning and emotional state.[202,203] In one survey of older women with one or more nutrients below the 80% RDA, 40% reported unexplained tiredness, joint pains, and shortness of breath, all of which could be mistaken symptoms of depression or anxiety.[204] Whanger and Wang[205] presented the results from a study of geriatric psychiatric patients which found that over 70% who were admitted to a state hospital had inadequate or barely adequate diets. Of these individuals, 50% had borderline or low levels of fólic acid, while 12% had borderline or low levels of vitamin B_{12}. Deficiencies in folacin have been associated with the following behaviors: apathy, weakness, irritability, forgetfulness, lack of motivation, hostility, paranoia, and depression.[206] Vitamin B_{12} deficiency may result in loss of vibratory/position sense, paresthesia (a burning, tingling sensation), poor muscle coordination, ataxia (inability to coordinate voluntary body movements), confusion, memory loss, agitation, depression, delusions, hallucinations, and paranoia.[207] Similarly, Whanger[122] has reviewed research on B vitamin deficiencies and has noted the resultant behavioral symptomology of irritability, emotional instability, and loss of recent memory. Vitamin B_6 (pyridoxine) deficiency has been associated with irritability, weakness, nervousness, insomnia, apathy, and depression.[208] Thiamin deficiency (vitamin B_1) has been linked to a loss of appetite, depression, irritability, confusion, memory loss, inability to concentrate, and a sensitivity to noise.[209] Dakshinamurti[210] has described the early signs of a niacin deficiency as headaches, irritability, apprehension, sleeplessness, and emotional instability. Advanced cases of niacin deficiency may result in confusional psychosis with delirium and catatonia. Kinsman and Hood[211] have found hysteria, depression, fatigue, and hypochondriasis associated with a vitamin C deficiency.

Several researchers have alerted professionals to vitamin deficiencies as a possible source of causation for changes in psychological functioning.[212-214] Symptoms typically associated with senile dementia have been sometimes found to be caused by serious vitamin B deficiencies.[213] Although the frequency of the total dementias caused by nutritional deficiencies remains unclear (it has been estimated to be only 1 to 2% by Schneck[215]), it is still an area that should be addressed by professionals searching for sources of causation of disturbed psychological functioning in elderly persons.

Supplementation of vitamin C in a group of chronic psychiatric patients showed significant decreases in depression and manic and paranoid symptomatology.[216] Supplementation of serum folate was reported by Carney and Sheffield[217] to be associated with significant improvements in functional and organic psychosis. Similarly, Taylor[218] has reported that supplementation of vitamin C and B complex improved physical and mental functioning in a group of chronically ill and hospitalized elderly persons. It appears that more research is needed to disentangle the relationship between vitamin deficiency and emotional states and what role vitamin supplementation may play in this relationship.

VIII. CONCLUSION

Psychological factors have been shown to influence eating and resultant nutritional status in the elderly. One's appetite is not solely influenced by sensations of hunger or a physical need for food, but additionally by sensory perception (e.g., smell, taste, hearing, vision), emotional state, attitudes toward food, previous eating habits, and the company one dines with.[53,219] All of these factors must be considered in the nutritional assessment of the elderly individual.

Sensory perception can be a key element in one's nutritional status, especially for older individuals. Foods that have "lost" their aroma or taste may cease to be of interest. In their place one may substitute items that are less nutritious but yet are more pleasing to the senses, such as extremely sweet or salty snack foods. One may also be more reluctant to prepare foods if the task is becoming more difficult or less pleasurable due to diminished sight or hearing.

Attitudes toward food and eating habits shaped over many years can also be major factors influencing one's nutritional status. It has been shown that the psychological context, both historical and present, plays a part in the foods one eats. If a given food was thought to contain certain properties, either beneficial or harmful, when one was first learning about food and nutrition, it is likely that information will be acted upon throughout life. The penchant of society in general for red meat may be an example of this idea. On the other hand, if one grew up in a situation or era where just obtaining a minimally sufficient amount of food was very often difficult, e.g., the Depression years, the tendency to overindulge in more plentiful times may be a common occurrence. The admonition often given children to "clean your plate" may also be a carryover from that earlier, more difficult period.

Additionally, it has been shown that one's perception of isolation, whether actually real or not, may significantly impact on eating behavior and nutrition. For the elderly who, almost by definition, find themselves at somewhat greater risk of experiencing some sort of isolation and loneliness, this linkage may be of heightened importance and concern. Loss of a lifetime mate or an increasing number of health problems which serve to close more tightly one's social range are two very real factors for many of the elderly.

The assessment of the nutritional status of a population or an individual is, at best, complex. Experts in physiology, medicine, and nutrition are not yet in agreement about what comprises good nutrition, either for the population as a whole or for a subgroup such as the elderly. What does seem clear, however, is that those interested in understanding and improving the nutritional status of any group or individual will need to consider the social and psychological as well as the chemical and physiological elements. This need to consider the nonorganic may be important, most of all, with the elderly.

IX. EPILOGUE: WHERE DO WE GO FROM HERE?*

From the vantage point of the social scientist, a wide range of issues implicated in the nutritional status of older people remain to be explored. Like so many other crucial topics in social gerontology, nutrition is demanding of an interdisciplinary focus which will draw expertise from the spectrum of social and behavioral sciences. An early item on the research agenda must be a question of focus. At a global level it is mandatory that we concentrate not only on the institutionalized or indigent population, but also on individuals from various social classes and all walks of life. To date, too strong an emphasis has been placed on low-income and disadvantaged elderly. Though certainly necessary, such a tendency can lead to the perpetuation of a new set of myths which may impede understanding and intervention for the vast majority of older Americans. At the same time there is still a need for a better understanding of the role of ethnicity in shaping nutritional patterns in later life. Normative analysis without trammeling on the basis of self-identity will be a difficult task but it is one which will remain pressing as long as there are distinct subcultures in our pluralistic society.

As a corollary, social class membership is also deserving of greater and more sensitive attention. At the very least, a composite index of SES is necessary. Merely measuring income or educational achievement will not suffice; neither will a simple ranking of occupational prestige. Social class is necessarily comprised of the three components taken together, and must be dealt with as such. The key element underlying the concept of social classes is the differential access to opportunity accorded to individuals belonging to one class category or another. In terms of nutrition the task is not merely to ascertain position in a hierarchical ranking scheme, but to determine the extent to which that position shapes eating behavior, differential knowledge, attitudes, and subjective views of appropriate nutrition. Perhaps most importantly, a theoretical paradigm to guide research must be adopted in place of what has, to date, been a piecemeal approach. As the two parts of this chapter

---

* Authors: Hendricks, Calasanti, Gibbs, and Turner.

should make clear, adequate nutrition results from a confluence of social-familial, personal-physiological, and financial resources. For the social scientist interested in nutrition the next step should be to integrate these three dimensions of personal resource inventories into a processual or transactional model.

Having formulated an overarching theoretical perspective, the research gaps can be addressed and steps taken to alleviate shortcomings. First, however, renewed efforts must also be undertaken to avoid confounding age, period, and cohort effects. The greatest share of the existing body of research has utilized cross-sectional methodologies, which, while they are parsimonious, do not discriminate age differences from age changes. Despite their expense, longitudinal studies should become the design of choice. Short of that, more methodologically precise cross-lagged investigations can be utilized. In addition, models for assessing change through linear structural analysis strategies need to be adopted in order to draw valid conclusions from observed changes. Finally, sampling problems have imposed serious limitations on our present state of knowledge. This is not to suggest that only scientifically acceptable survey research should be done — there are many problems which are more appropriately investigated using ethnographic methodologies — only to contend that more serious consideration be given to the generalizability of the findings.

As was noted in Part II of this chapter, psychologists working in the area of nutrition have tended to focus on situational and emotional factors linked primarily with eating disorders. In the future, research efforts must also be made to identify constitutive elements of normal everyday nutritional behavior. While the present discussion attempts to draw attention to those psychological and emotional factors hypothesized as being increasingly important in relation to dietary status over the course of adult life, little is yet known about baseline psychological variables which serve as a general foundation for nutrition. Of course, understanding the etiology of eating disorders will remain a priority but it would be a fallacy to assume that these same factors comprise a comprehensive psychology of everyday eating. Only when research has more fully addressed the normative psychological components of sound nutrition can age differences be interpreted in a valid and reliable manner.

Though a sizable body of literature documents changing psychomotor processes during adulthood, relatively little insight into how these affect actual food consumption has been forthcoming. Knowing, for example, that the savoriness of food influences eating and that losses in any, or all, of the senses may influence appetite does not reveal what actually occurs when sensory deficits are realized. In short, more research is needed to specifically determine what adjustments individuals make in their nutritional intake due to alterations in sensory acuity.

Those dealing with adult nutrition, either directly or indirectly, must have a wide-ranging interdisciplinary perspective if their efforts are to be brought to fruition. This chapter attempts to add to the breadth of focus in nutrition research. Future success, however, demands that more work be done to effectively and efficiently bring together the myriad physiological, social, psychological, and economic factors impinging on the elderly. Only then will we be able to understand and react to their nutritional desires and needs.

REFERENCES

1. **Sherwood, S.,** Gerontology and the sociology of food and eating, *Int. J. Aging Hum. Dev.*, 1, 61, 1970.
2. *America in Transition: An Aging Society*, Current Population Reports, Ser. P-23, No. 128, U.S. Bureau of the Census, Washington, D.C., 1983.
3. White House Conference on Aging, *Chartbook on Aging in America*, U.S. Government Printing Office, Washington, D.C., 1981.

4. National Center for Health Statistics, *Vital and Current Health Statistics*, Ser. 3, No. 22, Public Health Service, U.S. Department of Health and Human Services, Washington, D.C., 1982.
5. **Kitagawa, E. M. and Hauser, P. M.**, *Differential Mortality in the United States: A Study in Socioeconomic Epidemiology*, Harvard University Press, Cambridge, Mass., 1973.
6. **Gelfand, D.**, *Aging: The Ethnic Factor*, Little, Brown, Boston, 1982.
7. **Cowgill, D. O.**, The demography of age, in *Daily Needs and Interests of Older People*, Morris, W. W. and Bader, I. M., Eds., Charles C Thomas, Springfield, Ill., 1983, 28.
8. **Sinicropi, A. V.**, Economic security and income maintenance, in *Daily Needs and Interests of Older People*, Morris, W. W. and Bader, I. M., Eds., Charles C Thomas, Springfield, Ill., 1983, 53.
9. **Ehrenhalt, S. M.**, in "Older men left in job squeeze", Mike McNamee, *USA Today*, November 28, 1983.
10. U.S. Department of Commerce, Money Income and Poverty Status of Families and Persons in the United States: 1981, Current Population Reports, Ser. P-60, No. 134, Washington, D.C., 1982.
11. Social Security Administration, Relative importance of the aged, 1980, *Social Security Bull.*, 46, 9, 1983.
12. *Health: United States, 1979*, (PHS-801232), Public Health Service, U.S. Department of Health, Education and Welfare, Hyattsville, Md., 1979.
13. **Morrison, S. D.**, Nutrition and longevity, *Nutr. Rev.*, 41, 133, 1983.
14. **Fries, J. F.**, Aging, natural death, and the compression of morbidity, *N. Engl. J. Med.*, 303, 130, 1980.
15. **Hendricks, J. and Hendricks, D. C.**, *Aging in Mass Society: Myths and Realities*, 2nd ed., Little, Brown, Boston, 1981.
16. **Schneider, E. L. and Brody, J. A.**, Aging, natural death and the compression of morbidity: another view, *N. Engl. J. Med.*, 309, 854, 1983.
17. **Colvez, A. and Blanchet, M.**, Disability trends in the United States population 1966-76: analysis of reported causes, *Am. J. Public Health*, 71, 464, 1981.
18. **Katz, S., Branch, L. G., Branson, M. H., Papsidero, J. A., Beck, J. A., and Creer, D. S.**, Active life expectancy, *N. Engl. J. Med.*, 309, 1218, 1983.
19. **Grundy, E.**, Demography and old age, *J. Am. Geriatr. Soc.*, 31, 325, 1983.
20. **Davis, M. A. and Randall, E.**, Social change and food habits of the elderly, in *Aging and Society: Selected Reviews of Recent Research*, Riley, N. W., Hesas, B. B., and Bond, K., Eds., Lawrence Erlbaum Associates Publishers, Hillsdale, N.J., 1983, 199.
21. **Mead, M.**, *Food Habits Research: Problems of the 1960s*, National Academy of Sciences, Washington, D.C., 1964.
22. **Rux, J. M.**, Thoughts on culture, nutrition and the aged, *J. Nutr. Elderly*, 1, 15, 1981.
23. **Pliner, P.**, Family resemblance in food preferences, *J. Nutr. Educ.*, 15, 137, 1983.
24. **Boykin, L. S.**, Soul foods for some older Americans, *J. Am. Geriatr. Soc.*, 23, 38, 1975.
25. **Howell, S. C. and Loeb, M. B.**, Culture, myths, and food preferences among aged, *Gerontology*, 9 (3), 31, 1969.
26. **Bass, M. A., Wakefield, L., and Kolasa, K.**, *Community Nutrition and Individual Food Behavior*, Burgess, Minneapolis, 1979.
27. **Sanjur, D.**, *Sociocultural Perspectives of Nutrition*, Prentice-Hall, Englewood Cliffs, N.J., 1981.
28. **Foley, C., Hertzler, A., and Anderson H.**, Attitudes and food habits, *J. Am. Dietet. Assoc.*, 75, 13, 1979.
29. **Jalso, S. B., Burns, M. M., and Rivers, J. M.**, Nutrition beliefs and practices, *J. Am. Dietet. Assoc.*, 47, 263, 1965.
30. **Fewster, W. I., Bostian, L. R., and Powers, R. D.**, Measuring the connotative meanings of foods, *Home Econ. Res. J.*, 2, 44, 1973.
31. **Davis, A. K. and Davis, R. L.**, Food facts, fads, fallacies and folklore of the elderly, in *Handbook of Geriatric Nutrition: Principles and Applications for Nutrition and Diet in Aging*, Hsu, J. M. and Davis, R. L., Eds., Noyes, Park Rige, N.J., 1981, 328.
32. **Natow, A. B. and Heslin, J.**, *Geriatric Nutrition*, CBI, Boston, 1980.
33. **Elder, G. H.**, *Children of the Great Depression*, University of Chicago Press, Chicago, 1974.
34. **Miller, H. W.**, *Plan and Operation of the Health and Nutrition Examination Survey: United States, 1971-1973*, Ser. 1, No. 10a, National Center for Health Statistics, Vital and Health Statistics Public Health Service, U.S. Department of Health and Human Services, Washington, D.C., 1973.
35. Center for Disease Control, *Ten-State Nutrition Survey, 1968—1970*, Health Services and Mental Health Administration, Washington, D.C., 1972.
36. **Brody, E. M., Kleban, M. H., and Moles, E.**, What older people do about their day-to-day mental health and physical health symptoms, *J. Am. Geriatr. Soc.*, 31, 489, 1983.
37. **Blau, Z. S.**, *Old Age in a Changing Society*, Franklin Watts, New York, 1973.
38. **Stub, H. R.**, *The Social Consequences of Long Life*, Charles C Thomas, Springfield, Ill., 1982.
39. **Watkin, D. M.**, Nutrition for the aging and the aged, in *Modern Nutrition in Health and Disease*, Goodhart, R. S. and Shils, M. E., Eds., Lea & Febiger, Philadelphia, 1980, 781.

40. **McIntosh, W. A. and Shifflett, P. A.,** Interrelations among instrumental forms of social support and the dietary adequacy among the elderly, presented at Rural Sociological Society meetings, 1983.

41. **Grandjean, A. C., Kerth, L. L., Kara, G. C., Smith, J. L., and Schaefer, N. E.,** Nutritional status of elderly participants in a congregate meals program, *J. Am. Dietet. Assoc.,* 78, 324, 1981.

42. **Learner, R. M. and Kivett, V. R.,** Discriminators of perceived dietary inadequacy among rural elderly, *J. Am. Dietet. Assoc.,* 78, 327, 1981.

43. **Glick, P. C.,** Updating the family life cycle, *J. Marr. Fam.,* 39, 5, 1977.

44. **Hertzler, A. A. and Vaughan, C. E.,** The relationship of family structure and interaction to nutrition, *J. Am. Dietet. Assoc.,* 74, 23, 1979.

45. **Coughenour, C. M.,** Functional aspects of food consumption activity and family life cycle stages, *J. Marr. Fam.,* 34, 656, 1972.

46. **Cross, B., Herrmann, R. O., and Warland, R. H.,** Effect of family life-cycle stage on concerns about food selection, *J. Am. Dietet. Assoc.,* 67, 131, 1975.

47. **Cosper, B. A. and Wakerfield, L. M.,** Food choices of women, *J. Am. Dietet. Assoc.,* 66, 152, 1975.

48. **Singleton, N., Kirby, A. L., and Overstreet, M. H.,** Snacking patterns of elderly females, *J. Nutr. Elderly,* 2, 3, 1982.

49. **Osborn, M. O.,** Nutrition of the aged, in *Daily Needs and Interests of Older People,* Morris, W. W. and Bader, I. M., Eds., Charles C Thomas, Springfield, Ill., 1983, 170.

50. **Kalish, R. A.,** *Late Adulthood Perspectives on Human Development,* Brooks/Cole, Monterey, Calif., 1982.

51. **Gubrium, J. F.,** Being single in old age, in *Time, Roles, and Self in Old Age,* Gubrium, J. F., Ed., Human Sciences Press, New York, 1976.

52. **Shifflett, P. A. and McIntosh, W. A.,** Correlates and consequences of social isolation for the diet of the elderly, presented to the Rural Sociological Society meetings, 1983.

53. **Roe, D. A.,** *Geriatric Nutrition,* Prentice-Hall, Englewood Cliffs, N.J., 1983.

54. **Brown, E. L.,** Factors influencing food choices and intake, *Geriatrics,* September, 89, 1976.

55. **Yurkiw, M. A., Krondl, A., Krondl, M., and Coleman, P.,** Anthropometric and dietary assessment of select single-living urban elderly, *J. Nutr. Elderly,* 2, 3, 1983.

56. **Anon.,** Nutritional status of a healthy, middle class, elderly population, *Nutr. Rev.,* 41, 143, 1983.

57. **Edwards, R.,** *Contested Terrain: The Transformation of the Workplace in the Twentieth Century,* Basic Books, New York, 1979.

58. **Edwards, R., Reich, M., and Gordon, D., Eds.,** *Labor Market Segmentation,* D.C. Heath, Lexington, Mass., 1975.

59. **Hendricks, J. A. and McAllister, C. E.,** An alternative perspective on retirement: a dual economic approach, *Ageing Soc.,* 3, 279, 1983.

60. **Baran, P. and Sweezy, P.,** *Monopoly Capital,* Monthly Review Press, New York, 1966.

61. **Beck, E. M., Horan, P., and Tolbert, C.,** Social stratification in industrial society: further evidence for a structural alternative, *Am. Sociol. Rev.,* 45, 712, 1980.

62. **Dowd, J. J.,** *Stratification Among the Aged,* Brooks/Cole, Monterey, Calif., 1980.

63. **Myles, J.,** Income inequality and status maintenance: concepts, methods, and measures, *Res. Aging,* 3, 123, 1981.

64. **Roe, D. A.,** *Clinical Nutrition for the Health Scientist,* CRC Press, Boca Raton, Fla., 1979.

65. **Reaburn, J. A., Krondl, M., and Lau, D.,** Social determinants in food selection, *J. Am. Dietet. Assoc.,* 74, 1979.

66. **Grotowski, M. L. and Sims, L. S.,** Nutritional knowledge, attitudes, and dietary practices of the elderly, *J. Am. Dietet. Assoc.,* 72, 499, 1978.

67. **Caliendo, M. A. and Smith, J.,** Factors influencing the nutrition knowledge and dietary intake of participants in the Title III-C Meal Program, *J. Nutr. Elderly,* 1, 65, 1981.

68. **Rawson, I. G., Weinberg, E. I., Herold, J. A., and Holtz, J.,** Nutrition of rural elderly in southwestern Pennsylvania, *Gerontologist,* 18, 24, 1978.

69. **Caliendo, M. A. and Batcher, M.,** Factors influencing the dietary status of participants in the national nutrition program for the elderly. II. Relationships between dietary quality, program participation, and selected variables, *J. Nutr. Elderly,* 1, 41, 1980.

70. **Reid, D. L. and Miles, J. E.,** Food habits and nutrient intakes of noninstitutionalized senior citizens, *Can. J. Public Health,* 68, 154, 1977.

71. **Sherman, E. M. and Brittan, M. R.,** Contemporary food gatherers: a study of food shopping habits of an elderly urban population, *Gerontologist,* 13, 358, 1973.

72. **Slesinger, D. P., McDivitt, M., and O'Donnell, F. M.,** Food patterns in an urban population: age and sociodemographic correlates, *J. Gerontol.,* 35, 432, 1980.

73. **LeClerc, H. L. and Thornbury, M. E.,** Dietary intakes of Title III Meal Program recipients and non-recipients, *J. Am. Dietet. Assoc.,* 83, 573, 1983.

74. **Singleton, N., Overstreet, M. H., and Schilling, P. E.,** Dietary intakes and characteristics of two groups of elderly females, *J. Nutr. Elderly,* 1, 77, 1980.

75. **Schafer, R. B. and Keith, P. M.,** Social-psychological factors in the dietary quality of married and single elderly, *J. Am. Dietet. Assoc.*, 81, 30, 1982.

76. **Guseman, P. K., Sapp, S. G., and McIntosh, W. A.,** Cultural components of consumer demands for agricultural products, presented at Rural Sociological Society meetings, 1983.

77. **Natow, A. B. and Heslin, J. A.,** Understanding the cultural food practices of elderly observant Jews, *J. Nutr. Elderly*, 2, 49, 1982.

78. **Chen, P. N.,** Minority elderly: continuity/discontinuity of life patterns in nutrition programs, *J. Nutr. Elderly*, 1, 65, 1980.

79. **Munro, H. N.,** The status of the elderly: major gaps in nutrient allowances, *J. Am. Dietet. Assoc.*, 76, 137, 1980.

80. **Edwards, S. J.,** Nutrition and lifestyle, in *Nutrition in the Middle and Later Years*, Feldman, E. B., Ed., PSG, Boston, 1983, 1.

81. **Bazzarre, T. L., Yarhas, J. A., and Wu, S. L.,** Measures of food intake among rural elderly, *J. Nutr. Elderly*, 2, 3, 1983.

82. **Guthrie, H. A., Black, K., and Madden, J. P.,** Nutritional practices of elderly citizens in rural Pennsylvania, *Gerontologist*, 12, 330, 1972.

83. **Baird, P. C. and Schutz, H. G.,** Life style correlates of dietary and biochemical measures of nutrition, *J. Am. Dietet. Assoc.*, 76, 228, 1980.

84. **Sherwood, S.,** Sociology of food and eating: implications for action for the elderly, *Am. J. Clin. Nutr.*, 26, 1108, 1973.

85. **Ashford, N. and Holloway, F. M.,** Transportation patterns of older people in six urban centers, *Gerontologist*, 12, 43, 1972.

86. **Roundtree, J. L. and Tinklin, G. L.,** Food beliefs and practices of selected senior citizens, *Gerontologist*, 15, 537, 1975.

87. **Hertzler, A. A.,** Recipes and nutrition education, *J. Am. Dietet. Assoc.*, 83, 466, 1983.

88. **Sorenson, A. W. and Ford, M.,** Diet and health for senior citizens: workshops by the health team, *Gerontologist*, 21, 257, 1981.

89. **Clarke, M. and Wakefield, L. M.,** Food choices of institutionalized vs. independent-living elderly, *J. Am. Dietet. Assoc.*, 66, 600, 1975.

90. **Axelson, M. L. and Penfield, M. P.,** Food and nutrition related attitudes of elderly persons living alone, *J. Nutr. Educ.*, 15, 23, 1983.

91. **Fernandes, J.,** Undernutrition among the elderly, *J. Nutr. Elderly*, 1, 79, 1981.

92. **Todhunter, E. N.,** Life style and nutrient intake in the elderly, in *Dimensions of Aging*, Hendricks, J. and Hendricks, C. D., Eds., Winthrop Publishers, Cambridge, Mass., 1979, 113.

93. **Endres, J. M., Theobald, L., Galligos, C. R., and Sawicki, M.,** Energy and nutrient content of menus, foods served, and foods consumed by institutionalized elderly, *J. Nutr. Elderly*, 2, 3, 1982.

94. **Fusillo, A. E. and Beloian, A. M.,** Consumer nutrition knowledge and self-reported food shopping behavior, *Am. J. Public Health*, 67, 846, 1977.

95. **Sempos, C. T., Johnson, N. E., Elmer, P. J., Allington, J. K., and Matthews, M. E.,** A dietary survey of 14 Wisconsin nursing homes, *J. Am. Dietet. Assoc.*, 81, 35, 1982.

96. **Lefkowitz, L.,** Food demonstration as a social activity in a nursing home, *J. Nutr. Elderly*, 2, 31, 1982.

97. **McConnel, C. E. and Deljavan, F.,** Consumption patterns of the retired household, *J. Geriatr.*, 38, 480, 1983.

98. National Dairy Council, Nutrition of the elderly, Dairy Council Dig., 48, 1, 1977.

99. **Weimer, J. P.,** The nutritional status of the elderly, *J. Nutr. Elderly*, 2, 17, 1983.

100. **Krout, J. A.,** Knowledge and use of services by the elderly: a critical review of the literature, *Int. J. Aging Hum. Dev.*, 17, 153, 1983.

101. **Caliendo, M. A.,** Factors influencing the dietary status of participants in the national nutrition program for the elderly. I. Population characteristics and nutritional intakes, *J. Nutr. Elderly*, 1, 23, 1980.

102. **Burkhardt, J. E., Lago, A. M., and Blattenberger, L. B.,** Factors affecting the demand for congregate meals at nutrition sites, *J. Gerontol.*, 38, 614, 1983.

103. **Skelly, F. R.,** The attitudes of the consumer, *Nutr. Rev.*, (Suppl.), 35, 1982.

104. **Caliendo, M. A. and Smith, J.,** Preliminary observations on the dietary status of participants in the Title III-C meal program, *J. Nutr. Elderly*, 1, 21, 1981.

105. **Wong, H., Korndl, M., and Williams, J. I.,** Long-term effect of a nutrition intervention program for the elderly, *J. Nutr. Elderly*, 2, 31, 1982.

106. **Costantakos, C. M. and Schaeffer, B.,** Nutrition education for the elderly: a preprofessional experience for foods and nutrition students, *J. Nutr. Elderly*, 1, 17, 1981.

107. **Shannon, B. M., Smiciklas-Wright, H., Davis, B. W., and Lewis, C.,** A peer educator approach to nutrition for the elderly, *Gerontologist*, 23, 123, 1983.

108. **Ludman, E. K.,** The emergent curriculum: a tool for nutrition education programs for the elderly, *J. Nutr. Elderly*, 2, 17, 1983.

109. **Natow, A. B. and Heslin, J. A.,** Nutrition education in the later years, *J. Nutr. Elderly,* 1, 101, 1981.
110. **Kaplan, J.,** Three perspectives on the 1981 White House Conference on Aging: services, *Gerontologist,* 22, 127, 1982.
111. **Houseman, D. H.,** Food cooperatives and community gardens save money for the elderly, *Aging,* 337, 20, 1983.
112. **Minkler, M.,** Social support and social action organizing in a "grey ghetto": the Tenderloin Experience, *Int. Q. Comm. Health Ed.,* 3(1), 3, 1983.
113. **Vigersky, R. A., Ed.,** *Anorexia Nervosa,* Raven Press, New York, 1977.
114. **Lerner, R. M. and Hultsch, D. F.,** *Human Development: A Life-Span Perspective,* McGraw-Hill, New York, 1983.
115. **Bennett, R. and Eckman, J.,** Attitudes toward aging: a critical examination of recent literature and implications for future research, in *The Psychology of Adult Development and Aging,* Eisdorder, C. and Lawton, M. P., Eds., American Psychological Association, Washington, D.C., 1973, 575.
116. **Schwartz, A. N., Snyder, C. L., and Peterson, J. A.,** *Aging and Life: An Introduction to Gerontology,* 2nd ed., Holt, Rinehart & Winston, New York, 1984.
117. **Howell, S. C. and Loeb, M. B.,** Nutrition and aging. Culture, myths and food preferences among aged, *Gerontologist,* 9(3), 66, 1969.
118. **Freedman, M. L. and Marcus, D. L.,** Anemia and the elderly: is it physiological or pathology? *Am. J. Med. Sci.,* 280, 81, 1980.
119. **Shanas, E. and Maddox, G.,** Aging, health, and the organization of health resources, in *Handbook of Aging and the Social Sciences,* Binstock, R. and Shanas, E., Eds., Van Nostrand Reinhold, New York, 1976, 592.
120. **Maddox, G.,** Self-assessments of health status: a longitudinal study of selected elderly subjects, *J. Chronic Dis.,* 17, 449, 1964.
121. **Ferraro, K. F.,** Self-ratings of health among the old and the old-old, *J. Health Soc. Behav.,* 21, 377, 1980.
122. **Whanger, A. D.,** Nutrition, diet and exercise, in *Handbook of Geriatric Psychiatry,* Busse, E. W. and Blazer, D. B., Eds., Van Nostrand Reinhold, New York, 1980, 473.
123. **Neugarten, B. L.,** The awareness of middle age, in *Middle Age and Aging,* Neugarten, B. L., Eds., University of Chicago Press, Chicago, 1968, 93.
124. **Huyck, M. H.,** Sex, gender and aging, *Humanitas,* 13, 1977.
125. **Maddox, G. L. and Douglas, E. G.,** Self-assessment of health, *J. Health Soc. Behav.,* 14, 87, 1973.
126. **Barrows, C. H. and Roeder, L. M.,** Nutrition, in *Handbook of the Biology of Aging,* Finch, C. and Hayflick, L., Eds., Van Nostrand Reinhold, New York, 1977.
127. **Edwards, J. N. and Klemmack, D. L.,** Correlates of life satisfaction: a re-examination, *J. Gerontol.,* 28, 497, 1973.
128. **Medley, M. L.,** Satisfaction with life among persons 65 years and older, *J. Gerontol.,* 31, 443, 1976.
129. **Barfield, R. E. and Morgon, J. N.,** Trends in satisfaction with retirement, *Gerontologist,* 18, 19, 1978.
130. **Beck, S. H.,** Adjustment to and satisfaction with retirement, *J. Gerontol.,* 37, 616, 1982.
131. **Schafer, R. B.,** The self-concept as a factor in diet selection and quality, *J. Nutr. Educ.,* 11, 37, 1979.
132. **Crapo, P.,** Nutrition in the aged, in *Clinical Internal Medicine in the Aged,* Schrier, R. W., Ed., W. B. Saunders, Philadelphia, 1982, 167.
133. **Troll, L. E.,** *Early and Middle Adulthood,* Brooks/Cole, Monterey, Calif., 1975.
134. **Butler, R. N.,** Why are older consumers so susceptible? *Geriatrics,* 23, 83, 1968.
135. **Schafer, R. and Yetley, E. A.,** Social psychology of food faddism, *J. Am. Dietet. Assoc.,* 66, 129, 1975.
136. **Kuske, T.,** Quackery and fad diets, in *Nutrition in the Middle and Later Years,* Feldinen, E. G., Ed., John Wright, Boston, 1983, 291.
137. **Engen, T.,** Taste and smell, in *Handbook of the Psychology of Aging,* Birren, J. and Schaie, K. W., Eds., Van Nostrand Reinhold, New York, 1977, 554.
138. **Troll, L. E.,** *Continuations: Adult Development and Aging,* Brooks/Cole, Monterey, Calif., 1982.
139. **Busse, W. E.,** How mind, body, and environment influence nutrition in the elderly, *Postgrad. Med.,* 63, 118, 1978.
140. **Baker, K. A., Didcock, E. A., Kemm, J. R., and Patrick, J. M.,** Effect of age, sex, and illness on salt taste detection thresholds, *Age Ageing,* 12, 159, 1983.
141. **Grzegorczyk, P. B., Jones, S. W., and Mistretta, C. M.,** Age-related differences in salt taste acuity, *J. Gerontol.,* 34, 834, 1979.
142. **Moore, L. M., Nielsen, C. R., and Mistretta, C. M.,** Sucrose taste thresholds: age-related differences, *J. Gerontol.,* 37, 64, 1982.
143. **Schiffman, S. and Pasternak, M.,** Decreased discrimination of food odors in the elderly, *J. Gerontol.,* 34, 73, 1979.
144. **Schiffman, S.,** Food recognition by the elderly, *J. Gerontol.,* 32, 586, 1977.

145. **Schemper, T., Voss, S., and Cain, W. S.,** Odor identification in young and elderly persons: sensory and cognitive limitations, *J. Gerontol.,* 36, 446, 1981.
146. **Rovee, C. K., Cohen, R. Y., and Shlapak, W.,** Life-span stability in olfactory sensitivity, *Dev. Psychol.,* 11, 311, 1975.
147. **Lytle, L. D.,** Control of eating behavior, in *Nutrition and the Brain,* Vol. 2, Wurtman, R. J. and Wurtman, J. J., Eds., Raven Press, New York, 1977, 1.
148. **Huyck, M. H. and Hoyer, W. J.,** *Adult Development and Aging,* Wadsworth, Belmont, Calif., 1982.
149. **Fozard, J., Wolf, E., Bell, B., McFarland, R., and Podolsky, S.,** Visual perception and communication, in *Handbook of the Psychology of Aging,* Birren, J. and Schaie, K. W., Eds., Van Nostrand Reinhold, New York, 1977, 497.
150. **Gilbert, J. G.,** Age changes in color matching, *J. Gerontol.,* 12, 210, 1957.
151. **Corso, J. F.,** Auditory perception and communication, in *Handbook of the Psychology of Aging,* Birren, J. E. and Schaie, K. W., Eds., Van Nostrand Reinhold, New York, 1977, 535.
152. **Anderson, R. G., Simpson, K., and Roeser, R.,** Auditory dysfunction and rehabilitation, *Geriatrics,* 38, 101, 1983.
153. **Erikson, E.,** *Childhood and Society,* Norton, New York, 1950.
154. **Babcock, C. G.,** Food and its emotional significance, *J. Am. Dietet. Assoc.,* 24, 390, 1948.
155. **Menzies, I. E. P.,** Psychosocial aspects of eating, *J. Psychosom. Res.,* 14, 223, 1970.
156. **Savitsky, E.,** Psychological factors in nutrition of the aged, *Soc. Casework,* 34, 435, 1953.
157. **Bruch, H.,** The treatment of eating disorders, *Mayo Clin. Proc.,* 51, 266, 1976.
158. **Fathauer, G. H.,** Good habits - an anthropologists' view, *J. Am. Dietet. Assoc.,* 37, 335, 1960.
159. **Gift, H.,** *Nutrition, Behavior and Change,* Prentice-Hall, Englewood Cliffs, N.J., 1972.
160. **Robinson, C. H. and Lawler, M.,** *Normal and Therapeutic Nutrition,* 15th ed., Macmillan, New York, 1977.
161. **Berger, R.,** Nutritional needs of the aged, in *Nursing and the Aged,* Burnside, I. M., Ed., McGraw-Hill, New York, 1976.
162. **Pumpian-Mindlin, E.,** The meanings of food, *J. Am. Dietet. Assoc.,* 30, 576, 1954.
163. **Chappell, M.,** The language of food, *Am. J. Nurs.,* 72, 1294, 1972.
164. **Saxon, S. V. and Etten, M. J.,** Psychological aspects of nutrition in aging, in *Handbook of Geriatric Nutrition,* Hsu, J. M. and Davis, R. L., Eds., Noyes, Park Ridge, N.J., 1981, 19.
165. **Harrill, I., Erbes, C., and Schwartz, C.,** Observations on food acceptance by elderly women, *Gerontologist,* 16, 349, 1976.
166. **Weinberg, J.,** Psychological implications of the nutritional needs of the elderly, *J. Am. Dietet. Assoc.,* 60, 293, 1972.
167. **Cobb, S.,** Social support as a moderator of life stress, *J. Psychosom. Med.,* 38, 300, 1976.
168. **Lynch, J. J.,** *The Broken Heart,* Basic Books, New York, 1977.
169. **Brockington, F. and Lempert, S. M.,** *The Stockport Survey, The Social Needs of the Over 80's,* University Press, Manchester, England, 1976.
170. **Exton-Smith, A. N. and Caird, F. K.,** Eds., *Metabolic and Nutritional Disorders in the Elderly,* John Wright & Sons, Bristol, England, 1980.
171. **Bradburn, N. and Caplovits, D.,** *Reports on Happiness,* Aldine, Chicago, 1965.
172. **Posner, B. M.,** *Nutrition and the Elderly,* D.C. Heath, Lexington, Mass., 1979.
173. **Fowler, F. J. and McCalla, M. E.,** Need and Utilization of Services Among the Aged in Greater Boston, Report to the Administration on Aging, Washington, D.C., 1969.
174. **Postma, J. S.,** The Characteristics and Needs of the Eugene-Springfield Nutrition Congregate Meals Program Participation and their Perception of the Program's Effects and Operation, Doctoral dissertation, University of Oregon, Eugene, 1974.
175. **Kim, S., Schriver, J. E., and Campbell, K. M.,** Nutrition education for nursing home residents, *J. Am. Dietet. Assoc.,* 78, 362, 1981.
176. **Schaffer, C. B. and Donlon, P. T.,** Medical causes of psychiatric symptoms in the elderly, *Clin. Gerontol.,* 1, 3, 1983.
177. **Garetz, F. K.,** Breaking the dangerous cycle of depression and faulty nutrition, *Geriatrics,* 33, 73, 1976.
178. **Butler, R. N. and Lewis, M. I.,** *Aging and Mental Health,* C.V. Mosby, St. Louis, 1982.
179. **Brozek, J.,** Nutrition, malnutrition and behavior, *Ann. Rev. Psychol.,* 29, 157, 1978.
180. **Ernst, P., Beran, B., Safford, F., and Kleinhauz, M.,** Isolation and the symptoms of chronic brain syndrome, *Gerontologist,* 118, 468, 1978.
181. **Epstein, L. J.,** Symposium on age differentiation in depressive illness, depression in the elderly, *J. Gerontol.,* 31, 278, 1976.
182. **Moore, H. B.,** Psychologic facts and dietary fancies, *J. Am. Dietet. Assoc.,* 28, 789, 1952.
183. **Clark, A. N. G., Mankikas, G. D., and Gray, I.,** Diogenes syndrome, a clinical study of gross neglect in old age, *Lancet,* 1, 366, 1975.

184. **Exton-Smith, A. N.,** Physiological aspects of aging: relationship to nutrition, *Am. J. Clin. Nutr.,* 26, 853, 1972.
185. **Anderson, W. F., Cohen, C., Hyams, D. E., Millard, P. H., Ploweright, N. M., Woodford-Williams, E., and Berry, W. T. C.,** Clinical and subclinical malnutrition in old age, in *Symposia of the Swedish Nutrition Foundation X: Nutrition in Old Age,* Almqvist & Wiksell, Stockholm, 1972.
186. **Krehl, W. A.,** The influence of nutritional environment on aging, *Geriatrics,* 29, 65, 1974.
187. **Rao, D. B.,** Problems of nutrition of the aged, *J. Am. Geriatr. Soc.,* 21, 362, 1973.
188. **Pelcovits, J.,** Nutrition to meet the human needs of older Americans, *J. Am. Dietet. Assoc.,* 60, 297, 1972.
189. **Baum, S. K.,** Loneliness in elderly persons: a preliminary study, *Psychol. Rep.,* 50, 1317, 1982.
190. **Hale, W. D.,** Correlates of depression in the elderly: sex differences and similarities, *J. Clin. Psychol.,* 38, 253, 1982.
191. **Albanese, A. A.,** *Nutrition for the Elderly,* Alan R. Liss, New York, 1980.
192. **Conrad, E. H.,** Psychogenic obesity-effects of social rejection upon hunger, food craving, food consumption, and drive-reduction value of eating for obese vs. normal individuals, *Psychosom. Med.,* 32, 556, 1970.
193. **Glucksma, M. L.,** Psychiatric observations of obesity, *Adv. Psychosom. Med.,* 7, 194, 1972.
194. **Rowland, N. E. and Antelman, S. M.,** Stress-induced hyperphagia and obesity in rats: a possible model for understanding human obesity, *Science,* 191, 310, 1976.
195. **Schacter, S. and Gross, L.,** Manipulated time and eating behavior, *J. Per. Soc. Psych.,* 10, 98, 1968.
196. **Rodin, J.,** Causes and consequences of time perception differences in overweight and normal weight people, *J. Per. Soc. Psych.,* 31, 898, 1975.
197. **Roth, M.,** Diagnosis of senile and related forms of dementia, in *Alzheimer's Disease: Senile Dementia and Related Disorders,* Vol. 7, Katzman, R., Terry, R. D., and Bich, K. L., Eds., Raven Press, New York, 1978, 71.
198. **Stahelin, H. B., Hofer, H. O., Vogel, M., Held, C., and Seiler, W. O.,** Energy and protein consumption in patients with senile dementia, *Gerontologist,* 29, 145, 1983.
199. **Vincent, M. and Gibson, R. S.,** Dietary intake of a group of chronic geriatric psychiatric patients, *Gerontologist,* 28, 245, 1982.
200. **MacLennan, W. J., Martin, P., and Mason, B. J.,** Causes for reduced dietary intake in a long-stay hospital, *Age Ageing,* 4, 175, 1975.
201. **Nelson, F. L. and Farberow, N. L.,** Indirect self-destructive behavior in the elderly nursing home patient, *J. Gerontol.,* 35, 949, 1980.
202. **Howell, S. C. and Loeb, M. B.,** Diet and the nervous system: effects on behavior and emotions in the older adult, *Gerontologist,* 9(3), 53, 1969.
203. **Blass, J.,** Food selection in the aged, *Int. J. Obesity,* 4, 377, 1980.
204. **Kelley, L., Ohlson, M., and Harper, L.,** Food selection and well-being of aging women, *J. Am. Dietet. Assoc.,* 33, 466, 1957.
205. **Whanger, A. D. and Wang, H. S.,** Vitamin B_{12} deficiency in normal and aged and elderly psychiatric patients, in *Normal Aging II,* Palmore, E., Ed., Duke University Press, Durham, N.C., 1974.
206. **Kane, F. J., Jr. and Lipton, M.,** Folic acid and mental illness, *S. Med. J.,* 63, 603, 1970.
207. **Herbert, V.,** Folic acid and vitamin B_{12}, in *Modern Nutrition in Health and Disease: Dietotherapy,* 5th ed., Goodhart, R. S. and Shils, M. E., Eds., Lea & Febiger, Philadelphia, 1973.
208. **Bertolini, A. M.,** *Gerontologic Metabolism,* Charles C Thomas, Springfield, Ill., 1969.
209. **Cheraskin, E. and Ringsdorf, W. M., Jr.,** *Psychodietetics,* Stein & Day, New York, 1974.
210. **Dakshinamurti, K.,** B vitamins and nervous system function, in *Nutrition and the Brain,* Vol. 1, Wurtman, R. J. and Wurtman, J. J., Eds., Raven Press, New York, 1977.
211. **Kinsman, R. A. and Hood, J.,** Some behavioral effects of ascorbic acid deficiency, *Am. J. Clin. Nutr.,* 24, 455, 1971.
212. **Kalchthaler, T. and Rigor Tan, M. E.,** Anemia in institutionalized elderly patients, *J. Am. Geriatr. Soc.,* 28, 108, 1980.
213. **Eisdorfer, C. and Cohen, D.,** The cognitively impaired elderly: differential diagnosis, in *The Clinical Psychology of Aging,* Storandt, M. A., Siegler, I. C., and Elias, M. F., Eds., Plenum Press, New York, 1978.
214. **Gaitz, C. M.,** Identifying and treating depression in an older patient, *Geriatrics,* 38, 2, 1983.
215. **Schneck, S. A.,** Aging of the nervous system and dementia, in *Clinical Internal Medicine in the Aged,* Schrier, R. W., Ed., W. B. Saunders, Philadelphia, 1982, 41.
216. **Milner, G.,** Ascorbic acid in chronic psychiatric patients — a controlled trial, *Br. J. Psychiatr.,* 109, 294, 1963.
217. **Carney, M. W. P. and Sheffield, B. F.,** Associations of subnormal serum folate and vitamin B_{12} values and effects of replacement therapy, *J. Nerv. Ment. Dis.,* 150, 404, 1970.

218. **Taylor, G. F.,** A clinical survey of elderly people from a nutritional standpoint, in *Vitamins in the Elderly,* Exton-Smith, A. N. and Scott, D. L., Eds., John Wright & Sons, Bristol, England, 1968.

219. **Brosin, H. W.,** The psychology of appetite, in *Modern Nutrition in Health and Disease,* Wohl, M. G. and Goddhart, R. S., Eds., Lea & Febiger, Philadelphia, 1968.

Section III
Assessment of Nutritional Status

Chapter 5

DIETARY CHARACTERISTICS OF THE ELDERLY

Fudeko Maruyama

TABLE OF CONTENTS

I. INTRODUCTION

Food-related practices which determine nutrient intake and influence nutritional status are the subjects of this chapter. Patterns of food selection, use, and expenditures for food by the elderly are reviewed. Dietary supplements, misinformation, nutritional knowledge, and beliefs which affect nutritional well being are discussed. Intake and status of specific nutrients which are discussed in appropriate chapters elsewhere are not included. A review of dietary guidelines and nutritional advice for the elderly concludes this chapter.

II. PATTERNS OF FOOD CONSUMPTION

A. Meals Per Day

Studies indicate that the elderly eat more regularly and skip fewer meals than younger adults. Hunter and Linn[1] examined the daily food consumption patterns of 182 subjects over 65 years of age and found that 45% ate 3 full meals and 45% consumed 2 full meals. Davidson et al.[2] found that most of the 800 elderly subjects in their study ate the customary 3 meals a day although more than one third ate more frequently. Statistics vary because of differences in definition of a meal among studies. According to the U.S. Department of Agriculture Food Consumption Survey 1977—1978 (USDA-FCS),[3] adults 75 years and older had the highest proportion: 58%, eating 3 times during the day. Only 39% of individuals of all age groups ate 3 meals.

B. Breakfast

Breakfast is frequently described by elderly as their favorite meal. Hunter and Linn[1] found that 70% of their elderly subjects ate a full breakfast containing foods representing at least 3 of the 4 basic food groups. Other investigators have reported that breakfast was eaten regularly by 90% of the elderly and was the favorite meal for more than one third of the subjects.[4] A National Medical Care Expenditure Survey showed that 88% of white persons over 65 and 72% of black persons over 65 ate breakfast.[5] The USDA-FCS defined breakfast as an "eating occasion" rather than a full meal and found that 94 to 98% of persons over 65 ate breakfast while 86% of individuals of all age groups did so.[3]

The elderly averaged somewhat higher percentages of their day's energy and most nutrients from breakfast than most groups of younger adults.[3]

C. Snacks

Singleton et al.[6] examined the snacking patterns of 95 noninstitutionalized low-income families over 60 years of age; 44% of their subjects snacked in the previous 24 hr. This is consistent with the 47% of 65- to 74-year-old females and 40% of 75 and over females who reported snacks in the USDA-FCS,[3] but higher than the 37% of men and women over 60 years of age who reported eating between meals reported by Todhunter;[4] 58 to 61% of younger adult females in the USDA-FCS[3] reported eating snacks. A larger percentage of elderly males snacked (55% of 65- to 74-year-olds and 49% of those 75 and over) than did elderly females. Among younger age groups there was no difference in percent males and percent females who consumed snacks.[3]

Of the snacks consumed by elderly women in Singleton's study,[6] 60% were from the Four Basic Food Groups; 18% were cakes, cookies, and pastries. Snacks provided 5 to 10% of the mean intake of protein, 10 to 15% of the mean intake for calories and fat, and 31% of the sucrose. These figures are slightly higher than those reported by the USDA-FCS.[3]

Persons over 65 consumed potato chips and salty snacks less frequently than did younger adults.[7] The elderly thus snack less frequently than younger adults and choose more nutritious, less salty snacks.

D. Meals Away from Home

Elderly persons eat out less frequently than younger adults.[3,4,8] Pao et al.[8] found that 35% of males 65 to 74 years of age and 27% of males 75 years and over obtained and ate food away from home at least once during a 3-day survey period. Among females, 37% of women 65 to 74 years of age and 26% of women 75 and over ate away from home at least once during a 3-day period.[8]

The higher cost of restaurant meals relative to food prepared at home and the limited mobility of some elderly persons may be reasons for their eating more meals at home.

III. FOOD ACCEPTANCE

A. Milk and Dairy Products

Milk is liked and consumed by a large proportion of the elderly. A survey of 347 participants in a congregate meal program for the elderly revealed that 90% consumed milk as a beverage when a choice of whole, skim, chocolate, or buttermilk was offered.[9] Furthermore, milk was equally well accepted by Mexican-Americans, blacks, and Anglo-Saxons.[9] Harrill et al.[10] studied food acceptance by elderly women and found that 67% liked milk.

Lack of agreement in survey results are often due to variations in food groupings among different studies. Approximately one third of the persons over 65 years of age in the Health and Nutrition Examination Survey (HANES) reported that they seldom or never consumed whole milk,[11] but when dairy products such as ice cream, cheese, and pudding were included, total consumption of foods in the milk group increased. The USDA-FSC found that 93% of persons over 65 used milk and milk products[7] and that elderly females consumed more milk and milk products per person than middle-aged females.[7,8] The amounts of milk and milk products consumed per day by females as total calcium equivalents were as follows:[7]

- 35 to 50 years of age: 215 g (as whole milk)
- 51 to 64 years of age: 232 g (as whole milk)
- 65 to 74 years of age: 248 g (as whole milk)
- 75+ years of age: 274 g (as whole milk)

Between 32 and 40% of elderly women drank milk at least once a day in contrast to 25 to 27% of younger women.[8] For men 65 and older there was a slight increase in milk consumption over that of men 35 to 50 years of age:[8] 312 vs. 301 g.

Between 37 and 43% of elderly men drank milk at least once a day in contrast to 30 to 32% of younger men. Despite widespread use of milk and milk products, however, average intake as calcium equivalents is only slightly over one cup,[7] thus accounting for the inadequate intake of calcium in the U.S.

B. Meat, Poultry, and Fish

Meat, including beef, poultry, and pork, is well liked by the elderly.[10,11] Between 71 and 91% of subjects in Harrill's[10] study reported liking meat. Total consumption of meat, poultry, and fish by females over 65 is less than that of all individuals, however.[7] Females 65 to 74 years of age consumed 165 g of meat, poultry, and fish, and women 75+ years of age consumed only 148 g/day. The average for all individuals was 204 g/day.[7] Beef was the most popular meat and chicken was a distant second; 99% of persons over 65 reported eating meat, poultry, or fish.

C. Other Protein Foods

Eggs are another good source of protein. The USDA-FCS found an increase in per capita consumption of eggs among men with increasing age. Between 68 and 72% of elderly men used eggs while 58 to 59% of elderly women used eggs. The average for all individuals was 55%.[7]

The elderly consumed less cooked dried beans than did younger adults.[7,8]

D. Grain Products

Grain products, such as bread, rolls, and cereals, are convenient to use and liked by 67% of the elderly according to Harrill.[10] Grain products contribute a larger share of the food energy consumed by persons over 65 than of younger adults.[3] The elderly, especially those 75 and over, ate cooked cereals more frequently than any other age group.[8] Ready-to-eat cereals were more popular and were eaten more frequently by older adults than by young and middle-aged adults.[3,7,8] Older adults preferred ready-to-eat cereals with lower sugar levels while children and teenagers consumed the highly sugared varieties more frequently.[8]

With regard to bread and cereal products, the elderly consume more high fiber, wheat germ, bran, oatmeal, and multigrain breads than younger people, and the 75+ age group consumes more whole and cracked wheat breads than any other age group.[7] Thus, the elderly appear to be more nutrition conscious in their selection of grain products than younger adults.

E. Vegetables and Fruits

Consumption and popularity of vegetables increased with age, especially among women, according to the USDA-FCS. Per capita consumption of vegetables for all individuals was 198 g/day, while that of men and women 65 to 74 years of age was 224 to 256 g/day.[7]

The elderly are especially fond of fruit. Consumption of fruit, including juices, fresh and canned fruits, and dried fruits was 30 to 40% greater than that of younger adults.[7]

F. Other Food and Beverages

Consumption of coffee, tea, and alcoholic beverages declined with age while the consumption of sugar and sweets increased slightly with age.[3,7]

IV. EXPENDITURE FOR FOOD

A. Amount Spent

It is commonly believed that many elderly citizens eat inadequately because they lack the resources and mobility to purchase and prepare nourishing food.

In a Bureau of Labor Statistics study in 1972—1973, households in which the head was over 64 years of age spent an average of 21.5% of their income for food compared with 16.9% for households headed by people under 65,[12] reflecting the lower incomes of senior citizens.

They also spent their at-home food dollar differently, allocating more to fresh fruits and fresh vegetables and less to red meats, dairy products, beverages, and prepared foods than did younger households. The 65-and-over households spent 38.2% of their at-home food dollar on meat, poultry, fish, and eggs, while the 45- to 64-year-old households spent 40.0%.

Axelson and Penfield[13] used food records to study the expenditures for categories of foods by persons 60 years of age or older living alone. A large portion (78%) of the respondents' food dollar went for the purchase of foods from 6 categories: meat and meat alternates (25%), vegetables (12%), fruits (12%), milk and milk products (11%), beverages (12%), and flour products (6%). The average total expenditure for food was comparable to the cost suggested for the USDA Low-Cost Food Plan,[14] which was $59 to 66 per month at the time of the study for females living alone. Although direct comparisons are impossible because food categories are not the same in the Axelson study and USDA food plans, the allocation of food dollars agrees with the recommendations for well-balanced nutritious diets by the Consumer Nutrition Division of USDA,[15] which are 28% for meat, poultry, and fish, 23%

for vegetables and fruits, 17% for milk and milk products, 20% for grain products, 5% for beans, and 7% for fats, sweets, and beverages.

B. Shopping

Studies show that grocery shopping is a source of pleasure, recreation, and exercise for the elderly and that most shop at least once a week for food.[13,16] Food shopping is a diversionary activity for older people and many shop even when not intending to buy; this, however, is not unique to older people. Also, the elderly spend most of their food dollar for food at the grocery store and little on food away from home.[12,13]

V. USE OF DIETARY SUPPLEMENTS

A. Trends

Specially formulated nonfood preparations containing vitamins, minerals, protein, or a combination of these and other ingredients are widely used by the elderly, and appropriateness of supplements used is generally unrelated to need.[17,18] Recent studies have found that 50 to 72% of elderly use supplements.[19-22] This appears to be an upward trend from earlier studies which found that 35 to 50% of aging or older persons use vitamin/mineral supplements.[23-25] The USDA-FCS in 1965—1966[26] found that 34% of men and women aged 75 and over took vitamin/mineral supplements. Among men and women 65 to 74 years of age, supplement use was 27% in the 1965—1966 survey.

B. Type and Extent of Use

Read[20] studied the type and extent of use of dietary supplements by elderly men and women in relatively good health; 66% of the subjects indicated the use of supplements. Women were twice as likely as men to use supplements. Ascorbic acid and vitamin E were the most popular choices.

Multivitamins, calcium, potassium, B-complex, and vitamin A were taken by 14 to 27% of the users. Less common supplements included lecithin, alfalfa, kelp, iodine, garlic, and fennel seeds. Concern about "bone strength", doctor's advice, "energy", and "general health" were commonly given reasons for taking supplements.

Multivitamin/mineral supplements, vitamin C, and vitamin E were the most frequently used supplements reported by Gray et al.[21] Consumption of supplements was unrelated to dietary intake of nutrients.

Dietary and supplemental intakes were assessed by Garry et al.[22] from 3-day food records collected from 270 healthy men and women over 60 years of age; 57% of the men and 61% of the women reported routinely ingesting one or more vitamin or mineral supplements. Ascorbic acid was the vitamin consumed most, even though more than 90% of the people received at least 100% of the RDA for ascorbic acid from diet alone.

Although use of supplements by older persons is high, it is not out of line with the level of supplement consumption by adults in general. Schutz et al.[27] found that 67% of a sample of adults in 7 states used some form of food supplements with 40% consuming 1 to 3 supplements per day. The most frequently cited reason for using supplements was to "prevent colds" and other illnesses. The three most frequently used dietary supplements were multiple vitamins, vitamin C, and multiple vitamins plus iron. Results of FDA surveys have indicated that 55% of the U.S. population use food supplements.[28]

The elderly are especially susceptible to food fads and misrepresentations of so-called health food products.[29,30] The subject of food fads, and misinformation has been reviewed by Herbert and Barrett[31] and others,[32,33] and the American Dietetic Association has issued a position paper[34] on the subject, calling attention to the need to dispel popular myths concerning the nutritional value of health foods, diets promising quick weight loss, and excessive vitamin, mineral, and dietary supplementation.

It has been estimated that Americans are spending between 5 and 10 billion dollars each year on unneeded, useless, and frequently harmful food supplements and other forms of nutrition quackery.[30] No estimate of the amount spent by the elderly is available, but assuming that the elderly comprise 11% of the population and portioning expenditure evenly across all age groups, 11% of 5 to 10 billion dollars is roughly 500,000 to 1 billion dollars, which could be better spent for wholesome food.

A critical evaluation of dietary supplements and health aids has recently been reported by Dubick.[35-37] Research indicates conclusively that for healthy people, intakes of nutrients beyond the RDA do not improve health or provide greater resistance to disease.[35]

Vitamin B_{12}, or cyanocobalamin, has been popular for many years, particularly among the elderly as a quick "pick-me-up". Controlled studies show that vitamin B_{12} does not relieve tiredness.[35] Vitamin C is taken in high doses by many in the mistaken belief that it will prevent the common cold and influenza. Advertisements claiming that vitamin E retards aging are directed toward the elderly and vitamin A is receiving much publicity as a possible preventive agent for cancer. The Committee on Diet, Nutrition, and Cancer of the National Research Council specifically advises against taking vitamin A supplements as a cancer preventive.[38]

The American Medical Association has recommended that when daily vitamin supplements are indicated, the dosage should contain amounts no greater than the Recommended Dietary Allowance (RDA). Therapeutic dosage at levels of three to five times the RDA may be justified when there is a medical reason such as prolonged illness, chronic disease, and poor food intake over a long period of time. Megavitamin dosage, generally defined as ten times or more in excess of RDA, are not recommended; even the water-soluble vitamins in levels exceeding ten times the RDA can cause symptoms of toxicity.[30]

VI. FADS AND FALLACIES

A. Misinformation

Misinformation about food causes unnecessary concern and sometimes inadvisable food restrictions by the elderly. Citrus fruits are frequently omitted from the diet by older people because they mistakenly believe that oranges and grapefruit are "too acid". This is a misunderstanding because the stomach is normally acid due to the presence of hydrochloric acid, which is essential for proper digestion of food. Symptoms frequently attributed to an "acid stomach" are not caused by acid food.   Milk and cheese are believed to be constipating by many people and may be the reason some older people avoid milk. Constipation is a common complaint of older people and usually is due to a diet very low in fiber and high in processed carbohydrate foods as well as insufficient intake of fluids.

B. Weight Reduction

Since many older people are overweight or obese they are vulnerable targets of promoters of diet aids and weight loss plans. Appetite suppressants are widely advertised as an aid to weight loss and are frequently taken by people who are unaware of the possible dangers involved in their use. Elderly who need to lose weight because they have hypertension, heart disease, or kidney or thyroid problems should not use these drugs.[30]

Protein supplements in the form of liquids, powders, and tablets are available for weight reduction. Using them as the only source of energy is a dangerous practice; the FDA has reported 58 deaths associated with their use.[36] Since this negative publicity, protein supplements have reemerged in "diet plans" in combination with limited amounts of other foods, but their use is not recommended except under close medical supervision.[39,40] The FDA now requires a warning label on protein diet foods.

Elderly with impaired kidney function may suffer serious side effects from the low carbohydrate-high protein ketogenic diets for weight reduction. Many versions of this diet exist under various names such as the Stillman Diet, Doctors Quick Weight Loss Diet, Dr. Atkins Diet, and the Scarsdale Diet.[32,,36] Not only are these diets potentially harmful to the aging kidney, but they are not effective in controlling weight because they fail to improve eating habits.

VII. NUTRITIONAL KNOWLEDGE AND BELIEFS

Nutritional knowledge tests administered by Grotkowski and Sims[41] revealed a low level of knowledge among the noninstitutionalized elderly. Other studies, in which the nutritional knowledge of adults of different age groups was compared, have shown the elderly (over 60) group to have the lowest level of knowledge.[42-44]

Elderly subjects studied by Grotkowski and Sims[41] had a tendency to overrate the adequacy of their diets, yet had misconceptions about the need for vitamin/mineral supplements for energy. Although 95% of the senior citizens studied by Roundtree and Tinklin[45] believed that a balanced diet is composed of foods from the four food groups, many believed that everyone should take vitamin and mineral supplements to remain healthy and energetic. These misconceptions and beliefs may not be unique to older persons. Axelson and Penfield[46] used 97 belief statements to assess the food-related attitudes of subjects age 60 or older and found that they do not differ from the food-related attitudes of other age groups.

Sources of nutrition information most used by elderly have been reported as television, physicians, magazines, cookbooks, newspapers, and health food stores.[20,41,44] Recognized sources of reliable nutrition information, such as registered dietitians and Cooperative Extension Service agents, appear to have minimal impact.

VIII. NUTRITIONAL ADVICE FOR OPTIMAL NUTRITION OF THE ELDERLY

A. Energy

Information regarding requirements for many essential nutrients for the elderly is still lacking. There is evidence, however, to support the RDA for a protein level slightly greater than 0.8 g/kg/day (or 12% of total caloric needs) and reduced calories to compensate for decreases in lean body mass, resting metabolic rate, and physical activity.[48-50]

The Committee on Dietary Allowances of the National Research Council recommends that energy allowances for persons between 51 and 75 years of age be reduced to about 90% of the amount required as a young adult and for persons beyond age 75 years to about 75 to 80% of that amount.[47] The Committee on Dietary Allowances also recommends that the elderly receive 12% or more of their energy intake from protein in contrast to 10% for younger adults.[47] Based on these two recommendations the energy and protein intakes of older people should be

Age	Male	Female
51—75	2400 kcal, 72 g	1800 kcal, 54 g
76+	2050 kcal, 62 g	1600 kcal, 48 g

Energy and protein are the only RDAs affected by aging at the present time. It should be noted that the increased allowance for protein is discussed in the text, however, the RDA table[47] does not list the increased protein allowances for persons over 51 years of age.

B. Protein

Nutritional advice regarding protein intake for the elderly range from no increase (0.8 g mixed protein per kilogram body weight)[51,52] to 2.0 g/kg/day by Rao,[53] who believes 20 to

25% of total calories should be derived from protein. Kohrs[54] endorses the recommendation of the National Research Council that the elderly consume 12% of their energy needs as protein, which is slightly more than 0.8 g/kg/day, because studies have shown that the elderly are one of the few groups in the U.S. that is likely to consume inadequate amounts of protein.[55] Also, older persons may be on medications or have chronic illnesses which increase protein need. Following the recommendation would help to avoid excessive amounts of protein that the aging kidney may not be able to handle.[54]

Food guides for the elderly[56-58] suggest two to three servings per day of protein-rich foods such as meat, poultry, fish, and beans. One serving would include 2 to 3 oz of cooked meat, chicken, or fish, 2 eggs, 1 cup of cooked dry beans, 4 tablespoons of peanut butter, or $1/_2$ cup of nuts. Milk and cheese also provide protein and can be used to increase the value of other protein sources such as cereals, beans, and nuts.

C. Fat

Moderation in consumption of fat is now a dietary guideline[59] for all adults, but it is especially important for the elderly because of their reduced need for calories. Kohrs[54] recommends a modest reduction of fat to 35% of total calories from fat (the average is 40.3%[3]). She advised elderly persons to omit all unnecessary fat from the diet by trimming fat from meat, limiting intake of salad dressings made from fat, omitting fried foods, and cooking without fat. Winick[51] recommends 25 to 30% of total daily calories from fat while Rao[51] states that sound nutrition for the elderly should include only 20% of total daily calories derived from fat. Such a reduction would be difficult to achieve without drastic changes in food selection and preparation methods.

The ratio of saturated fat to polyunsaturated fat and the amount of cholesterol are two additional considerations related to fat intake. Cholesterol is of special concern to the elderly because of its perceived relationship to heart disease.[60] Dietary intervention trials to study the effect of the kind of fat and amount of cholesterol in the diet on cardiovascular disease have found that the older subjects do not benefit from alterations in diet.[61] Thus, it may not be prudent to suggest increasing polyunsaturated fats or reducing cholesterol intake in the diet for the elderly.

D. Carbohydrate and Fiber

If an elderly person consumed 12% of total calories as protein and 35% as fat, 53% of the energy needs should be consumed as carbohydrate. Those who need to restrict caloric intake because of a weight problem should reduce consumption of simple carbohydrates such as sweet desserts and candy. Whole grain cereals and fruits and vegetables are good carbohydrate sources for the elderly. Consumption of fruits, vegetables, whole grain cereals, and legumes will also increase the intake of dietary fiber. Adequate fiber in the diet can alleviate constipation, the most common chronic digestive problem in the elderly. Constipation is 2.5 times more prevalent in people over 65 than in people 45 to 64 years of age.[62]

E. Daily Food Guide

The USDA Daily Food Guide[63,64] is a useful tool for providing the elderly, with little knowledge of the scientific basis of nutrition, basic advice for obtaining a nutritionally adequate diet. The Daily Food Guide is intended as a starting point. After initial exposure and for continuing efforts more sophisticated approaches which address specific nutritional problems or concerns of the elderly are needed. Weight control, dietary supplements, fluid intake, and making economical food purchases are among the concerns of the elderly.[60,65,66]

F. Food Plans at Different Levels of Cost

The USDA has prepared guides for selecting nutritious diets at different levels of cost. Such guides or food plans are revised from time to time to take into account new information about nutritional needs, nutritive values of food, food consumption, and food prices. Most recent revisions were done in 1983,[67,68] and prior to that in 1974—1975. The USDA Human Nutrition Information Service in Hyattsville, Md. publishes estimates for the cost of food at home based on food plans at four cost levels. For a couple 51 years of age and over the cost for food for 1 week in October 1983 was $33.30 on the thrifty plan, $42.30 on the low-cost plan, $51.90 on the moderate-cost plan, and $61.80 on the liberal plan. These figures serve as a guide to food expenditure. The USDA Food Plans[67,68] also list the quantities of food included per week for males and females of different age groups. Unfortunately, the elderly are part of the 51 years and over group and the quantities of food for them may be somewhat excessive. Nevertheless, the food plans are a useful tool when planning or evaluating food needed by the elderly.

REFERENCES

1. **Hunter, K. I. and Linn, M. W.,** Cultural and sex differences in dietary patterns of the urban elderly, *J. Am. Geriatr. Soc.,* 27, 359, 1979.
2. **Davidson, C. S., Livermore, J., Anderson, P., and Kaufman, S.,** The nutrition of a group of apparently healthy aging persons, *Am. J. Clin. Nutr.,* 10, 181, 1962.
3. Food and Nutrient Intakes of Individuals in 1 Day in the United States, Spring 1977, 1977 to 1978, Preliminary Report No. 2, Nationwide Food Consumption Survey, U.S. Department of Agriculture, Washington, D.C., 1980.
4. **Todhunter, N. E.,** Lifestyle and nutrient intake in the elderly, in *Nutrition and Aging,* Winick, M., Ed., John Wiley & Sons, New York, 1976, 119.
5. **Manuel, R. C. and Berk, M. L.,** A look at similarities and differences in older minority populations, *Aging,* 339, 21, 1983.
6. **Singleton, N., Kirby, A. L., and Overstreet, M. H.,** Snacking patterns of elderly females, *J. Nutr. Elderly,* 2(2), 3, 1982.
7. Food Intakes: Individuals in 48 States, Year 1977 to 1978, Report No. 1-1, Nationwide Food Consumption Survey, U.S. Department of Agriculture, Washington, D.C., 1983.
8. **Pao, E. M., Fleming, K. H., Guenther, P. M., and Mickle, S. J.,** Foods Commonly Eaten by Individuals: Amount Per Day and Per Eating Occasion, Home Economics Research Report No. 44, U.S. Department of Agriculture, Washington, D.C., 1982.
9. **Marrs, D. C.,** Milk drinking by the elderly of three races, *J. Am. Dietet. Assoc.,* 72, 495, 1978.
10. **Harrill, I., Erbes, C., and Schwartz, C.,** Observations on food acceptance by elderly women, *Gerontologist,* 16, 349, 1976.
11. Food Consumption Profiles of White and Black Persons Aged 1-74 Years: United States, 1971-74, DHEW Publication No. (PHS) 79-1658, U.S. Department of Health, Education and Welfare, Washington, D.C., 1979.
12. **Gallo, A. E., Salathe, L. E., and Boehm, W. T.,** Senior Citizens: Food Expenditure Patterns and Assistance, Agricultural Economic Report No. 426, U.S. Department of Agriculture, Economics, Statistics and Cooperative Extension, Washington, D.C., 1979.
13. **Axelson, M. L. and Penfield, M. P.,** Factors associated with food expenditures of elderly persons living alone, *Home Econ. Res. J.,* 12 (Suppl.2), 228, 1983.
14. **Peterkin, B., Chassy, J., and Kerr, R.,** The Low Cost Food Plan, U.S. Department of Agriculture, Consumer and Food Economics Institute, Washington, D.C., 1975.
15. **Peterkin, B. B.,** Making food dollars count, *Family Econ. Rev.,* 4, 23, 1983.
16. **Mason, J. B. and Bearden, W. O.,** Profiling the shopping behavior of elderly consumers, *Gerontologist,* 18, 454, 1978.
17. **Natow, A. and Heslin, J. A.,** *Geriatric Nutrition,* CBI, Boston, 1980, 189.
18. **Osborn, M. O.,** Nutrition of the aged, in *Hoffman's Daily Needs and Interests of Older People,* Morris, W. W. and Bader, I. M., Eds., Charles C Thomas, Springfield, Ill., 1983, 170.
19. **Yearick, E. S., Wang, M. L., and Pisias, S. J.,** Nutritional status of the elderly: dietary and biochemical findings, *J. Gerontol.,* 35, 663, 1980.
20. **Read, M. H. and Graney, A. S.,** Food supplement usage by the elderly, *J. Am. Dietet. Assoc.,* 80 (3), 250, 1982.

21. **Gray, G. E., Paganini-Hill, A., and Ross, R. K.,** Dietary intake and nutrient supplement use in a Southern California retirement community, *Am. J. Clin. Nutr.,* 38(1), 122, 1983.
22. **Garry, P. J., Goodwin, J. S., Hunt, W. C., Hooper, E. M., and Leonard, A. G.,** Nutritional status in a healthy elderly population: dietary and supplemental intakes, *Am. J. Clin. Nutr.,* 36, 319, 1982.
23. **Steinkamp, R. C., Cohen, N. L., and Walsh, H. E.,** Resurvey of an aging population — fourteen year follow-up, *J. Am. Dietet. Assoc.,* 46, 103, 1965.
24. **Bovit, C. L.,** The food of older persons living at home, *J. Am. Dietet. Assoc.,* 46, 285, 1965.
25. **Davidson, C. S., Livermore, J., Anderson, P., and Kaufman, S.,** The nutrition of a group of apparently healthy aging persons, *Am. J. Clin. Nutr.,* 10, 181, 1962.
26. Food Nutrient Intake of Individuals in the United States, Household Food Consumption Survey, 1965-66, Rep. No. 11, U.S. Department of Agriculture, Agricultural Research Service, 1967.
27. **Schutz, H. G., Read, M., Bendel, R., Bhalla, V. S., Harrill, I., Monagle, J. E., Sheehan, E. T., and Standal, B. R.,** Food supplement use in seven Western states, *Am. J. Clin. Nutr.,* 36, 897, 1982.
28. Consumer Nutrition Knowledge Survey, Report I, 1973-1974 and Report II, 1975, DHEW Publ. No. 76-2058, Food and Drug Administration, Department of Health, Education and Welfare, Washington, D.C., 1976.
29. **Osborn, M. D.,** Nutrition of the aged, in *Hoffman's Daily Needs and Interests of Older People,* Morris, W. W. and Bader, I. M., Charles C Thomas, Springfield, Ill., 1983, 170.
30. **Davis, A. K. and Davis, R. L.,** Food facts, fads, fallacies, and folklore of the elderly, in *Handbook of Geriatric Nutrition,* Hsu, J. and Davis, R., Eds., Noyes Publications, Park Ridge, N.J., 1981, 328.
31. **Herbert, V. and Barrett, S.,** *Vitamins and "Health" Foods: The Great American Hustle,* George F. Stickley, Philadelphia, 1981.
32. **White, P. L.,** Nutrition misinformation and food faddism, *Nutr. Rev. Suppl.,* 1, 32, 1974.
33. **Kuske, T. T.,** Quackery and fad diets, in *Nutrition in the Middle and Later Years,* Feldman, E. B., Ed., PSG, Boston, 1983.
34. American Dietetic Association, Position paper on food and nutrition misinformation on selected topics, *J. Am. Dietet. Assoc.,* 66, 277, 1975.
35. **Dubick, M. A. and Rucker, R. B.,** Dietary supplements and health aids — a critical evaluation. I. Vitamins and minerals, *J. Nutr. Educ.,* 15, 47, 1983.
36. **Dubick, M. A.,** Dietary supplements and health aids — a critical evaluation. II. Macronutrients and fiber, *J. Nutr. Educ.,* 15, 88, 1983.
37. **Dubick, M. A.,** Dietary supplements and health aids — a critical evaluation. III. Natural and miscellaneous products, *J. Nutr. Educ.,* 15, 123, 1983.
38. Committee on Diet, Nutrition and Cancer, National Research Council, *Diet, Nutrition and Cancer,* National Academy Press, Washington, D.C., 1982.
39. Food and Drug Administration, Cambridge diet update, *FDA Talk Paper,* December 30, 1982.
40. Fed. Regist., April 6, 1984.
41. **Grotkowski, M. L. and Sims, L. S.,** Nutritional knowledge, attitudes, and dietary practices of the elderly, *J. Am. Dietet. Assoc.,* 72, 499, 1978.
42. **Young, C. M., Waldner, B. G., and Berresford, K.,** What the homemaker knows about nutrition. II. Level of nutrition knowledge, *J. Am. Dietet. Assoc.,* 32, 218, 1956.
43. **Abelson, H., Shrayer, S., and Genzelman, S.,** Food and Nutrition Knowledge and Beliefs. A Nationwide Study Among Shoppers, Division of Consumer Studies, Bureau of Foods, Food and Drug Administration, Department of Health, Education and Welfare, Washington, D.C., 1974.
44. **Jalso, S. B., Burns, M. M., and Rivers, J. M.,** Nutritional beliefs and practices, *J. Am. Dietet. Assoc.,* 47, 263, 1965.
45. **Roundtree, J. L. and Tinklin, G. L.,** Food beliefs and practices of selected senior citizens, *Gerontologist,* 15, 537, 1975.
46. **Axelson, M. L. and Penfield, M. P.,** Food and nutrition-related attitudes of elderly persons living alone, *J. Nutr. Educ.,* 15, 23, 1983.
47. Recommended Dietary Allowances, *9th rev. ed., National Research Council, National Academy of Sciences,* 1980.
48. **Harper, A. E.,** Recommended dietary allowances for the elderly, *Geriatrics,* 33(Suppl.5), 73, 1978.
49. **Harper, A. E.,** Nutrition, aging and longevity, *Am. J. Clin. Nutr.,* 36, 737, 1982.
50. **Bazzarre, T. L.,** Nutritional requirements of the elderly, in *Medical Care of the Elderly: A Practical Approach,* McCue, J. D., Ed., D.C. Heath, Indianapolis, 1983, 335.
51. **Winick, M.,** Nutrition for the elderly, in *Nutrition and Health,* Vol. 1, Suppl. no. 6, Institute of Human Nutrition, Columbia University, New York, 1979.
52. **Posner, B. M.,** *Nutrition and the Elderly,* D.C. Heath, Indianapolis, 1979, 35.
53. **Rao, D. B.,** Problems of nutrition in the aged, *J. Am. Geriatr. Soc.,* 21, 362, 1973.
54. **Kohrs, M. B.,** A rational diet for the elderly, *Am. J. Clin. Nutr.,* 36, 796, 1982.

55. **O'Hanlon, P. and Kohrs, M. B.,** Dietary studies of older Americans, *Am. J. Clin. Nutr.*, 31, 1257, 1978.
56. Society for Nutrition Education, *A Guide for Food and Nutrition in Later Years*, Berkeley, Calif., 1976.
57. Food Guide for Older Folks, Home and Garden Bull. No. 17, U.S. Department of Agriculture, Washington, D.C., 1971.
58. U.S. Department of Health and Human Services, National Institute on Aging, Food: Staying Healthy After 65, *Age Page*, 1980.
59. Dietary Guidelines for Americans, *U.S. Department of Agriculture and U.S. Department of Health and Human Services*, 1980.
60. **Elwood, T. W.,** Nutritional concerns of the elderly, *J. Nutr. Educ.*, 7(Suppl.2), 50, 1975.
61. **Hazzard, W. R.,** Aging and atherosclerosis: interactions with diet, heredity and associated risk factors, in *Nutrition, Longevity, and Aging*, Rockstein, M. and Sussman, M. L., Eds., Academic Press, New York, 1976, 143.
62. **Shank, R. E.,** Nutritional characteristics of the elderly — an overview, in *Nutrition, Longevity, and Aging*, Rockstein, M. and Sussman, M. L., Eds., Academic Press, New York, 1976, 9.
63. Guide to a Better Diet, Extension Service Leaflet No. 567, U.S. Department of Agriculture, Washington, D.C., 1980.
64. Food, Home and Garden Bull. No. 228, U.S. Department of Agriculture, Washington, D.C., 1980.
65. **Shaefer, A. E.,** Nutrition policies for the elderly, *Am. J. Clin. Nutr.*, 36, 819, 1982.
66. **Osborn, M. O.,** Nutrition of the aged, in *Hoffman's Daily Needs and Interests of Older People*, Morris, W. W. and Bader, I. M., Eds., Charles C Thomas, Springfield, Ill., 1983, 170.
67. **Cleveland, L. E. and Peterkin, B. B.,** USDA family food plans, *Family Econ. Rev.*, 2, 12, 1983.
68. **Kerr, R. L., Peterkin, B. B., Blum, A. J., and Cleveland, L. E.,** USDA 1983 thrifty food plan, *Family Econ. Rev.*, 1, 18, 1984.

Chapter 6

METHODS FOR THE ASSESSMENT OF NUTRITIONAL STATUS

Howerde E. Sauberlich

TABLE OF CONTENTS

I. INTRODUCTION

The elderly in the U.S. population has increased in both proportion and absolute number. Approximately two thirds of the elderly population are suffering from some form of chronic disease or illness that may influence their nutrient needs. A poor nutritional status may contribute to or exacerbate chronic and acute diseases, delay recovery from illness, or accelerate the development of degenerative diseases associated with aging. Thus, an important health aspect in the elderly is their nutritional status.

Nutritional assessment should provide a characterization of the body compartments in a static or functional sense so as to determine the need or non-need for nutritional repletion or modification. Biochemical studies make possible definitive appraisal of the nutritional status of an individual or population. Biochemical tests may provide pre- or subclinical information on the nutritional status of an individual. Biochemical procedures used to evaluate nutritional status are based largely on nutrient levels and the presence of abnormal metabolic products in blood and urine, alterations in activities of certain blood enzymes, nutrient metabolites in urine, and saturation or load tests.

In addition to biochemical, dietary, and clinical information, anthropometric measurements can be helpful in the diagnosis of malnutrition.[1-5] Unfortunately, anthropometric measurements have been less developed for the elderly as compared to age groups under 65 years of age. With aging, there is a decrease in body weight and height and a change in body composition. Lean body mass decreases with a relative increase in body weight due to fat. Simple anthropometric measurements, such as tricep skinfold thickness and midarm muscle circumference, must be evaluated with caution and uncertainty with the elderly. Reference standards for the elderly have not been adequately developed. Both the 1959 and 1983 Metropolitan reference weights were derived from data on people age 25 to 59 years of age.

Some of the difficulties and complexities of assessing the nutritional status in the elderly have been reviewed and summarized previously.[1,6-9]

II. LABORATORY PROCEDURES

A. Vitamins

1. Vitamin A

Vitamin A (retinol) deficiency has seldom been observed in elderly people, although low dietary intakes of the vitamin for this age group have been reported.[10-12] However, few studies have examined the occurrence of night blindness and dark adaptation in the elderly. Vitamin A deficiency primarily affects infants and children. In these groups, a severe vitamin A deficiency frequently results in blindness. Hypovitaminosis A continues to be an important nutritional problem in several areas of the world, particularly in South Asian countries such as Indonesia, Sri Lanka, and Bangladesh. Recently, a renewed interest in vitamin A status for all age groups has occurred, with the observation that low serum vitamin A (retinol) levels may be associated with an increased risk of cancer.[13-15] In addition to impaired night blindness and dark adaptation, vitamin A deficiency in the adult may result in skin lesions (follicular hyperkeratosis) on the arms and thighs as well as in an impaired resistance to infections. Impaired dark adaptation may occur in subjects with chronic alcoholism or with hepatic and gastrointestinal diseases.[16]

Although measurements of dark adaptation can serve as a sensitive functional assessment of vitamin A status, the procedure is too involved for routine use. The majority of the vitamin A reserves in the body are present in the liver,[17,18] but seldom would liver biopsy samples be available for a direct assessment of vitamin A status. Consequently, measurements of serum retinol levels are the only practical means of assessing vitamin A status. Clinical signs of a vitamin A deficiency are associated with low retinol levels in the serum.[17-19]

Malabsorption conditions, infection, fever, and liver disease may also result in low serum levels of retinol. Serum vitamin A levels of 30 μg/ℓ or higher are desirable (Table 1). Abnormally elevated serum retinol (vitamin A) levels are usually a reflection of excessive or toxic intakes of retinol.

Retinol can be measured in serum with the use of microtechniques that employ either fluorometric, spectrophotometric, trifluoracetic acid colorimetric, or high performance liquid chromatography (HPLC) procedures.[17,19-27] The results obtained from the use of the various methods are quite comparable, but each requires careful execution of the analysis. In addition, samples for analysis must be properly handled.[28] For those laboratories that have HPLC systems, this procedure may be the method of choice.[22,23,25,26]

The measurement of plasma retinol-binding protein levels permits another means of evaluating vitamin A nutritional status.[29] The plasma-binding protein serves to transport retinol in the body and correlates closely with plasma retinol levels.[29-32] Only 2 to 10 μℓ of plasma are required to perform this measurement. The measurement commonly utilizes commercial assay kits. Otherwise the procedure is difficult to establish.

2. Vitamin D and Calcium

Osteoporosis is a bone disease common in older men and in women after menopause.[33-35] The demineralized bone condition is associated with decreased calcium absorption and an increased calcium loss in the urine, which leads to thinning of bones and bone fractures. Osteomalacia is a bone condition associated with a deficiency of vitamin D and calcium that also may occur in the elderly.[33-39] Its frequency is uncertain since both osteoporosis and osteomalacia may occur in the same subject. Osteomalacia results in a softening of the bones due to impaired mineralization and may be accompanied with pain, muscular weakness, tenderness, anorexia, and loss of weight. In the elderly, calcium is frequently less efficiently absorbed and thus they may have an increased dietary requirement for calcium in order to achieve calcium balance. In contrast, calcium intakes in the elderly may fall because of their lower intakes of milk, dairy products, and other sources of calcium. Intakes of calcium as high as 1400 mg/day have been recommended for the elderly.[33] Moreover, exposure to sunlight tends to be lower in the elderly and thereby results in a reduced synthesis of vitamin D (cholecalciferol; vitamin D_3) in their skin.[36] In addition, the ability to synthesize vitamin D in the skin of the elderly is decreased.[40]

Vitamin D nutritional status may be evaluated directly through the measurement of serum levels of 25-hydroxycholecalciferol (principal circulating metabolite of vitamin D_3) and of 1,25-dihydroxycholecalciferol (most active form).[41] Competitive protein binding assays are commonly used for their measurement.[41] Recently, radioimmunoassay procedures have been developed for the determination of 1,25-dihyroxycholecalciferol in serum.[41] HPLC has enhanced the ability to measure vitamin D metabolites in serum. Serum measurements of 25-hydroxycholecalciferol provide an excellent index of vitamin D stores, while levels of 1,25-dihydroxycholecalciferol are less reliable for this purpose. With rickets or osteomalacia, serum levels of 25-dihydroxycholecalciferol may or may not change.[41-44] Markedly elevated levels of 25-hydroxycholecalciferol may indicate excessive intakes of vitamin D.[41] Although the measurements of 25-hydroxycholecalciferol and 1,25-dihydroxycholecalciferol are technically difficult to perform, they are particularly useful in clinical research and with subjects suspected of having a defect in vitamin D metabolism. Low serum levels of 25-hydroxycholecalciferol are frequently encountered in the elderly, particularly in late winter or early spring.

Serum calcium levels and serum alkaline phosphatase activity may be used as an indirect measurement of vitamin D status.[19] The increase in serum alkaline phosphatase that may occur with vitamin D deficiency is not specific and may occur with metastatic carcinomas and a number of other conditions.[19] Similarly, the blood calcium level is controlled very

Table 1
SOME LABORATORY MEASUREMENTS USED FOR ASSESSING NUTRITIONAL STATUS[a]

Nutrient	Test	Adult normal or acceptable values[b]
Vitamin A	Serum vitamin A	30 μg/dℓ
	Serum retinol-binding protein	30—40 μg/mℓ
Vitamin D	Serum 25-hydroxycholecalciferol	20 ng/mℓ
	Serum 1,25-dihydroxycholecalciferol	3 ng/dℓ
	Serum alkaline phosphatase	40 King-Armstrong units
		15 Bodansky units
Vitamin E	Plasma vitamin E level	0.7 mg/dℓ
	Erythrocyte hemolysis test	10% Hemolysis
Vitamin K	Prothrombin test	11—18 sec
Vitamin C	Serum ascorbic acid level	0.30 mg/dℓ
	Leukocyte ascorbic acid level	15 mg/dℓ cells (20 μg/10⁸ cells)
Thiamin (B₁)	Erythrocyte TK-TPP stimulation	15%
	Urinary thiamin excretion	65 μg/g Creatinine
Riboflavin (B₂)	Erythrocyte glutathione reductase stimulation coefficient	1.20
	Urinary riboflavin excretion	80 μg/g Creatinine
Niacin	Urinary 2-pyridone/N'-methylnicotin-amide ratio	1.0
Vitamin B₆	Erythrocyte GOT activity	1.1 IU/mℓ cells
	Erythrocyte GOT-PLP stimulation coefficient	2.0
	Plasma pyridoxal phosphate	5.0 ng/mℓ
	Xanthurenic acid excretion following a tryptophan load	50 mg/24-hr urine collection
	Urinary vitamin B₆ excretion	20 μg/g creatinine
	Urinary 4-pyridoxic acid excretion	1.0 mg/24 hr
Folic acid	Serum folic acid level	6.0 ng/mℓ
	Erythrocyte folic acid level	160 ng/mℓ cells
Vitamin B₁₂	Serum vitamin B₁₂ level	200 pg/mℓ
Biotin	Urinary biotin excretion	30—60 μg/24 hr
	Blood biotin levels	120—240 ng/dℓ
Pantothenic acid	Urinary pantothenate excretion	2.0 mg/g creatinine
Iron	Hemoglobin (blood)	Men: 14 g/dℓ
		Women: 12 g/dℓ
	Hematocrit	Men: 44%
		Women: 38%
	Serum transferrin	170—250 mg/dℓ
	Serum iron	50 μg/dℓ
	Total iron binding capacity	225—400 μg/dℓ
	Serum transferrin saturation	15%
	Serum ferritin	10 ng/mℓ
	Erythrocyte protoporphyrin	70 μg/dℓ
Zinc	Urinary zinc	400—600 μg/24 hr
	Serum zinc level	80—120 μg/dℓ
	Hair zinc level	100 μg/g
Copper	Serum copper level	75—125 μg/dℓ
Selenium	Plasma selenium level	100 μg/ℓ
Iodine	Serum protein-bound iodine	4—8 μg/dℓ
	Urinary iodine excretion	50 μg/g creatinine
Sodium	Serum sodium level	130—155 mEq/ℓ
Potassium	Serum potassium level	3.5—5.5 mEq/ℓ
Chloride	Serum chloride level	100—106 mEq/ℓ

Table 1 (continued)
SOME LABORATORY MEASUREMENTS USED FOR ASSESSING
NUTRITIONAL STATUS[a]

Nutrient	Test	Adult normal or acceptable values[b]
Phosphorus	Serum phosphorus level	3—4 mg/dℓ
Calcium	Serum calcium level	9—11 mg/dℓ (4.5—5.5 mEq/ℓ)
Magnesium	Serum magnesium level	1.50—2.75 mg/dℓ (1.25—2.25 mEq/ℓ)
Essential fatty acid deficiency	Serum triene/tetraene ratio	0.4
Lipids	Serum triglycerides	150 mg/dℓ
	Serum total cholesterol	Men: 115—215 mg/dℓ Women: 120—240 mg/dℓ
Immunocompetence tests	Total lymphocyte count	1500 Cells/mm³
	Rosette-forming cells	50%
	Skin antigen tests	5-mm Skin induration
Protein	Serum proteins	6.5 g/dℓ
	Serum albumin	3.5 g/dℓ
	Serum thyroxine-binding prealbumin	20—50 mg/dℓ
	Serum fibronectin	30—40 mg/dℓ
	Urine creatinine (24 hr)	Men: 20—28 mg/kg body weight Women: 14—22 mg/kg body weight

[a] Adapted from Sauberlich, H. E., *Surv. Synth. Pathol. Res.*, 2, 120, 1983.
[b] Normal or acceptable values may vary somewhat depending upon the method used. Normal values for infants, children, pregnant women, and the elderly may differ from those listed.

closely over a wide range of intakes so that blood calcium determinations do not provide a suitable evaluation of vitamin D status or of dietary calcium intakes. Blood calcium values outside of normal would be suspect of pathological problems before that of a nutritional nature.[34] However, excess intakes of vitamin D may cause hypercalcemia. Similarly, prolonged low dietary intakes of calcium in children have been reported to result in hypocalcemia and in elevated serum alkaline phosphate levels.[45]

3. Vitamin E

Although vitamin E (tocopherol) is an essential nutrient for the human, deficiencies of the nutrient are relatively rare.[19,46-50] Low plasma vitamin E levels have been observed in newborn infants (particularly with premature infants), in cystic fibrosis patients, with protein-energy malnutrition, in certain malabsorption syndromes, with high intakes of polyunsaturated fat, and in certain other diseases or genetic abnormalities. Low plasma tocopherol levels are usually not associated with any clinical symptomatology. However, depressed plasma tocopherol levels, such as with premature infants, are associated with an increased susceptibility to in vitro hemolysis of the erythrocytes. This observation serves as the basis of the erythrocyte hemolysis test. For the test, washed erythrocytes are incubated with hydrogen peroxide under specified conditions and the hemolysis produced during the incubation period is measured.[19] The test is easily performed and requires only a small amount of blood.

Vitamin E nutritional status can be evaluated by direct measurements of the tocopherol levels in plasma.[19,48-51] Various methods are available for the measurement of tocopherols in plasma and serum, including gas chromatography and fluorometric and colorimetric procedures.[19,27,52,53] At present, HPLC is the method of choice because of its sensitivity, specificity, and simplicity.[22,23,25,26,53-56] It should be noted, however, that plasma levels of tocopherol appear to be associated with the plasma lipid concentration, and specifically with the plasma β-lipoprotein content.[57-59] Consequently, low plasma lipid concentrations are associated with low plasma vitamin E concentrations. Thus, to evaluate vitamin E status properly on the basis of plasma tocopherol levels, information on the plasma lipid status is necessary. If such information is available, plasma levels of tocopherol may be expressed in terms of milligrams per gram of plasma lipids. When plasma lipid concentrations are normal, tocopherol levels above 0.70 mg/ℓ of plasma are considered acceptable.

4. Vitamin K

A dietary deficiency of vitamin K is rare in the adult since the vitamin is fairly abundant in the diet and may be synthesized by the intestinal flora.[19,60-66] A deficiency may occur in newborn infants whose body stores of the vitamin are frequently low. Deficiencies may be noted in the elderly, and in patients with diseases affecting the absorption or metabolism of vitamin K, and in patients receiving certain drugs such as oral antibiotics, which interfere with the intestinal synthesis of the vitamin, or anticoagulants, which interfere with prothrombin synthesis.

Vitamin K is essential for the carboxylation of specific glutamic acid residues in prothrombin to form γ-carboxyglutamic acid. The resulting activated prothrombin is then capable of binding with calcium to permit the formation of thrombin. Prothrombin time measurements can serve as a practical indirect indication of vitamin K status. Unless severe liver disease is present, a low prothrombin activity due to a vitamin K deficiency should respond rapidly to vitamin K administration. Although direct plasma measurements of vitamin K are feasible, the current methods are not practical for the routine assessment of vitamin K status.

Vitamin K is also essential for the synthesis in the bone of osteocalcin, the most abundant γ-carboxyglutamic acid-containing protein in the body.[62] In the human, γ-carboxyglutamic acid is not metabolized but excreted quantitatively in the urine. Serum osteocalcin can be measured by radioimmunoassay and the urinary level of γ-carboxyglutamic acid by resin chromatography.[62] The urinary levels of γ-carboxyglutamic acid are increased in osteoporosis[63] and depressed with anticoagulation treatment, such as with warfarin.[64] Whether or not measurements of urinary γ-carboxyglutamic acid excretion or of serum osteocalcin levels will serve as indicators of vitamin K status or of calcium metabolism remains to be established.

5. Vitamin C

A deficiency of vitamin C (ascorbic acid) in the diet results in a condition referred to as "scurvy". Detailed clinical descriptions of the syndrome are available elsewhere.[67-69] In brief, some of the signs and symptoms of scurvy include petechiae, ecchymoses, coiled hairs, hyperkeratosis, gum changes, and congested follicles. Scorbutic patients usually complain of lethargy, aching of the legs, and weakness. A deficient intake of vitamin C results in low plasma ascorbate levels and little or no urinary excretion of the vitamin. With the ready availability of dietary sources of vitamin C, subjects with scurvy are seldom observed in the U.S. Nevertheless, vitamin C deficiency is a pediatric problem in certain areas of the world, in alcoholics, and in neglected elderly patients.[70-73] Elderly subjects with dental problems frequently avoid eating vegetables and fruits that would serve as a source of vitamin C. Lower blood ascorbate levels have been observed in the elderly than in younger age groups.[74-80] Vitamin C supplements enhanced the blood levels of the elderly.[74,79] Behavioral changes were observed in experimental vitamin C deficiency before the presence of clinical signs.[81]

Vitamin C nutritional status is assessed through the use of indirect methods since a biochemical functional test is not available.[19,82-85] For this purpose, determinations of ascorbic acid levels in plasma (serum), erythrocytes, whole blood, leukocytes, and urine have been used. The easiest and most commonly used procedure to evaluate vitamin C status has been the measurement of plasma (serum) ascorbic acid levels.[82] Human studies have demonstrated a linear relationship between vitamin C intake and serum ascorbate level. In patients with scurvy, little or no ascorbic acid would be present in the serum. Serum ascorbic acid levels above 0.30 mg/dℓ (1.65 μmol/ℓ) are considered acceptable.

Ascorbic acid levels in erythrocytes or whole blood are somewhat less sensitive indicators of vitamin C status than serum levels of the vitamin. Vitamin C levels in erythrocytes never fall to the low levels observed in serum with low intakes of ascorbic acid or in the scorbutic patient. Leukocyte levels of ascorbic acid provide an indication of body stores of the vitamin.[19,82,86] The determination of vitamin C in leukocytes is somewhat more difficult because of the need to isolate the cells.[87,88] Urinary excretion of ascorbic acid declines to low or undetectable levels in scurvy or in subjects with low intakes of vitamin C. However, urinary levels of ascorbic acid have not been a reliable indicator of vitamin C status due to analytical difficulties and to the instability of the vitamin in urine.[19,82]

Numerous methods are available for the measurement of vitamin C in serum and blood samples.[82,83,85,87] The 2,6-dichloroindophenol and 2,4-dinitrophenylhydrazine colorimetric methods, the α, α'-dipyridyl method, and the *o*-phenylenediamine fluorometric method are the most commonly used procedures. Each method of measurement provides comparable results. More recently, HPLC techniques have been applied to the measurement of ascorbic acid in serum (plasma), leukocytes, and urine.[82,85,88-92] Regardless of the method used, it should be recognized that vitamin C is highly unstable in biological samples and requires prompt stabilization with trichloroacetic acid, meta-phosphoric acid, or perchloric acid.

6. Thiamin

Dietary survey studies suggest that thiamin intakes are less than desirable in a significant number of the elderly population.[12,93-99] However, clinical cases of a thiamin deficiency in the form of beriberi are rare in the U.S.[100] A deficiency is often observed among chronic alcoholics.[99,101] Occasionally, signs of a thiamin deficiency are observed in patients with a history of heavy consumption of "junk" foods, sweets, carbohydrates, and carbonated beverages.[102,103] High carbohydrate intakes increase the requirement for thiamin.[104] A thiamin deficiency results in an impairment of the functions of the heart and nervous system.[69,100]

Thiamin nutritional status can be evaluated through the use of erythrocyte transketolase activity measurements.[19,102,104-107] Transketolase is a thiamin pyrophosphate (TPP)-requiring enzyme that catalyzes two reactions in the pentose phosphate pathway of the erythrocyte and, hence, serves as a functional measurement of thiamin adequacy. The test provides an early indication of thiamin insufficiency prior to the onset of clinical manifestations.[19] Both manual and automated procedures are available to measure unstimulated and TPP-stimulated erythrocyte transketolase activities.[19,104,107-112]

Thiamin levels in the urine reflect dietary intakes of the vitamin.[19,104,106] However, these measurements are not as reliable an indicator of thiamin status as erythrocyte transketolase activity values.[19,93] Urinary levels of thiamin are influenced by recent intakes of the vitamin and therefore may not provide an accurate evaluation of the thiamin status of an individual. However, under controlled experimental conditions, urinary thiamin excretion levels and erythrocyte transketolase activities reflect thiamin intakes.[104,112] Thiamin can be readily measured in urine with the use of the thiochrome fluorometric method or by microbiological assay with *Lactobacillus viridescens* as the assay organisms.[19] More recently, thiamin has been determined in urine with the use of HPLC.[113] Direct measurement of thiamin in blood has been difficult and unreliable. New methods for measuring blood thiamin levels, including HPLC, may prove adequate for use in the assessment of thiamin nutritional status.[114-116] Other laboratory procedures exist for assessing thiamin nutrition, but the procedures are less sensitive and, therefore, seldom used.[19,106]

7. Riboflavin

Riboflavin deficiency is relatively common in many areas of the world.[19] Populations with low intakes of dairy products and animal protein sources are particularly prone to inadequate intakes of riboflavin. A high prevalence of riboflavin deficiency has been reported for some areas in India[107] and Thailand.[118] Evidences of inadequate intakes of riboflavin have been reported for high school students,[119] alcoholics,[120] pregnant women,[121] and the elderly.[10,12,77,97,122-125]

A riboflavin deficiency is associated with ocular, skin, and mucous membrane changes.[126,127] The deficiency is commonly characterized by corneal vascularization, cheilosis, glossitis, angular stomatitis, and seborrheic dermatitis about the scrotum, face, and nose. Some of these signs may be observed with other deficiencies, such as vitamin C, niacin, vitamin B_6, or iron. Clinical or biochemical manifestations of a riboflavin deficiency usually respond promptly with administration of the vitamin. Riboflavin nutriture can be assessed easily with erythrocyte glutathione reductase measurements.[19,108,119,120,128-133] This riboflavin-dependent enzyme provides for a sensitive functional test of riboflavin status. The enzyme is quite stable and the procedures used to measure its activity are reliable and simple. The enzyme activities are usually expressed in terms of activity coefficients, representing the degree of stimulation resulting from the in vitro addition of flavin adenine dinucleotide.[19,120]

Urinary riboflavin levels can also provide a reasonable evaluation of riboflavin status since clinical signs of a deficiency can be closely related to riboflavin intakes and its urinary excretion.[19,127] Riboflavin levels in the urine tend to reflect the recent dietary intake of riboflavin and, hence, may be prone to day-to-day variations for a given person. However, these effects are minimal in individuals subsisting on marginal or inadequate intakes of riboflavin whose body stores are depleted. When 24-hr urine collections are not practical or reliable, as may often occur with the elderly, random urine samples may be used and analyzed for riboflavin and the results expressed in terms of urine creatinine level.[19,120] Riboflavin levels in the urine may be measured by microbiological or fluorometric methods.[19,120] Recently, a radioassay[134] and a protein-binding assay[135] were reported that could be used for the measurement of riboflavin in urine. HPLC offers promise as a simple and rapid procedure to determine riboflavin levels in the urine.[136,137] These newer analytical procedures require further evaluation to establish their reliability. Other laboratory procedures exist for assessing riboflavin nutritional status, but are less sensitive or reliable and, therefore, seldom used.[19,120]

8. Niacin

Niacin (nicotinic acid) deficiency, although common in the Southern U.S. prior to 1950, is now relatively rare. Occasionally, niacin deficiency, or pellagra, is observed in association with alcoholism. However, pellagra remains an important nutritional problem in certain maize-eating areas of the world, particularly in South Africa and India. In India, pellagra occurs among both populations consuming maize as their primary dietary staple and those whose staple is jowar (variety of millet). Pellagra has been frequently described as the disease of dermatitis, diarrhea, and dementia.[69,138]

Little recent information is available as to the niacin nutritional status of the elderly.[11,12,97,139] The information available would indicate that niacin deficiency is not a serious problem in the healthy, noninstitutionalized elderly. However, low urinary levels of N-methylnicotinamide have been observed in elderly patients in the U.K.[11]

A functional biochemical test is not available for assessing body reserves of niacin. The only practical procedure for evaluating niacin nutriture is the measurement of the urinary levels of the two major niacin metabolites: 2-pyridone and N^1-methylnicotinamide.[119,140] The ratio of these two values is used as an index of niacin nutritional status. With a niacin deficiency, the urinary excretion of 2-pyridone falls more profoundly than that of N^1-methylnicotinamide. A ratio of less than 1.0 is indicative of a latent niacin deficiency. Normally, a ratio of 1.3 to 4 should be observed. Previously, the measurement of 2-pyridone was laborious and unreliable.[19] With the use of HPLC, procedures are now available that permit both metabolites to be measured easily and quickly in urine.[141-143]

Although other approaches have been investigated for the evaluation of niacin status, none has been satisfactory. Thus, for example, the measurement of niacin compounds in blood or its components has not proven reliable or practical for this purpose. Similarly, nicotinic acid is present in only small amounts in the urine, and its excretion is relatively uninfluenced by dietary intakes of niacin. Thus, in contrast to riboflavin and thiamin, nicotinic acid levels in the urine cannot serve as a useful indicator of niacin intakes.

9. Vitamin B_6

An overt clinical deficiency of vitamin B_6 (pyridoxine) is rather rare, but a subclinical deficiency may be fairly widespread.[144-147] Thus, evidence of a vitamin B_6 deficiency may occur during pregnancy, with the use of oral contraceptive agents, in alcoholic patients, and with the use of certain medications such as isoniazid in the treatment of tuberculosis, anticonvulsants, and prolonged use of penicillamine.[146] In the adult human, a deficiency of vitamin B_6 is displayed in abnormal electroencephalograms, depression, hyperirritability, seizures, seborrheic dermatitis, eczema about the eyes, ears, mouth, and nose, and angular stomatitis, cheilosis, and glossitis. The lesions of the face are not specific to a deficiency of vitamin B_6, since similar lesions may occur with deficiencies of other vitamins.

Both dietary intake studies and biochemical assessments indicate that vitamin B_6 status in the elderly is often inadequate.[123,124,148-153] Based on abnormal erythrocyte aminotransferase activities, 19 to 56% of the elderly studied had an inadequate vitamin B_6 status. Supplements of vitamin B_6 were able to correct the condition.[123,151] The abnormal tryptophan metabolism observed in elderly subjects was also corrected with supplements of vitamin B_6.[152]

Vitamin B_6 deficiency is rather difffficult to diagnose on the basis of clinical symptoms and signs. However, a number of useful laboratory procedures are available to evaluate vitamin B_6 nutritional status.[19,145-147,154] These include determination of plasma pyridoxal phosphate levels, erythrocyte aminotransferase activity measurements, tryptophan load tests, and urinary excretion of 4-pyridoxic acid or intact vitamin B_6. Measurements of erythrocyte aspartate aminotransferase (AST, GOT) and erythrocyte alanine aminotransferase (ALT, GPT) activities along with the in vitro stimulation effect of pyridoxal-5-phosphate have been commonly used for the assessment of vitamin B_6 status.[145,154] The activities of the erythrocyte aminotransferases decrease markedly in a vitamin B_6 deficiency and are also markedly stimulated by the in vitro addition of pyridoxal-5-phosphate. Coupled enzyme spectrophotometric procedures or colorimetric methods are generally employed to measure the activities.[108,155-158]

The dietary intake of vitamin B_6 is reflected in the urinary excretion of the vitamin and of its metabolite, 4-pyridoxic acid.[19,145,154,158-160] The levels of the compounds in urine fall rapidly with low or inadequate intakes of vitamin B_6.[159-161] At present, microbiological assays are the only practical means of measuring vitamin B_6 in urine.[159] Although HPLC techniques are under development, the current procedures are not practical for the assessment of vitamin B_6 status.[158] However, the determination of 4-pyridoxic acid in urine can be easily performed by the use of HPLC.[160-163] The methods previously available were rather tedious and involved.

Plasma pyridoxal-5-phosphate levels can also indicate a depletion of vitamin B_6 reserves. Enzymatic assays using either apotyrosine decarboxylase or apotryptophase are generally used to measure pyridoxal-5-phosphate levels in plasma.[158,159,164-168]

The measurement of the urinary excretion of xanthurenic acid following a tryptophan load is a sensitive and useful functional test for detecting or evaluating a vitamin B_6 deficiency.[19,144-147,154,169] In this test a subject is given 2 to 5 g of L-tryptophan orally and the xanthurenic acid excretion is measured in the following 24-hr urine collection. Normal subjects will excrete less than 50 mg of xanthurenic acid following the load test. Alternative procedures that have been used on occasion are the methionine load tests[170] or the more expensive kynurenine load test.[169,171,172] The urinary excretion of cystathionine increases substantially following a methionine load in the vitamin B_6-deficient subject.[170,173]

10. Folacin

Folacin (folic acid, folate) intakes for most free-living elderly persons have been considered adequate.[174,175] Folate status, based on biochemical assessments, was also adequate for the majority of U.S. free-living elderly. However, folic acid deficiency has been observed in certain segments of the elderly population, such as those with low incomes, and those who are institutionalized or hospitalized. In some instances, a deficiency may result from the use of certain medications (e.g., anticonvulsants), impaired absorption, or alcoholism.[176,177] Colman et al.[176,178] observed an increased incidence of folate deficiency in the elderly age 61 years and above. Other reports have also provided evidence of folate deficiencies in the elderly, particularly among the handicapped and institutionalized aged.[153,179-185] Widespread folacin deficiency was observed in low-income urban black elderly persons.[181] In most instances, the low folate blood levels could be associated with poor dietary habits and low dietary folacin intakes.

Folacin functions in the transfer of methyl groups essential for the synthesis of nucleic acids and in the metabolism of certain amino acids. A deficiency of the vitamin results in a megaloblastic anemia that may be accompanied by a decrease in the number of white blood cells (leukopenia).

Folate nutritional status is usually evaluated through measurement of folic acid levels in serum and erythrocytes.[19,176,186,187] However, the erythrocyte folate level has been regarded as a more accurate and less variable quantitative index than serum folate as to the severity of folacin deficiency in a subject.[19,176] Microbiological assays are commonly used for the measurement of total folic acid activity in serum and erythrocytes. *Lactobacillus casei* is the organism of choice since its growth response is nearly equal for all the folate forms present in serum and prepared erythrocytes.[19,176,188] However, microbiological assays for folacin must be conducted with care to avoid analytical errors and problems of bacterial contamination and of antibiotics or antifolates present in the samples.

Recently, various radioassay procedures have been developed for measuring folic acid levels in serum and erythrocytes.[189-192] For convenience and easy use, a number of commercial radioassay kits have become available.[193] The radioassay methods have been useful with serum samples but have been less reliable with erythrocyte specimens.[191,193-195]

Various other less-used procedures are available to assess folacin nutritional status. The procedures include neutrophil hypersegmentation measurements,[176,196,197] deoxyuridine suppression tests,[176,198,199] and the determination of urinary excretion of formiminoglutamic acid (FIGLU).[19,176] However, these tests are not specific for folacin deficiency, but may also be applied to evaluate vitamin B_{12} status. Consequently, subjects with a suspected folacin deficiency should be evaluated further to eliminate a possible dietary vitamin B_{12} deficiency or pernicious anemia.

11. Vitamin B_{12}

A deficiency of vitamin B_{12} (cobalamin) due to low dietary intakes is relatively rare. However, it may occur in vegans who maintain a strict vegetarian diet devoid of eggs, meat, and dairy products.[200] Most cases of vitamin B_{12} deficiency observed in the U.S. are the result of an impaired absorption of the vitamin due to a lack of the intrinsic factor (pernicious anemia). In some studies, latent pernicious anemia was observed in the elderly.[179,182,201] In Sweden and the U.K., pernicious anemia occurs in nearly 1% of the people over the age of 60 years.[200] However, vitamin B_{12} absorption has not been reported to decline with age.[202] The dietary intake of vitamin B_{12} was found to be adequate in an elderly Nordic population.[153] A deficiency in vitamin B_{12} will result in a sore tongue, paresthesias, weakness, and neurologic changes.[200] A deficiency in either vitamin B_{12} or folic acid will result in a morphologically identical macrocytic anemia, megaloblastic bone marrow changes, and hypersegmented polymorphonuclear neutrophils. Because of the close metabolic interrelationship that exists between vitamin B_{12} and folacin, vitamin B_{12} nutritional status must also be evaluated in terms of folacin nutrition.[176]

Vitamin B_{12} nutritional status may be assessed from information on the serum levels of the vitamin.[19,200] Low body contents of vitamin B_{12} are associated with low serum levels of the vitamin. The measurement may be performed with either microbiological assays or radioassay techniques.[19,200] Earlier, some of the vitamin B_{12} radioassay procedures failed to detect certain patients with pernicious anemia.[203-205] Recent modifications in the commercial radioassay kits appear to have eliminated this problem. With the ready availability of the commercial kits, the radioassay for vitamin B_{12} has proven convenient, simple, and reliable. Commercial radioassay kits are also available that allow for the simultaneous measurement of both vitamin B_{12} and folic acid in serum samples. As noted in the section on folacin, neutrophil hypersegmentation measurements, deoxyuridine suppression tests, or formiminoglutamic acid excretion levels may assist in vitamin B_{12} evaluations.[19,176,198,199] These procedures are not entirely specific to a vitamin B_{12} deficiency and, therefore, have received limited application. When a vitamin B_{12} deficiency has been indicated, a Schilling test or plasma vitamin B_{12} uptake test is necessary to establish the cause.

12. Biotin

Biotin deficiency in the adult human seldom occurs. In a few instances, a biotin deficiency has been induced in the human experimentally or as the result of the ingestion of diets that included large amounts of raw eggs or raw egg whites.[206-209] In a few instances a biotin deficiency has been associated with parenteral nutrition.[210] Symptoms of a deficiency included anorexia, nausea, dry scaly dermatitis, depression, glossitis, pallor, and muscle pain.[209] Because of the rareness with which biotin deficiency occurs, few studies have been devoted to evaluating biotin nutritional status in the human.[207,208,211] The existing reports would indicate that measurement of the biotin level in blood or urine may provide some information as to the nutritional status of the vitamin.[19] Microbiological assays are the only practical means to measure the biotin level in biological samples.[19,212]

13. Pantothenic Acid

Although a pantothenic acid deficiency produces characteristic deficiency syndromes in animal species, an uncomplicated clinical case of pantothenic acid deficiency has not been recognized for the human.[213-215] The burning-feet syndrome has been associated with a deficiency of pantothenic acid as well as of riboflavin and niacin. Hence, in some instances, this syndrome responds to pantothenic acid treatment. A characteristic of the syndrome is a burning sensation of the soles of the feet and occasionally in the palms of the hands. Blood and urine levels of pantothenic acid are below normal. Occasionally, alcoholic patients are encountered with low blood pantothenic acid levels. Otherwise, few studies have been conducted on the pantothenic acid status in human subjects.[19,215-220] Recently, Srinivasan et al.[218,221] studied pantothenic acid nutritional status in the elderly by measuring pantothenic acid levels in blood and urine by both radioimmunoassay and microbiological assay. The radioimmunoassay had equal or greater sensitivity than the microbiological assay. The microbiological procedures may use either *Saccharomyces uvarum*, *Lactobacillus casei*, or *L. plantarum*.[19]

Over the age range of 65 to 90 years of age, the urinary excretion of pantothenic acid was not age dependent.[218] This and other studies indicate that urinary excretions of pantothenic acid correlated with dietary intakes of the vitamin. However, blood pantothenic acid levels showed a less reliable relationship with dietary intakes of the vitamin.[19,218]

B. Minerals

1. Iron

A dietary deficiency of iron will result in a hypochromic, microcytic anemia. A deficiency may also be associated with fatigue, listlessness, palpitation on exertion, angular stomatitis, sore tongue, dyspnea, and tingling or numbness. These observations, however, are not specific to an iron deficiency, since many may occur with other nutrient deficiencies.

Reports on iron nutritional status in the elderly are limited.[181,182,222-233] Some studies indicate that the iron requirements of the elderly are lower than those of younger age groups and, hence, the elderly should be less prone to a dietary iron deficiency.[222,223] In some studies with the elderly no evidence of an iron deficiency was observed.[181,222] Aside from vitamin B_{12} and folacin deficiencies occurring in the adult, the presence of anemia must be evaluated in terms of being caused by occult gastrointestinal bleeding (e.g., due to gastric or duodenal ulcers, gastritis, bleeding diverticula, etc.), medications, infections, malabsorption, cancer, and other diseases.[222,223,230,234] Thus, it cannot be assumed that anemia in the elderly can be equated with an iron deficiency.

Various laboratory procedures are available to evaluate iron nutritional status.[19,235] Hemoglobin, hematocrit, and reticulocyte count determinations can establish the presence of anemia, but additional measurements are required to establish the presence of an iron deficiency. For this purpose, measurement of serum ferritin levels has become exceedingly useful. Serum ferritin can be measured with the use of radioassay methods that require only a small sample of blood. With the availability of commercial radioassay kits, the analysis is simple and convenient to perform. Low serum ferritin levels are associated with depleted body stores of iron. Approximately 1 μg of ferritin per liter of serum is equivalent to 8 mg of storage iron.[236]

Additional laboratory tests widely used for the evaluation of iron nutritional status include serum transferrin, percent transferrin saturation, serum iron and total iron-binding capacity, erythrocyte protoporphyrin, mean corpuscular volume, and mean corpuscular hemoglobin concentration. Elevated total iron-binding capacity, elevated erythrocyte protoporphyrin levels, lowered serum iron levels, and a low percent saturation are indicative of an iron deficiency. Guidelines used to evaluate iron nutritional status in the elderly are the same as those used for younger age groups (Table 1). Evidence suggests that normal hematological values for the elderly may differ from that of the younger age groups.[222] A need exists for an evaluation and establishment of normal values for the older age groups.

2. Zinc

Zinc is essential for the human for the maintenance of normal function of numerous enzymes.[237-241] Thus, zinc participates in the function of enzymes such as alcohol dehydrogenase, alkaline phosphatase, superoxide dismutase, carboxypeptidase A, carbonic anhydrases, RNA polymerase, and DNA polymerase. In young males, zinc deficiency was associated with poor appetite, growth retardation, hypogonadism, mental lethargy, and skin changes.[237,240] In some instances, zinc supplementation promoted wound healing and improved taste acuity (hypogeusia).[241]

Few studies have been conducted on the zinc nutritional status of the elderly.[230,241-246] Dietary intake information suggests that zinc intakes by the elderly may be low.[241,242] The elderly had low blood and hair zinc concentrations and impaired taste acuity, suggesting that this population is at risk for a zinc deficiency.[241-245,247,248] Undoubtedly, the risk of zinc deficiency is increased in elderly subjects with disease states, such as alcoholic liver cirrhosis, malabsorption, and chronic renal failure.

At present, no single laboratory test can be used to reliably evaluate zinc nutritional status.[237,239,240] Procedures available that are used include serum or plasma zinc levels, erythrocyte zinc, hair zinc, salivary zinc, urinary zinc, and taste acuity measurements.[238,239] The determination of serum or plasma zinc by atomic absorption spectrophotometry is the most frequently used test for the evaluation of zinc nutriture.[238,249] Although the analytical procedure is simple and reliable, care must be maintained to avoid exogenous zinc contamination of the samples or from hemolysis. Erythrocytes contain over ten times more zinc than does plasma.[238] The zinc levels in the erythrocytes and hair may be used also for the evaluation of zinc status. Because of the slow zinc turnover rates in these tissues, the measurements can reflect long-term zinc status. Hair samples must be used with caution because of their susceptibility to environmental contamination.[238,239] Although more tedious to perform, neutrophil zinc levels have been proposed as a useful measurement of zinc nutritional status as has the alkaline phosphatase activity of the neutrophils.[250] With zinc depletion, the urinary excretion of zinc decreases. Thus, measurements of 24-hr urinary excretions of zinc have been useful in diagnosing a zinc deficiency.

3. Copper

Although a dietary copper deficiency may occur in the human, it is relatively rare.[230,238,239,244,246,251,252] Nevertheless, diets in the U.S. are often marginal in terms of providing the Recommended Dietary Allowance (RDA) for copper.[251-253] Copper nutriture has been assessed by the measurement of copper levels in serum or plasma. Serum copper levels fall with a deficiency of copper and increase again with copper supplementation. Copper levels in biological specimens are commonly determined with the use of atomic absorption spectrometry. Hair copper and urinary copper levels have not proven reliable for evaluating copper nutritional status.

4. Chromium

Evidence exists suggesting that chromium deficiency may contribute to abnormal glucose and fat metabolism in the elderly in the U.S.[230,244,254-258] Chromium supplements have been reported to improve glucose tolerance and lower blood cholesterol levels in the elderly.[254-258] The body stores of chromium appear to deplete gradually with aging.[258] At present, no satisfactory test is available to assess the chromium status of an individual.[258,259] In a lengthy and expensive procedure, a person's responses to chromium supplementation can serve in a retrospective evaluation.

5. Selenium

The association of selenium deficiency with Keshan disease of China has stimulated interest in this trace element.[251,260,261] Various procedures have been employed to evaluate selenium status including urinary selenium levels, whole blood, erythrocyte, and plasma selenium levels, selenium levels in hair, toenails and fingernails, and glutathione peroxidase activities of platelets or erythrocytes.[249,260-265] No one method is entirely satisfactory for assessing selenium status; hence, several techniques are employed.

6. Other Trace Elements

Manganese, molybdenum, and fluorine are also considered essential for man. However, no satisfactory assessment methods, standards, or criteria of adequacy have been established for these nutrients.[230,249,251]

7. Iodine

With the use of iodized table salt, the incidence of goiter is low in the U.S. Urinary excretions of iodine of less than 50 μg/g of creatinine are considered indicative of a deficient intake of iodine.[19,265,266] Also, with a dietary deficiency of iodine, serum protein-bound iodine is decreased while radioiodine uptake is increased.

8. Sodium, Potassium, and Chloride

It is difficult to attach nutritional significances to changes in serum levels of sodium, potassium, or chloride which may be due to certain pathological states. Consequently, serum levels of these nutrients do not always reflect the body content of these elements because of the influence of factors such as hydration, dehydration, or chronic renal failure. Nevertheless, true deficiencies can occur with inadequate intakes, or with excessive losses due to extreme sweating, diarrhea, vomiting, and certain disease conditions.[267] In addition to serum measurements of these elements, urinary excretion levels may also be helpful.[267,268] For this purpose, 24-hr urine collections over a period of several days should be used to obtain an estimate of an individual's intake. Low dietary intakes of potassium and the use of drugs, such as laxatives or diuretics, increase the risk of potassium depletion in the elderly.

9. Phosphorus

The average normal range for serum phosphorus in the adult appears to be 2.5 to 4.0 mg/dℓ, while levels of 1.0 mg/dℓ or lower may have serious clinical consequences. From a nutritional viewpoint, interpretation of serum phosphorus levels is difficult because of the number of factors observed to influence serum phosphorus levels. Reduced serum levels occur with hyperparathyroidism, rickets, sprue, insulin treatment, vomiting, or diarrhea, while elevated serum values occur with renal disease, hypoparathyroidism, diabetes, and healing fractures. Hypophosphatemia requires prompt diagnosis of cause and treatment with the administration of inorganic or organic phosphates.[269]

10. Calcium

Serum calcium levels are closely controlled by the body over a wide range of intakes. Consequently, serum or blood calcium values outside of normal would be suspect of a pathological problem before that of a nutritional nature. At present, a suitable laboratory technique for monitoring the adequacy of calcium intake does not exist. See the section on vitamin D.

11. Magnesium

A deficiency in magnesium due solely to a low dietary intake is seldom reported for the adult. A deficiency is usually associated with malabsorption syndromes, diarrhea, renal tubular dysfunction, and alcoholism.[266,270] In a magnesium deficiency, serum magnesium values are markedly reduced. Urine magnesium levels are also ordinarily very low. Atomic absorption spectroscopy appears to be the easiest means to measure magnesium levels in serum and urine specimens.[271]

C. Lipids

A dietary deficiency of essential fatty acids is rare in the adult human, but it has been encountered in prolonged fat-free parenteral nutrition.[272-276] An essential fatty acid deficiency can be diagnosed at an early stage by gas-liquid chromatography analysis of the fatty acids present in the plasma or red cell membranes.[19,273] Most widely used for evaluating essential fatty acid status is the ratio of 20:3 ω 9/20:4 ω6 (triene to tetraene ratio). For serum, ratios below 0.4 are considered acceptable, although a ratio as low as 0.2 has also been suggested.[273,274] Other aspects of lipid metabolism are frequently measured. This may include serum total cholesterol, high- and low-density lipoproteins, phospholipids, triglycerides, or total lipids.[266,277,278]

D. Protein-Energy Malnutrition

Dietary intakes of protein generally fall with increasing age. The protein content or lean body mass in the adult also diminishes with age.[279] However, it is uncertain that the protein requirements of the elderly per kilogram of body weight are any different from that of younger adults.[280] Although protein-energy malnutrition (protein-calorie malnutrition) is often observed in children in developing countries in the form of kwashiorkor or as marasmus, the condition is rarely noted in children or adults in the U.S.[19,281,282] However, a high incidence of protein-energy malnutrition is observed in hospitalized patients.[282-286]

Various laboratory procedures are available for evaluating protein nutritional status in the adult.[1,2,19] Because of the ease of determination, serum albumin assessments are frequently performed. Serum albumin levels of 4.5 g/dℓ are observed in the normal subject, while the serum albumin level may fall to less than 2.5 g/dℓ in the severely protein-depleted individual. Serum albumin levels have been reported to fall slightly with aging.[1] Total serum protein determinations, although useful, are a less sensitive indicator of protein status than serum albumin measurements. Serum transferrin levels[287] and total iron-binding capacity have also been useful in evaluating protein nutritional status. Due to their short biological half-life, thyroxine-binding prealbumin, retinol-binding prealbumin, and fibronectin appear to be sensitive and early indicators of protein nutritional status.[2,287,288-293]

Urinary excretion levels of creatinine, although variable and influenced by age, can provide some information as to the protein nutritional status.[1,6,282] Lymphocyte counts under 1800 mm³ are an indication of protein malnutrition.[294]

E. Immunocompetence Tests

Numerous studies have shown an association of energy and impaired immunocompetence with an increase in morbidity and mortality from infectious disease.[1,2,295,296] The impaired cell-mediated immune function appears to be associated with protein-energy malnutrition,[294,297] although deficiencies in other nutrients, including zinc and vitamin A, may cause impairment. With advancing age, however, the immune function may decline. Nevertheless, evaluation of immunocompetence can provide an indirect measure of nutritional status.[296,298-300] Cell-mediated immune function may be evaluated by the use of recall antigens, usually *Candida*, mumps, streptokinase-streptodornase, coccidioidin, and tuberculin skin test antigens.[301,302] Other less well-established measures of immunocompetence have been proposed that include serum concentrations of complement component C3, leukocyte terminal transferase concentration, secretory IgA levels in body secretions, and bactericidal capacity of neutrophils.[298,299] Further investigations as to the reliability and usefulness of immunocompetence tests as a measure of nutritional status in the elderly are needed.[303]

III. DIETARY STUDIES

Dietary assessments are difficult to perform and no single procedure is universally accepted.[304-315] Dietary assessment may be conducted by (1) 24-hr recall of food intake, (2) diet diary, (3) diary-interview, (4) food frequency, (5) diet history, (6) food purchases, and (7) actual food intake. The success of any dietary methodology is highly dependent upon the use of trained dietitians. Numerous forms and questionnaires are available for use in the dietary interviews.

The 24-hr recall procedure is most commonly used, but the intake data are often inaccurate due to omission of foods, errors in estimates of amounts eaten, and lack of representation of the food usually eaten. If cooperation can be maintained, 7-day (or longer) diet diaries provide a more reliable index of the nutrient intakes and dietary habits of an individual than the 24-hr recall. Occasionally, 3- or 5-day diet diaries are used. Because of the dependency upon memory, dietary recall methodologies may have disadvantages for use with the elderly.

The food frequency technique provides qualitative data on food intakes since information as to specific exact portion sizes is not usually requested of the participant.

Dietary intake information usually requires the conversion of the data to daily intakes of nutrients with the use of food composition tables. The limitations of these food composition tables should be recognized since for some nutrients, such as folic acid, the analytical data are either lacking or of questionable accuracy.[314] The dietary nutrient intakes calculated are commonly evaluated or compared to recommended dietary allowances (RDA) of the Food and Nutrition Board of the National Research Council.[66,304,308,310] Dietary intakes of nutrients less than the RDA cannot be considered to indicate the presence of deficiencies, but rather that a risk may exist. Biochemical data, however, can assess the presence of a deficiency.

In evaluating dietary intake data, a knowledge of the requirements is essential. At present, our knowledge of the actual nutrient requirements of the elderly is uncertain. To some extent this could be overcome by evaluating the consistency between prevalence of inadequate intakes and prevalence of biochemical indications of inadequacy.

Dietary assessments provide no direct measurement of nutritional status. With adequate data, however, dietary intakes may correlate with certain biochemical assessment.[312] Usually, dietary intake relationships are tenuous and cannot be relied upon to predict nutritional status. Nevertheless, reliable food consumption data may be useful to predict an increased risk to health from a particular pattern of food or nutrient intake.

IV. SUMMARY

Inadequacies exist in the laboratory procedures available for assessing the nutritional status of the elderly. Guidelines and criteria for evaluating biochemical nutritional data have been developed almost exclusively for younger age groups and may not be entirely applicable to the elderly. Until suitable criteria have been developed, biochemical nutritional data obtained on the elderly will need to be evaluated and interpreted with care. Of particular need are additional studies where attempts are made to correct abnormal biochemical findings with suitable nutrition intervention to accurately establish the presence of nutritional inadequacies.

ACKNOWLEDGMENTS

This work was supported in part by USPHS/NIH grant 5 PO1-CA28103 (Clinical Nutrition Research Unit). Additional support from the Susan Mott Webb Charitable Trust is also gratefully acknowledged. The author wishes to thank Ms. Cynthia Nobles for her secretarial assistance in the preparation of the manuscript.

REFERENCES

1. Assessing the Nutritional Status of the Elderly — State of the Art, Report of the Third Ross Roundtable on Medical Issues, Ross Laboratories, Columbus, Ohio, 1982.
2. **Levenson, S. M., Ed.,** Nutritional Assessment — Present Status, Future Directions and Prospects, Report of the Second Ross Conference on Medical Research, Ross Laboratories, Columbus, Ohio, 1981.
3. **Gray, G. E. and Gray, L. K.,** Validity of anthropometric norms used in the assessment of hospitalized patients, *J. Parenteral Enteral Nutr.,* 3, 366, 1979.
4. **Gray, G. E. and Gray, L. K.,** Anthropometric measurements and their interpretations: principles, practices, and problems, *J. Am. Dietet. Assoc.,* 77, 534, 1980.
5. **Pike, R. L. and Brown, M. D.,** *Nutrition: An Integrated Approach,* 3rd ed., Wiley, New York, 1984, 678.
6. **Bowman, B. B. and Rosenberg, I. H.,** Assessment of the nutritional status of the elderly, *Am. J. Clin. Nutr.,* 35, 1142, 1982.

7. **Harper, A. E. and Simopoulos, A. P.**, Summary, conclusions, and recommendations, *Am. J. Clin. Nutr.*, 35, 1098, 1982.
8. **Munro, H. N.**, Nutrition and the elderly. An introductory overview, *Bibl. "Nutr. Dieta"*, 33, 1, 1983.
9. **Exton-Smith, A. N.**, Epidemiological studies in the elderly: methodological considerations, *Am. J. Clin. Nutr.*, 35, 1273, 1982.
10. **Macleod, C. C., Judge, T. G., and Caird, F. I.**, Nutrition of the elderly at home. II. Intakes of vitamins, *Age Ageing*, 3, 209, 1974.
11. **Morgan, A. G., Kelleher, J., Walker, B. E., Losowsky, M. S., Droller, H., and Middleton, R. S.**, A nutritional survey in the elderly: blood and urine vitamin levels, *Int. J. Vitam. Nutr. Res.*, 45, 448, 1975.
12. **Harrill, I. and Cervone, N.**, Vitamin status of older women, *Am. J. Clin. Nutr.*, 30, 431, 1977.
13. **Kark, J. D., Smith, A. H., Switzer, B. R., and Hames, C. G.**, Serum vitamin A (retinol) and cancer incidence in Evans County, Georgia, *J. Natl. Cancer Inst.*, 66, 7, 1981.
14. **Peto, R., Doll, R., Buckley, J. D., and Sporn, M. B.**, Can dietary beta-carotene materially reduce human cancer rates? *Nature (London)*, 290, 201, 1981.
15. **Kark, J. D., Smith, A. H., and Hames, C. G.**, Serum retinol and the inverse relationship between serum cholesterol and cancer, *Br. Med. J.*, 284, 152, 1982.
16. **Carney, E. A. and Russell, R. M.**, Correlation of dark adaptation test results with serum vitamin A levels in diseased adults, *J. Nutr.*, 110, 552, 1980.
17. World Health Organization/USAID Meeting, Vitamin A deficiency and xerophthalmia, Technical Report Series in World Health Organization, No. 590, WHO, Geneva, 1976.
18. **Sauberlich, H. E., Hodges, R. E., Wallace, D. L., Kolder, H., Canham, J. E., Hood, J., Raica, N., Jr., and Lowry, L. K.**, Vitamin A metabolism and requirements in the human studied with the use of labeled retinol, *Vitam. Horm.*, 33, 251, 1975.
19. **Sauberlich, H. E., Dowdy, R. P., and Skala, J. H.**, *Laboratory Tests for the Assessment of Nutritional Status*, CRC Press, Cleveland, 1974.
20. **Garry, P. J.**, Vitamin A. Symp. Lab. Assess. Nutr. Status, Labbé, R. F., Ed., appearing in *Clin. Lab. Med.*, 1, 699, 1981.
21. **Underwood, B. A.**, The determination of vitamin A and some aspects of its distribution, mobilization and transport in health and disease, *World Rev. Nutr. Diet.*, 19, 123, 1974.
22. **Catignani, G. L. and Bieri, J. G.**, Simultaneous determination of retinol and α-tocopherol in serum or plasma by liquid chromatography, *Clin. Chem.*, 29, 708, 1983.
23. **Bieri, J. G., Tolliver, T. J., and Catignani, G. L.**, Simultaneous determination of α-tocopherol and retinol in plasma or red cells by high pressure liquid chromatography, *Am. J. Clin. Nutr.*, 32, 2143, 1979.
24. **Neeld, J. B., Jr. and Pearson, W. N.**, Macro- and micromethods for the determination of serum vitamin A using trifluoroacetic acid, *J. Nutr.*, 79, 454, 1963.
25. **Driskell, W. J., Neese, J. W., Bryant, C. C., and Bashor, M. M.**, Measurement of vitamin A and vitamin E in human serum by high-performance liquid chromatography, *J. Chromatogr.*, 231, 439, 1982.
26. **De Leenheer, A. P., De Bevere, V., De Ruyter, M. G. M., and Claeys, A. E.**, Simultaneous determination of retinol and α-tocopherol in human serum by high-performance liquid chromatography, *J. Chromatogr.*, 162, 408, 1979.
27. **Thompson, J. N., Erdody, P., and Maxwell, W. B.**, Simultaneous fluorometric determination of vitamins A and E in human serum and plasma, *Biochem. Med.*, 8, 403, 1973.
28. **Mejiá, L. A. and Arroyave, G.**, Determination of vitamin A in blood. Some practical considerations on the time of collection of the specimens and the stability of the vitamin, *Am. J. Clin. Nutr.*, 37, 147, 1983.
29. **Goodman, D. S.**, Plasma retinol-binding protein, *Ann. N.Y. Acad. Sci.*, 348, 378, 1980.
30. **Smith, F. R.**, Serum vitamin A, retinol-binding protein, and prealbumin concentrations in protein-calorie malnutrition. I. A functional defect in hepatic retinol releases, *Am. J. Clin. Nutr.*, 26, 973, 1973.
31. **Parviainen, M. T. and Ylitolo, P.**, Immunonephelometric determination of retinol-binding protein in serum and urine, *Clin. Chem.*, 29, 853, 1983.
32. **Goodman, D. S.**, Retinol-binding protein, prealbumin and vitamin A transport, *Prog. Clin. Biol. Res.*, 5, 313, 1976.
33. **Allen, L. H.**, The role of nutrition in the onset and treatment of metabolic bone diseases, in *Nutrition Update*, Vol. 1, Weininger, J. and Briggs, G. M., Eds., John Wiley & Sons, New York, 1983, 263.
34. **Thomas, W. C., Jr. and Howard, J. E.**, Disorders of calcium metabolism, in *CRC Handbook of Nutrition and Food*, Section E, *Nutritional Disorders*, Vol. 3, *Effects of Nutrient Deficiencies in Man*, Rechcigl, M., Ed., CRC Press, West Palm Beach, Fla., 1978, 139.
35. **Heaney, R. P., Gallagher, J. C., Johnston, C. C., Neer, R., Parfitt, A. M., Chir, B., and Whedon, G. D.**, Calcium nutrition and bone health in the elderly, *Am. J. Clin. Nutr.*, 36, 986, 1982.
36. **Anwar, M.**, Nutritional hypovitaminosis-D and the genesis of osteomalacia in the elderly, *J. Am. Geriatr. Soc.*, 26, 309, 1978.
37. **Holmes, R. P. and Kummerow, F. A.**, The vitamin D status of elderly Americans, *Am. J. Clin. Nutr.*, 38, 335, 1983.

38. **Parfitt, A. M., Gallagher, J. C., Heaney, R. P., Johnston, C. C., Neer, R., and Whedon, G. D.,** Vitamin D and bone health in the elderly, *Am. J. Clin. Nutr.,* 36, 1014, 1982.

39. **Omdahl, J. L., Garry, P. J., Hunsaker, L. A., Hunt, W. C., and Goodwin, J. S.,** Nutrition status in a healthy elderly population: vitamin D, *Am. J. Clin. Nutr.,* 36, 1225, 1982.

40. **Holick, M. F. and MacLaughlin, J. A.,** Aging significantly decreases the capacity of the human epidermis to produce vitamin D₃, *Clin. Res.,* 29, 408A, 1981.

41. **Duncan, W. E. and Haddad, J. G.,** Vitamin D assessment. The assay and their applications; Symp. Lab. Assess. Nutr. Status, Labbé, R. F., Ed., appearing in *Clin. Lab. Med.,* 1, 713, 1981.

42. **Stanbury, S. W.,** Vitamin D: metamorphosis from nutrient to hormonal system, *Proc. Nutr. Soc.,* 40, 179, 1981.

43. **Smith, R.,** Rickets and osteomalacia, *Hum. Nutr. Clin. Nutr.,* 36C, 115, 1982.

44. **Mawer, E. B.,** Clinical implications of measurements of circulating vitamin D metabolites, *Clin. Endocrinol. Metab.,* 9, 63, 1980.

45. **Pettifor, J., Ross, P., Moodley, G., and Shuenyane, E.,** Calcium deficiency in rural black children in South Africa: a comparison between rural and urban communities, *Am. J. Clin. Nutr.,* 32, 2477, 1979.

46. **Horwitt, M. K.,** Therapeutic uses of vitamin E in medicine, *Nutr. Rev.,* 38, 105, 1980.

47. **Roberts, H. J.,** Perspective on vitamin E as therapy, *JAMA,* 246, 129, 1981.

48. **McWhirter, W. R.,** Plasma tocopherol in infants and children, *Acta Paediatr. Scand.,* 64, 446, 1975.

49. **Underwood, B. A.,** Vitamin deficiency signs in man: vitamin E, in *CRC Handbook of Nutrition and Food,* Section E, *Nutritional Disorders,* Vol. 3, *Effect of Nutrient Deficiencies in Man,* Rechcigl, M., Ed., CRC Press, West Palm Beach, Fla., 1978, 123.

58. **Horwitt, M. K., Harvey, C. C., and Dahn, C. H., Jr.,** Relationship between levels of blood lipids, vitamins C, A, and E, serum copper compounds, and urinary excretions of tryptophan metabolites in women taking oral contraceptive therapy, *Am. J. Clin. Nutr.,* 28, 403, 1975.

59. **Brubacher, G., Stähelin, H. B., and Vuilleumier, J. P.,** Beziehung zwischen β-Lipoproteidgehalt des Serums und Plasma — Vitamin-E-Gehalt, *Int. J. Vitam. Nutr. Res.,* 44, 521, 1974.

60. **Rossi, E. C.,** Symposium on hemorrhagic disorders, *Med. Clin. N. Am.,* 56, 1, 1972.

61. **Quick, A. J.,** Bleeding problems, in *Clinical Medicine,* W. B. Saunders, Philadelphia, 1970.

62. **Gundberg, C. M., Lian, J. B., and Gallop, P. M.,** Measurement of γ-carboxyglutamate and circulating osteocalcin in normal children and adults, *Clin. Chim. Acta,* 128, 1, 1983.

63. **Lian, J. B., Steinberg, J., Gundberg, G., and Gallop, P. M.,** γ-Carboxyglutamate excretion in osteoporosis and Paget's disease, *Trans. Orthoped. Res. Soc.,* 5, 70, 1980.

64. **Levy, R. J. and Lian, J. B.,** γ-Carboxyglutamate excretion and warfarin therapy, *Clin. Pharm. Ther.,* 25, 562, 1979.

65. **Olsen, R. E., Vitamin K.,** in *Modern Nutrition in Health and Disease,* 6th ed., Goodhart, R. S. and Shils, M. E., Eds., Lea & Febiger, Philadelphia, 1980, 170.

66. Recommended Dietary Allowances, 9th ed., National Academy of Sciences, Washington, D.C., 1980.

67. **Hodges, R. E., Hood, J., Canham, J. E., Sauberlich, H. E., and Baker, E. M.,** Clinical manifestations of ascorbic acid deficiency in man, *Am. J. Clin. Nutr.,* 24, 432, 1971.

68. **Hodges, R. E.,** Ascorbic acid, in *Modern Nutrition in Health and Disease,* 6th ed., Goodhart, R. S. and Shils, M. E., Eds., Lea & Febiger, Philadelphia, 1980, 259.

69. **Sandstead, H. H.,** Clinical manifestations of certain classical deficiency diseases, in *Modern Nutrition in Health and Disease,* 6th ed., Goodhart, R. S. and Shils, M. E., Eds., Lea & Febiger, Philadelphia, 1980, 685.

70. **Henderson-Smart, D. J.,** Scurvy: a continuing paediatric problem, *Med. J. Aust.,* 2, 876, 1972.

71. **Vobecky, J. S., Vobecky, J., and Blanchard, R.,** Vitamin E and C levels in infants during the first year of life, *Am. J. Clin. Nutr.,* 29, 766, 1976.

72. **Clow, C. L., Laberge, C., and Scriver, C. R.,** Neonatal hypertyrosinemia and evidence for deficiency of ascorbic acid in arctic and subarctic people, *Can. Med. Assoc. J.,* 113, 624, 1975.

73. **Wilson, T. S., Weeks, M. M., Mukherjee, S. K., Murrell, J. S.,** and Andrews, C. S., A study of vitamin C levels in the aged and subsequent mortality, *Gerontol. Clin. (Basel),* 14, 17, 1972.

74. **Burr, M. L., Hurley, R. J., and Sweetnam, P. M.,** Vitamin C supplementation of old people with low blood levels, *Gerontol. Clin.,* 17, 236, 1975.

75. **Burr, M. L., Elwood, P. C., Hole, D. J., Hurley, R. J., and Hughes, R. E.,** Plasma and leukocytes ascorbic acid levels in the elderly, *Am. J. Clin. Nutr.,* 27, 144, 1974.

76. **Brook, M. and Grimshaw, J. J.,** Vitamin C concentration of plasma and leukocytes as related to smoking habit, age, and sex of humans, *Am. J. Clin. Nutr.,* 21, 1254, 1968.

77. **Salvatore, J. E., Vinton, P. W., and Rapuano, J. A.,** Nutrition in the aged: review of the literature, *J. Am. Geriatr. Soc.,* 17, 790, 1969.

78. **Irwin, M. I. and Hutchins, B. K.,** A conspectus of research on vitamin C requirements of man, *J. Nutr.,* 106, 823, 1976.

79. **Schorah, C. J., Newill, A., Scott, D. L., and Morgan, D. B.,** Clinical effects of vitamin C in elderly inpatients with low blood-vitamin-C levels, *Lancet,* 1, 403, 1979.

80. **Roine, P., Koivula, L., and Pekkarinen, M.,** Plasma vitamin C level and erythrocyte transketolase activity compared with vitamin intakes among old people in Finland, in *Proceedings of the 9th International Congress on Nutrition,* Vol. 4, S. Karger, Basel, 1972, 116.

81. **Kinsman, R. A. and Hood, J.,** Some behavioral effects of ascorbic acid deficiency, *Am. J. Clin. Nutr.,* 24, 455, 1971.

82. **Sauberlich, H. E.,** Ascorbic acid (Vitamin C). Symp. Lab. Assess. Nutr. Status, Labbé, R. F., Ed., appearing in *Clin. Lab. Med.,* 1, 673, 1981.

83. **Sauberlich, H. E.,** Vitamin C status: methods and findings, *Ann. N.Y. Acad. Sci.,* 258, 438, 1975.

84. **Sauberlich, H. E.,** Ascorbic acid, in *Present Knowledge in Nutrition,* 5th ed., Nutrition Foundation, New York, in press, 1984, 260.

85. **Sauberlich, H. E., Green, M. D., and Omaye, S. T.,** Determination of ascorbic acid and dehydroascorbic acid, in *Ascorbic Acid: Chemistry, Metabolism, and Uses,* Advances in Chemistry Series, Vol. 200, Seib, P. A. and Tolbert, B. M., Eds., American Chemical Society, Washington, D.C., 1982, 199.

86. **Turnbull, J. D., Sudduth, J. H., Sauberlich, H. E., and Omaye, S. T.,** Depletion and repletion of ascorbic acid in the Rhesus monkey: relationship between ascorbic acid concentration in blood components with total body pool and liver concentration of ascorbic acid, *Int. J. Vitam. Nutr. Res.,* 51, 47, 1981.

87. **Omaye, S. T., Turnbull, J. D., and Sauberlich, H. E.,** Selected methods for the determination of ascorbic acid in animal cells, tissues, and fluids, *Meth. Enzymol.,* 62, 3, 1972.

88. **Lee, W., Hamernyik, P., Hutchinson, M., Raisys, V. A., and Labbé, R. F.,** Ascorbic acid in lymphocytes: cell preparation and lipid-chromatographic assay, *Clin. Chem.,* 28, 2165, 1982.

89. **Wagner, E. S., Lindley, B., and Coffin, R. D.,** High-performance liquid chromatographic determination of ascorbic acid in urine, *J. Chromatogr.,* 163, 225, 1979.

90. **Tsao, C. S. and Salimi, S. L.,** Ultramicromethod for the measurement of ascorbic acid in plasma and white blood cells by high-performance liquid chromatography with electrochemical detection, *J. Chromatogr.,* 224, 477, 1981.

91. **Tsao, C. S. and Salimi, S. L.,** Differential determination of L-ascorbic acid and D-isoascorbic acid by reversed-phase high-performance liquid chromatography with electrochemical detection, *J. Chromatogr.,* 245, 355, 1982.

92. **Tsao, C. S. and Salimi, S. L.,** Influence of erythorbic acid on ascorbic acid retention and elimination in the mouse, *Int. J. Vit. Nutr. Res.,* 53, 258, 1983.

93. **Brown, P. T., Bergan, J. G., Parsons, E. P., and Krol, I.,** Dietary status of elderly people, *J. Am. Dietet. Assoc.,* 71, 41, 1977.

94. **Guthrie, H. A., Black, K., and Madden, J. P.,** Nutritional practices of elderly citizens in rural Pennsylvania, *Gerontology,* 12, 330, 1972.

95. **Pao, E. M. and Hill, M. M.,** Diets of the elderly — nutrition labeling and nutrition education, *J. Nutr. Educ.,* 6, 96, 1974.

96. **Todhunter, N. E. and Darby, W. J.,** Guidelines for maintaining adequate nutrition in old age, *Geriatrics,* 33, 49, 1978.

97. Ten-State Nutrition Survey 1968 to 1970. I. Historical development. II. Demographic data. III. Clinical, anthropometry, dental. IV. Biochemical. V. Dietary and highlights, U.S. Department of Health, Education and Welfare Publication No. (HSM) 72, 8130, 1972.

98. **Vir, S. C. and Love, A. H. G.,** Thiamin status of institutionalised and non-institutionalised aged, *Int. J. Vitam. Nutr. Res.,* 47, 325, 1977.

99. **Iber, F. L., Blass, J. P., Brin, M., and Leevy, C. M.,** Thiamin in the elderly — relation to alcoholism and to neurological degenerative disease, *Am. J. Clin. Nutr.,* 36, 1067, 1982.

100. **Platt, S.,** Thiamine deficiency in human beriberi and in Wernicke's encephalopathy, in *Thiamine Deficiency,* Wolstenholme, G. E. W. and O'Connor, M., Eds., Little, Brown, Boston, 1967, 135.

101. **Hoyumpa, A. M.,** Mechanisms of thiamin deficiency in chronic alcoholism, *Am. J. Clin. Nutr.,* 33, 2750, 1980.

102. **Lonsdale, D. and Shamberger, R. J.,** Red cell transketolase as an indicator of nutritional deficiency, *Am. J. Clin. Nutr.,* 33, 205, 1980.

103. **Kawai, C., Wakabayashi, A., Matsumura, T., and Yui, Y.,** Reappearance of beriberi heart disease in Japan. A study of 23 cases, *Am. J. Med.,* 69, 383, 1980.

104. **Sauberlich, H. E., Herman, Y. F., Stevens, C. O., and Herman, R. H.,** Thiamin requirement of the adult human, *Am. J. Clin. Nutr.,* 32, 2237, 1979.

105. **Warnock, L. G.,** Transketolase activity of blood hemolysate, a useful index for diagnosing thiamin deficiency, *Clin. Chem.,* 21, 432, 1975.

106. **Sauberlich, H. E.,** Biochemical alterations in thiamine deficiency. Their interpretation, *Am. J. Clin. Nutr.,* 20, 528, 1967.

107. **Brin, M.,** Transketolase (sedoheptulose-7-phosphate: D-glyceraldehyde-3-phosphate dihydroxyacetone-transferase, EC2.2.1.1) and the TPP effect in assessing thiamine adequacy, *Meth. Enzymol.*, 18A, 125, 1970.

108. **Bayoumi, R. A. and Rosalki, S. B.,** Evaluations of methods of coenzyme activation of erythrocyte enzymes for detection of deficiency of vitamin B_1, B_2, B_6, *Clin. Chem.*, 22, 327, 1976.

109. **Waring, P. P., Fisher, D., McDonnell, J., McGowan, E. L., and Sauberlich, H. E.,** A continuous-flow (Auto Analyzer II) procedure for measuring erythrocyte transketolase activity, *Clin. Chem.*, 28, 2206, 1982.

110. **Boni, L., Kieckens, L., and Hendrikx, A.,** An evaluation of a modified erythrocyte transketolase assay for assessing thiamine nutritional adequacy, *J. Nutr. Sci. Vitaminol. (Tokyo)*, 26, 507, 1980.

111. **Smeets, E. H. J., Muller, H., and De Wael, J. A.,** NADH-Dependent transketolase assay in erythrocyte hemolysates, *Clin. Chim. Acta*, 33, 379, 1971.

112. **Wood, B., Gijsbers, A., Goode, A., Davis, S., Mulholland, J., and Breen, K.,** A study of partial thiamin restriction in human volunteers, *Am. J. Clin. Nutr.*, 33, 848, 1980.

113. **Roser, R. L., Andrist, A. H., Harrington, W. H., Naito, H. K., and Londsale, D.,** Determination of urinary thiamine by high-performance liquid chromatography utilizing the thiochrome fluorescent method, *J. Chromatogr.*, 146, 43, 1979.

114. **Warnock, L. G., Prudhomme, C. R., and Wagner, C.,** The determination of thiamin pyrophosphate in blood and other tissues, and its correlation with erythrocyte transketolase activity, *J. Nutr.*, 108, 421, 1979.

115. **Kimura, M., Fujita, T., and Itokawa, Y.,** Liquid-chromatographic determination of the total thiamin content of blood, *Clin. Chem.*, 28, 29, 1982.

116. **Schrijver, J., Speek, A. J., Klosse, J. A., Van Rijn, H. J., and Schreurs, W. H.,** A reliable semiautomated method for the determination of total thiamin in whole blood by the thiochrome method with high-performance liquid chromatography, *Ann. Clin. Biochem.*, 19, 52, 1982.

117. **Bamji, M. A., Rameshwar Sarma, K. V., and Radhaiah, G.,** Relationship between biochemical and clinical indices of B-vitamin deficiency. A study in rural school boys, *Br. J. Nutr.*, 41, 431, 1979.

118. **Thurnham, D. I., Migasena, P., Vudhivai, N., and Supawan, V.,** A longitudinal study on dietary and social influences on riboflavin status in pre-school children in Northeast Thailand, *Southeast Asian J. Trop. Med. Public Health*, 2, 552, 1971.

119. **Sauberlich, H. E., Judd, J. H., Nichoalds, G. E., Broquist, H. P., and Darby, W. J.,** Application of the erythrocyte glutathione reductase assay in evaluating riboflavin nutritional status in a high school student population, *Am. J. Clin. Nutr.*, 25, 756, 1972.

120. **Nichoalds, G. E.,** Riboflavin, Symp. Lab. Assess. Nutr. Status, Labbé, R. F., Ed., appearing in *Clin. Lab. Med.*, 1, 685, 1981.

121. **Clarke, H. C.,** The riboflavin deficiency syndrome of pregnancy, *Surg. Forum*, 22, 394, 1971.

122. **Vir, S. C. and Love, A. H. G.,** Riboflavin status of institutionalized and non-institutionalized aged, *Int. J. Vitam. Nutr. Res.*, 47, 336, 1977.

123. **Hoorn, R. K., Flikweert, J. P., and Westerink, D.,** Vitamin B_1, B_2, and B_6 deficiencies in geriatric patients, measured by coenzyme stimulation of enzyme activities, *Clin. Chim. Acta*, 61, 151, 1975.

124. **Chen, L. H. and Fan-Chiang, W. L.,** Biochemical evaluation of riboflavin and vitamin B_6 status of institutionalized and non-institutionalized elderly in central Kentucky, *Int. J. Vitam. Nutr. Res.*, 51, 232, 1981.

125. **Garry, P. J., Goodwin, J. S., and Hunt, W. C.,** Nutritional status in a healthy elderly population: riboflavin, *Am. J. Clin. Nutr.*, 36, 902, 1982.

126. **Matts, S. G. F.,** Riboflavin in *Vitamins in Medicine*, Vol. 1, 4th ed., Barker, B. M. and Bender, D. A., Eds., William Heinemann, London, 1980, 398.

127. **Horwitt, M. K.,** Riboflavin, in *Modern Nutrition in Health and Disease*, 6th ed., Goodhart, R. S. and Shils, M. E., Eds., Lea & Febiger, Philadelphia, 1980, 197.

128. **Bamji, M. S.,** Enzymic evaluation of thiamin, riboflavin and pyridoxine status of parturient women and their newborn infants, *Br. J. Nutr.*, 35, 259, 1976.

129. **Tillotson, M. S. and Baker, E. M.,** An enzymatic measurement of the riboflavin status in man, *Am. J. Clin. Nutr.*, 25, 425, 1972.

130. **Glatzle, D., Vuilleumier, J. P., Weber, F., and Decker, K.,** Glutathione reductase test with whole blood, a convenient procedure for the assessment of the riboflavin status in humans, *Experientia*, 30, 665, 1974.

131. **Nichoalds, G. E., Lawence, J. D., and Sauberlich, H. E.,** Assessment of status of riboflavin nutriture by assay of glutathione reductase activity, *Clin. Chem.*, 20, 624, 1974.

132. **Thurnham, D. I. and Rathakette, P.,** Incubation of NAD (P) H_2: glutathione oxidoreductase (EC 1.6.4.2) with flavin adenine dinucleotide for maximal stimulation in the measurement of riboflavin status, *Br. J. Nutr.*, 48, 459, 1982.

133. **Böni, H., Wassmer, A., Brubacher, G., and Ritzel, G.,** Assessment of the vitamin B_2 status by means of the glutathione test, *Nutr. Metabl.,* 21(Suppl. 1), 20, 1977.

134. **Tillotson, J. A. and Bashor, M. M.,** Fluorometric apoprotein titration of urinary riboflavin, *Anal. Biochem.,* 107, 214, 1980.

135. **Lotter, S. E., Miller, M. S., Bruch, R. C., and White, H. B.,** III, Competitive binding assay for riboflavin and riboflavin-binding protein, *Anal. Biochem.,* 125, 110, 1982.

136. **Smith, M. D.,** Rapid method for determination of riboflavin in urine by high-performance liquid chromatography, *J. Chromatogr.,* 182, 285, 1980.

137. **Gatautis, V. J. and Naito, H. K.,** Liquid-chromatographic determination of urinary riboflavin, *Clin. Chem.,* 27, 1672, 1981.

138. **Bender, D. A.,** Niacin, in *Vitamins in Medicine,* Vol. 1, 4th ed., Barker, B. M. and Bender, D. A., Eds., William Heinmann, London, 1980, 315.

139. **Banerjee, A. K., Brocklehurst, J. C., Wainwright, H., and Swindell, R.,** Nutritional status of long-stay geriatric inpatients: effects of a food supplement, *Age Ageing,* 7, 237, 1978.

140. **deLange, D. J. and Joubert, C. P.,** Assessment of nicotinic acid status of population groups, *Am. J. Clin. Nutr.,* 15, 169, 1964.

141. **Sandhu, J. S. and Fraser, D. R.,** Measurement of niacin metabolites in urine by high performance liquid chromatography, a simple, sensitive assay of niacin nutritional status, *Int. J. Vitam. Nutr. Res.,* 51, 139, 1981.

142. **Carter, E. G. A.,** Quantitation of urinary niacin metabolites by reversed-phase liquid chromatography, *Am. J. Clin. Nutr.,* 36, 926, 1982.

143. **Kutnink, M. A., Vannucchi, H., and Sauberlich, H. E.,** A simple high performance liquid chromatography procedure for the determination of N^1-methylnicotinamide in urine, *J. Liquid Chromatogr.,* 7, 969, 1984.

144. **Sauberlich, H. E. and Canham, J. E.,** Vitamin B_6, in *Modern Nutrition in Health and Disease,* 6th ed., Goodhart, R. S. and Shils, M. E., Eds., Lea & Febiger, Philadelphia, 1980, 209.

145. **Sauberlich, H. E.,** Vitamin B_6 status assessment: past and present, in *Methods in Vitamin B-6 Nutrition,* Leklem, J. E. and Reynolds, R. D., Eds., Plenum Press, New York, 1981, 203.

146. **Human Vitamin B-6 Requirements:** Proceedings of a Workshop, National Academy of Sciences, Washington, D.C., 1978.

147. **Barker, B. M. and Bender, D. A.,** Vitamin B_6, in *Vitamins in Medicine,* Vol. 1, 4th ed., Barker, B. M. and Bender, D. A., Eds., William Heinemann, London, 1980, 348.

148. **Rose, C. S., György, P., Butler, M., Andres, R., Norris, A. H., Shock, N. W., Tobin, J., Brin, M., and Spiegel, H.,** Age differences in vitamin B_6 status of 617 men, *Am. J. Clin. Nutr.,* 29, 847, 1976.

149. **Vir, S. C. and Love, A. H. G.,** Vitamin B_6 status of institutionalised and non-institutionalised aged, *Int. J. Vitam. Nutr. Res.,* 47, 364, 1977.

150. **Vir, S. C. and Love, A. H. G.,** Nutritional evaluation of B groups of vitamins in institutionalised aged, *Int. J. Vitam. Nutr. Res.,* 47, 211, 1977.

151. **Vir, S. C. and Love, A. H.,** Vitamin B_6 status of the hospitalized aged, *Am. J. Clin. Nutr.,* 31, 1383, 1978.

152. **Crepaldi, G., Allegri, G., De Antoni, A., Costa, C., and Muggeo, M.,** Relationship between tryptophan metabolism and vitamin B_6 and nicotinamide in aged subjects, *Acta Vitaminol. Enzymol. (Milano),* 29, 140, 1975.

153. **Elsborg, L., Nielsen, J. A., Bertram, U., Helms, P., Nielsen, K., and Rosenquist, A.,** The intake of vitamins and minerals by the elderly at home, *Int. J. Vitam. Nutr. Res.,* 53, 321, 1983.

154. **Sauberlich, H. E., Canham, J. E., Baker, E. M., Raica, N., Jr., and Herman, Y. F.,** Biochemical assessment of the nutritional status of vitamin B_6 in the human, *Am. J. Clin. Nutr.,* 25, 629, 1972.

155. **Skala, J. H., Waring, P. P., Lyons, M. F., Rusnak, M. G., and Alletto, J. S.,** Methodology for determination of blood aminotransferases, in *Methods in Vitamin B-6 Nutrition: Analysis and Status Assessment,* Leklem, J. E. and Reynolds, R. D., Eds., Plenum Press, New York, 1980, 171.

156. **Leinert, J., Simon, I., and Hötzel, D.,** Evaluation of methods to determine the vitamin B_6 status of humans. I. α-EGOT: methods and validation, *Int. J. Vitam. Nutr. Res.,* 51, 145, 1981.

157. **Leinert, J., Simon, I., and Hötzel, D.,** Methods and their evaluation in the determination of vitamin B_6 status in man. II. α-EGOT: reliability of the parameter, *Int. J. Nutr. Res.,* 52, 24, 1982.

158. **Sauberlich, H. E.,** New methods for assessing nutriture of selected B-complex vitamins, *Annu. Rev. Nutr.,* 4, 377, 1984.

159. **Shultz, T. D. and Leklem, J. E.,** Urinary 4-pyridoxic acid, urinary vitamin B-6 and plasma pyridoxal phosphate as measures of vitamin B-6 status and dietary intake in adults, in *Methods in Vitamin B-6 Nutrition: Analysis and Status Assessment,* Leklem, J. E. and Reynolds, R. D., Eds., Plenum Press, New York, 1981, 297.

160. **Simon, I., Leinert, J., and Hötzel, D.,** Methods and their evaluation in estimating vitamin B_6 status in humans. III. Determination of 4-pyridoxic acid in urine, *Int. J. Vitam. Nutr. Res.,* 52, 280, 1982.

161. **Vannucchi, H., Kutnink, M. A., Kretsch, M. J., and Sauberlich, H. E.,** Measurement of urinary 4-pyridoxic acid excretion by vitamin B_6 depleted-repleted young women, *Proceedings Western Hemisphere Nutrition Congress VII,* in press.

162. **Gregory, J. F., III and Kirk, J. R.,** Determination of urinary 4-pyridoxic acid using high performance liquid chromatography, *Am. J. Clin. Nutr.,* 32, 879, 1979.

163. **Arend, R. A. and Brown, R. R.,** Comparison of analytical methods for urinary 4-pyridoxic acid, *Am. J. Clin. Nutr.,* 34, 1984, 1981.

164. **Li, T.-K. and Lumeng, L.,** Plasma PLP as indicator of nutrition status: relationship to tissue vitamin B_6 content and hepatic metabolism, in *Methods in Vitamin B-6 Nutrition: Analysis and Status Assessment,* Leklem, J. E. and Reynolds, R. D., Eds., Plenum Press, New York, 1981, 289.

165. **Lumeng, L., Lui, A., and Li, T.-K.,** Microassay of pyridoxal phosphate using tyrosine apodecarboxylase, in *Methods in Vitamin B-6 Nutrition: Analysis and Status Assessment,* Leklem, J. E. and Reynolds, R. D., Eds., Plenum Press, New York, 1981, 57.

166. **Bhagawan, H. N., Koogler, J. M., Jr., and Coursin, D. B.,** Enzymatic microassay of pyridoxal-5'-phosphate using L-tyrosine apodecarboxylase and L-(1-^{14}C) tyrosine, *Int. J. Vitam. Nutr. Res.,* 46, 160, 1976.

167. **Camp, V. M., Chipponi, J., and Faraj, B. A.,** Radioenzymatic assay for direct measurement of plasma pyridoxal-5'-phosphate, *Clin. Chem.,* 29, 642, 1983.

168. **Shin, Y. S., Rasshofer, R., Friedrich, B., and Endres, W.,** Pyridoxal-5'-phosphate determination by a sensitive micromethod in human blood, urine and tissues; its relation to cystathioninuria in neuroblastoma and biliary atresia, *Clin. Chim. Acta,* 127, 77, 1983.

169. **Brown, R. R.,** The tryptophan load test as an index of vitamin B-6 nutrition, in *Methods in Vitamin B-6 Nutrition: Analysis and Status Assessment,* Leklem, J. E. and Reynolds, R. D., Eds., Plenum Press, New York, 1981, 321.

170. **Linkswiler, H. M.,** Methionine metabolite excretion as affected by a vitamin B-6 deficiency, in *Methods in Vitamin B-6 Nutrition: Analysis and Status Assessment,* Leklem, J. E. and Reynolds, R. D., Eds., Plenum Press, New York, 1981, 373.

171. **Bender, D. A.,** Effects of oestradiol and vitamin B_6 on tryptophan metabolism in the rat: implications for the interpretation of the tryptophan load test for vitamin B_6 nutritional status, *Br. J. Nutr.,* 50, 33, 1983.

172. **Wolfe, H., Brown, R. R., and Arend, R. A.,** The kynurenine load test, an adjunct to the tryptophan load test, *Scand. J. Clin. Invest.,* 40, 9, 1980.

173. **Leklem, J. E., Linkswiler, H. M., Brown, R. R., Rose, D. P., and Anand, C. R.,** Metabolism of methionine in oral contraceptive users and control women receiving controlled intakes of vitamin B_6, *Am. J. Clin. Nutr.,* 30, 1122, 1977.

174. **Rosenberg, I. H., Bowman, B. B., Cooper, B. A., Halsted, C. H., and Lindenbaum, J.,** Folate nutrition in the elderly, *Am. J. Clin. Nutr.,* (Suppl.) 36, 1060, 1982.

175. **Justice, C., Howe, J., and Clark, H.,** Dietary intakes and nutritional status of elderly patients, *J. Am. Dietet. Assoc.,* 65, 639, 1974.

176. **Colman, N.,** Laboratory assessment of folate status, Symp. Lab. Assess. Nutr. Status, Labbé, R. F., Ed., appearing in *Clin. Lab. Med.,* 1, 775, 1981.

177. **Bonjour, J. P.,** Vitamins and alcoholism. II. Folate and vitamin B_{12}, Int. J. Vitam. Nutr. Res., *50,* 96.

178. **Colman, N., Barker, E. A., Barker, M., Green, R., and Metz, J.,** Prevention of folate deficiency by food fortification. IV. Identification of target groups in addition to pregnant women in an adult rural population, *Am. J. Clin. Nutr.,* 28, 471, 1975.

179. **Magnus, E. M., Bache-Wiig, J. E., Aanderson, T. R., and Melbostad, E.,** Folate and vitamin B_{12} (cobalamin) blood levels in elderly persons in geriatric homes, *Scand. J. Haematol.,* 28, 360, 1982.

180. **Thornton, W. E. and Thornton, B. P.,** Geriatric mental function and serum folate: a review and survey, *S. Med. J.,* 70, 919, 1977.

181. **Bailey, L. B., Wagner, P. A., Christakis, G. J., Aranjo, P. E., Appledorf, H., Davis, C. G., Masteryanni, J., and Dinning, J. S.,** Folacin and iron status and hematological findings in predominately black elderly persons from urban low-income households, *Am. J. Clin. Nutr.,* 32, 2346, 1979.

182. **Harant, Z. and Goldberger, J. V.,** Treatment of anemia in the aged: a common problem and challenge, *J. Am. Geriat. Soc.,* 23, 127, 1975.

183. **Raper, C. G. and Choudhury, M.,** Early detection of folic acid deficiency in elderly patients, *J. Clin. Pathol.,* 31, 44, 1978.

184. **Jägerstad, M.,** Folate intake and blood folate in elderly subjects, a study using the double sampling portion technique, *Nutr. Metab.,* 21(Suppl. 1), 29, 1977.

185. **Webster, S. G. and Leeming, J. T.,** Erythrocyte folate levels in young and old, *J. Am. Geriatr. Soc.,* 27, 451, 1979.

186. **Chanarin, I.,** The folates, in *Vitamins in Medicine,* Vol. 1, 4th ed., Barker, B. M. and Bender, D. A., Eds., William Heinemann, London, 1980, 247.

187. **Sauberlich, H. E.,** Detection of folic acid deficiency in populations, in *Folic Acid: Biochemistry and Physiology in Relation to the Human Nutrition Requirement,* National Academy of Sciences, Washington, D.C., 1977, 213.

188. **Scott, J. M., Ghanta, V., and Herbert, V.,** Trouble-free microbiological serum and red cell folate assays, *Am. J. Med. Technol.,* 40, 125, 1974.

189. **Brown, R. D., Robin, H., and Kronenberg, H.,** Folate assays — an alternative to microbiological assays and commercial kits, *Pathology,* 14, 449, 1982.

190. **Longo, D. L. and Herbert, V.,** Radioassay for serum and red cell folate, *J. Lab. Clin. Med.,* 87, 138, 1976.

191. **Theobald, R. A., Batchelder, M., and Sturgeon, M. F.,** Evaluation of pteroylglutamic acid and N^5-methyltetrahydrofolic acid reference standards in a new "No-Boil" radioassay for serum folate, *Clin. Chem.,* 27, 553, 1981.

192. **Waxman, S. and Schreiber, C.,** Determination of folate by use of radioactive folate and binding proteins, *Meth. Enymol.,* 66, 448, 1980.

193. **McGown, E. L., Lewis, C. M., Dong, M. H., and Sauberlich, H. E.,** Results with commercial radioassay kits compared with microbiological assay of folate in serum and whole-blood, *Clin. Chem.,* 24, 2186, 1978.

194. **Gutcho, S. and Bakke, O. M.,** Source of error in determination of erythrocyte folate by competitive binding radioassay — comments, *Clin. Chem.,* 24, 388, 1978.

195. **Shane, B., Tamura, T., and Stokstad, E. L. R.,** Folate assay: a comparison of radioassay and microbiological methods, *Clin. Chim. Acta,* 100, 13, 1980.

196. **Bills, T. and Spatz, L.,** Neutrophilic hypersegmentation as an indicator of incipient folic acid deficiency, *Ann. J. Clin. Pathol.,* 68, 263, 1977.

197. **Pietrzik, K., Urban, G., and Hötzel, D.,** Biochemical and hematological measures for determination of folate status in humans. I. Relation between serum folate and segmentation of neutrophil granuloycytes, *Int. J. Vitam. Nutr. Res.,* 48, 391, 1979.

198. **Wickramasinge, S. N.,** The deoxyuridine suppression test: a review of its clinical and research applications, *Clin. Lab. Haematol.,* 3, 1, 1981.

199. **Das, K. C., Manusselis, C., and Herbert, V.,** Simplifying lymphocyte culture and the deoxyuridine suppression test by using whole blood (0.1 mℓ) instead of separated lymphocytes, *Clin. Chem.,* 26, 72, 1980.

200. **Chanarin, I.,** The cobalamins (vitamin B_{12}), in *Vitamins in Medicine,* Vol. 1, 4th ed., Barker, B. M. and Bender, D. A., Eds., William Heinemann, London, 1980, 172.

201. **Elsborg, L., Lung, V., and Bastrup-Madsen, J.,** Serum vitamin B_{12} levels in the aged, *Acta. Med. Scand.,* 200, 309, 1976.

202. **McEvoy, A. W., Fenwick, J. D., Boddy, K., and James, O. F.,** Vitamin B_{12} absorption from the gut does not decline with age in normal elderly humans, *Age Ageing,* 11, 180, 1982.

203. **Kolhouse, J. F., Kondo, H., Allen, N. C., Podell, E., and Allen, R. H.,** Cobalamin analogues are present in human plasma and can mask cobalamin deficiency because current radioisotope dilution assays are not specific for the cobalamin, *N. Engl. J. Med.,* 299, 785, 1978.

204. **Cooper, B. A. and Whitehead, V. M.,** Evidence that some patients with pernicious anemia are not recognized by radio dilution assay for cobalamin in serum, *N. Engl. J. Med.,* 299, 816, 1978.

204a. **Cohen, K. L. and Donaldson, R. M., Jr.,** Unreliability of radiodilution assays as screening tests for cobalamin (vitamin B_{12}) deficiency, *JAMA,* 244, 1942, 1980.

205. **LeFebvre, R. J., Virji, A. S., and Mertens, B. F.,** Erroneously low results due to high nonspecific binding encountered with a radioassay kit that measures "true" serum vitamin B_{12}, *Am. J. Clin. Pathol.,* 74, 209, 1980.

206. **Mistry, S. P.,** Biotin, in *Vitamins in Medicine,* Vol. 1, 4th ed., Barker, B. M. and Bender, D. A., Eds., William Heinemann, London, 1980, 381.

207. **Bonjour, J. P.,** Biotin in man's nutrition and therapy — a review, *Int. J. Vitam. Nutr. Res.,* 47, 107, 1977.

208. **Roth, K. S.,** Biotin in clinical medicine — a review, *Am. J. Clin. Nutr.,* 34, 1967, 1981.

209. **Baugh, C. M., Malone, J. H., and Butterworth, C. E., Jr.,** Human biotin deficiency: a case history of biotin deficiency induced by raw egg consumption in a cirrhotic patient, *Am. J. Clin. Nutr.,* 21, 173, 1968.

210. **Mock, D. M., DeLorimer, A. A., Liebman, W. M., Sweetman, L., and Baker, H.,** Biotin deficiency: an unusual complication of parenteral alimentation, *N. Engl. J. Med.,* 304, 820, 1981.

211. **Brubacher, G. and Stransky, M.,** A nutritional survey of 38 single elderly persons with free choice of food, *Soz. Praeventivmed.,* 21, 197, 1976.

212. **Baker, H. and Frank, O.,** *Clinical Vitaminology,* John Wiley & Sons, New York, 1968, 22.

213. **Sauberlich, H. E.,** Pantothenic acid, in *Modern Nutrition in Health and Disease,* 6th ed., Goodhart, R. S. and Shils, M. E., Eds., Lea & Febiger, Philadelphia, 1980, 209.

214. **Hodges, R. E., Bean, W. B., Ohlson, M. A., and Bleiler, R. E.,** Factors affecting human antibody response. V. Combined deficiencies of pantothenic acid and pyridoxine, *Am. J. Clin. Nutr.,* 11, 187, 1962.

215. **Fry, P. C., Fox, H. M., and Tao, H. G.,** Metabolic response to a pantothenic acid deficiency diet in humans, *J. Nutr. Sci. Vitaminol.,* 22, 339, 1976.

216. **Markkanen, T.,** The metabolic significance of panthothenic acid as assessed by its levels in serum in various clinical and experimental conditions, *Int. J. Vitam. Nutr. Res.,* 43, 302, 1973.

217. **McCurdy, P. R.,** Is there an anemia responsive to pantothenic acid? *J. Am. Geriatr. Soc.,* 21, 88, 1973.

218. **Srinivasan, V., Christensen, N., Wyse, B. W., and Hansen, R. G.,** Pantothenic acid nutritional status in the elderly — institutionalized and non-institutionalized, *Am. J. Clin. Nutr.,* 34, 1736, 1981.

219. **Raghavan, P., Pimparkar, B. D., and Kulkarni, B. S.,** Pantothenic acid levels in blood and urine of patients with duodenal ulcer, *J. Postgrad. Med.,* 17, 57, 1971.

220. **Ishiguro, K.,** Aging effect of blood pantothenic acid content in females, *Tohoku J. Exp. Med.,* 107, 367, 1972.

221. **Wyse, B. W., Wittwer, C. W., and Hansen, R. G.,** Radioimmunoassay for pantothenic acid in blood and other tissues, *Clin. Chem.,* 25, 108, 1979.

222. **Lynch, S. R., Finch, C. A., Monsen, E. R., and Cook, J. D.,** Iron status of elderly Americans, *Am. J. Clin. Nutr.,* 36, 1032, 1982.

223. **Hallberg, L.,** Iron balance and iron deficiency in the elderly, *Bibl. "Nutr. Dieta",* 33, 107, 1983.

224. **Ventura, A., Senin, U., and Mannarino, E.,** Iron pathology in the elderly, *Bibl. "Nutr. Dieta",* 33, 113, 1983.

225. **Matzner, Y., Levy, S., Grossowicz, N., Izak, G., and Hershko, C.,** Prevalence and causes of anemia in elderly hospitalized patients, *Gerontology,* 25, 113, 1979.

226. **Kalchthaler, T. and Tan, M. E.,** Anemia in institutionalized elderly patients, *J. Am. Geriatr. Soc.,* 28, 108, 1980.

227. **Adler, S. S.,** Anemia in the aged: causes and considerations, *Geriatrics,* 35, 49, 1980.

228. **Takkunen, H. and Seppänen, R.,** Iron deficiency in the elderly population in Finland, *Scand. J. Soc. Med. (Suppl).,* 14, 151, 1977.

229. **Loria, A., Hershko, C., and Konijn, A. M.,** Serum ferritin in an elderly population, *J. Gerontol.,* 34, 521, 1979.

230. **Nordstrom, J. W.,** Trace mineral nutrition in the elderly, *Am. J. Clin. Nutr.,* 36, 788, 1982.

231. Preliminary Findings of the First Health and Nutritional Examination Survey, United States, 1971—72; Dietary and Biochemical Findings, HRA Publ. No. 74-1219-1, National Center for Health Statistics, Washington, D.C., 1974.

232. **Gershoff, S. N., Brusis, O. A., Nino, H. V., and Huber, A. M.,** Studies of the elderly in Boston. I. The effects of iron fortification on moderately anemic people, *Am. J. Clin. Nutr.,* 30, 226, 1977.

233. **Nordstrom, J. W., Abrahams, O. G., and Kohrs, M. B.,** Anemia among non-institutionalized white elderly, *Nutr. Rep. Int.,* 25, 97, 1982.

234. **Baker, S. J. and DeMaeyer, E. M.,** Nutritional anemia: its understanding and control with special reference to the work of the World Health Organization, *Am. J. Clin. Nutr.,* 32, 368, 1979.

235. **Sayers, M. H.,** Iron, Symp. Lab. Assess. Nutr. Status, Labbé, R. F., Ed., appearing in *Clin. Lab. Med.,* 1, 729, 1981.

236. **Walters, G. D., Miller, F. M., and Worwood, M.,** Serum ferritin and iron stores in normal subjects, *J. Clin. Pathol.,* 26, 770, 1973.

237. **Prasad, A.,** Clinical, biochemical and nutritional spectrum of zinc deficiency in human subjects: an update, *Nutr. Rev.,* 41, 197, 1983.

238. **Jacob, R. A.,** Zinc and copper, Symp. Lab. Assess. Nutr. Status, Labbé, R. F., Ed., appearing in *Clin. Lab. Med.,* 1, 743, 1981.

239. **Solomons, N. W.,** On the assessment of zinc and copper nutriture in man, *Am. J. Clin. Nutr.,* 32, 856, 1979.

240. **Prasad, A. S.,** Clinical and biochemical spectrum of zinc deficiency in human subjects, in *Clinical, Biochemical, and Nutritional Aspects of Trace Elements,* Prasad, A. S., Ed., Alan R. Liss, New York, 1982, 4.

241. **Sandstead, H. H., Henriksen, L. K., Greger, J. L., Prasad, A. S., and Good, R. A.,** Zinc nutriture in the elderly in relation to taste acuity, immune response, and wound healing, *Am. J. Clin. Nutr.,* (Suppl.)36, 1046, 1982.

242. **Hutton, C. W. and Hayes-Davis, R. B.,** Assessment of the zinc nutritional status of selected elderly subjects, *J. Am. Dietet. Assoc.,* 82, 148, 1983.

243. **Wagner, P. A., Krista, M. L., Bailey, L. B., Christakis, G. J., Jernigan, J. A., Aranjo, P. E., Appledorf, H., Davis, C. G., and Dinning, J. S.,** Zinc status of elderly black Americans from urban low-income households, *Am. J. Clin. Nutr.,* 33, 1771, 1980.

244. **Hsu, J. M.,** Current knowledge on zinc, copper, and chromium in aging, *World Rev. Nutr. Diet.,* 33, 44, 1979.

245. **Greger, J. L.,** Dietary intake and nutritional status in regard to zinc of institutionalized aged, *J. Gerontol.*, 32, 549, 1977.
246. **Vir, S. C. and Love, A. H. G.,** Zinc and copper status of the elderly, *Am. J. Clin. Nutr.*, 32, 1472, 1979.
247. **Greger, J. L. and Sciscoe, B. S.,** Zinc nutriture of elderly participants in an urban feeding program, *J. Am. Dietet. Assoc.*, 70, 37, 1977.
248. **Greger, J. L. and Geissler, A. H.,** Effect of zinc supplementation on taste acuity in the aged, *Am. J. Clin. Nutr.*, 31, 633, 1978.
249. **Johnson, H. L. and Sauberlich, H. E.,** Trace element analysis in biological samples, in *Clinical, Biochemical, and Nutritional Aspects of Trace Elements*, Prasad, A. S., Ed., Alan R. Liss, New York, 1982, 405.
250. **Prasad, A. S., Abbasi, A. A., Rabbani, P., and DuMouchelle, E.,** Effect of zinc supplementation on serum testosterone level in adult male sickle cell anemia subjects, *Am. J. Hematol.*, 10, 119, 1981.
251. **Gibson, R. S., Anderson, B. M., and Sabry, J. H.,** The trace metal status of a group of post-menopausal vegetarians, *J. Am. Dietet. Assoc.*, 82, 246, 1983.
252. **Williams, D. M.,** Clinical significance of copper deficiency and toxicity in the world population, in *Clinical, Biochemical, and Nutritional Aspects of Trace Elements*, Prasad, A. S., Ed., Alan R. Liss, New York, 1982, 277.
253. **Milne, D. B., Schnakenberg, D. D., Johnson, H. L., and Kuhl, G. L.,** Trace mineral intake of enlisted military personnel, *J. Am. Dietet. Assoc.*, 76, 41, 1980.
254. **Glinsmann, W. H. and Mertz, W.,** Effect of trivalent chromium on glucose tolerance, *Metabolism*, 15, 510, 1966.
255. **Levine, R. A., Streeten, D. H. P., and Doisy, R. J.,** Effect of oral chromium supplementation on the glucose tolerance of elderly human subjects, *Metabolism*, 17, 114, 1968.
256. **Liu, V. J. K. and Morris, I. S.,** Relative chromium response as an indicator of chromium status, *Am. J. Clin. Nutr.*, 31, 972, 1978.
257. **Offenbacher, E. G. and Pi-Sunyer, F. X.,** Beneficial effect of chromium-rich yeast on glucose tolerance and blood lipids in elderly subjects, *Diabetes*, 29, 919, 1980.
258. **Mertz, W.,** Clinical and public health significance of chromium, in *Clinical, Biochemical, and Nutritional Aspects of Trace Elements*, Prasad, A. S., Ed., Alan R. Liss, New York, 1982, 315.
259. **Vir, S. C. and Love, A. H. G.,** Chromium status of the aged, *Int. J. Vitam. Nutr. Res.*, 48, 402, 1978.
260. **Robinson, M. F.,** Clinical effects of selenium deficiency and excess, in *Clinical, Biochemical, and Nutritional Aspects of Trace Elements*, Prasad, A. S., Ed., Alan R. Liss, New York, 1982, 325.
261. **Levander, O. A.,** Selenium: biochemical actions, interactions, and some human health implications, in *Clinical, Biochemical, and Nutritional Aspects of Trace Elements*, Prasad, A. S., Ed., Alan R. Liss, New York, 1982, 345.
262. **Lane, H. W., Warren, D. C., Taylor, B. J., and Stool, E.,** Blood selenium and glutathione peroxidase levels and dietary selenium of free-living and institutionalized elderly subjects, *Proc. Soc. Exp. Biol. Med.*, 173, 87, 1983.
263. **Csallany, A. S., Zaspel, B. J., and Ayaz, K. L.,** Selenium and aging, in *Selenium in Biology and Medicine*, Spallholz, J. E., Martin, J. L., and Ganther, H. E., Eds., AVI, Westport, Conn., 1981, 118.
264. **Hojo, Y.,** Evaluation of the expression of urinary selenium level as ng Se/mg creatinine and the use of single-void urine as a sample for urinary selenium determination, *Bull. Environ. Contam. Toxicol.*, 27, 213, 1981.
265. **Kein, C. L. and Ganther, H. E.,** Manifestations of chronic selenium deficiency in a child receiving total parenteral nutrition, *Am. J. Clin. Nutr.*, 37, 319, 1983.
265a. **Allegrini, M., Pennington, J. A. T., and Tanner, J. T.,** Total diet study: determination of iodine intake by neutron activation analysis, *J. Am. Dietet. Assoc.*, 83, 18, 1983.
266. **Tietz, N. W., Ed.,** *Clinical Guide to Laboratory Tests*, W. B. Saunders, Philadelphia, 1983, 338.
267. **Isaksson, B.,** Requirements and appropriate intakes of electrolytes, *Bibl. "Nutr. Dieta"*, 33, 42, 1983.
268. **Pickering, G.,** Salt intake and essential hypertension, *Cardiovasc. Rev. Rep.*, 1, 13, 1980.
269. **Knochel, J. P.,** The pathophysiology and clinical characteristics of severe hypophosphatemia, *Arch. Intern. Med.*, 137, 203, 1977.
270. **Rude, R. K. and Singer, F. R.,** Magnesium deficiency and excess, *Ann. Rev. Med.*, 32, 245, 1981.
271. *Analytical Methods for Atomic Absorption Spectrophotometry*, Perkin-Elmer Corporation, Norwalk, Conn., 1971.
272. **Hutchinson, M. L. and Clemans, G. G.,** Essential fatty acids, Symp. Lab. Assess. Nutr. Status, Labbé, R. F., Ed., appearing in *Clin. Lab. Med.*, 1, 665, 1981.
273. **Rivers, J. P. W. and Frankel, T. L.,** Essential fatty acid deficiency, *Br. Med. Bull.*, 37, 59, 1981.
274. **Holman, R. T.,** Essential fatty acid deficiency in humans, in *CRC Handbook of Nutrition and Food, Section E, Nutritional Disorders*, Vol. 3, Rechcigl, M., Ed., CRC Press, West Palm Beach, Fla., 1978, 335.

275. **Richardson, T. J. and Sgoutas, D.,** Essential fatty acid deficiency in four adult patients during total parenteral nutrition, *Am. J. Clin. Nutr.,* 28, 258, 1975.
276. **Fleming, C. R., Smith, L. M., and Hodges, R. E.,** Essential fatty acid deficiency in adults receiving total parenteral nutrition, *Am. J. Clin. Nutr.,* 29, 976, 1976.
277. **Weinsier, R. L. and Butterworth, C. E., Jr.,** *Handbook of Clinical Nutrition,* C. V. Mosby, St. Louis, 1981.
278. **Feldman, E. B.,** Nutritional factors in cardiovascular disease, in *Nutrition in the Middle and Later Years,* Feldman, E. B., Ed., PSG, Littleton, Mass., 1983, 107.
279. **Forbes, G. B. and Halloran, E.,** The adult decline in lean body mass, *Hum. Biol.,* 48, 162, 1976.
280. **Munro, H. N.,** Protein nutriture and requirement in elderly people, *Bibl. "Nutr. Dieta",* 33, 61, 1983.
281. **Waterlow, J.,** Classification and definition of protein-calorie malnutrition, *Br. Med. J.,* 3, 566, 1972.
282. **Phinney, S.,** The assessment of protein nutrition in the hospitalized patient, Symp. Lab. Assess. Nutr. Status, Labbé, R. F., Ed., appearing in *Clin. Lab. Med.,* 1, 767, 1981.
283. **Bistrian, B. R., Blackburn, G. L., Hallowell, E., and Heddle, R.,** Protein status of general hospitalized patients, *JAMA,* 230, 858, 1974.
284. **Prevost, E. A. and Butterworth, C. E., Jr.,** Nutritional care of hospitalized patients, *Clin. Res.,* 22, 579A, 1974.
285. **Hill, G. L., Blackett, R. L., Pickford, I., Burkinshaw, L., Young, G. A., Warren, J. V., Schorah, C. J., and Morgan, D. B.,** Malnutrition in surgical patients. An unrecognized problem, *Lancet,* 1, 689, 1977.
286. **Weinsier, R. L., Hunker, E. M., Krumdieck, C. L., and Butterworth, C. E., Jr.,** Hospital malnutrition: a prospective evaluation of general medical patients during the course of hospitalization, *Am. J. Clin. Nutr.,* 32, 418, 1979.
287. **Eckman, I., Robbins, J. B., Van den Hamer, C. J. A., Lentz, J., and Scheinberg, I. H.,** Automation of a quantitative immunochemical microanalysis of human serum transferrin: a model system, *Clin. Chem.,* 16, 588, 1970.
288. **Shetty, P. A., Watrasiewicz, K. E., Jung, R. T., and James, W. P. T.,** Rapid-turnover transport proteins: an index of subclinical protein-energy malnutrition, *Lancet,* 2, 230, 1979.
289. **Ogle, C. K. and Alexander, J. W.,** The relationship of serum levels of prealbumin and retinol binding protein to bacteremia in burn patients, *J. Burn Care Rehabil.,* 3, 338, 1982.
290. **Scott, R. L., Sohmer, P. R., and MacDonald, M. G.,** The effect of starvation and repletion on plasma fibronectin in man, *JAMA,* 248, 2025, 1982.
291. **Young, G. A. and Hill, G. L.,** Assessment of protein-calorie malnutrition in surgical patients from plasma proteins and anthropometric measurements, *Am. J. Clin. Nutr.,* 31, 429, 1978.
292. **Smith, F. R., Suskin, R., Thanangkul, O., Leitzmann, C., Goodman, D. S., and Olsen, R. E.,** Plasma vitamin A, retinol-binding protein and prealbumin concentrations in protein-calorie malnutrition. III. Response to varying dietary treatments, *Am. J. Clin. Nutr.,* 28, 732, 1975.
293. **Ingenbleek, Y., Van Den Schrieck, H. G., De Nayer, P., and De Visscher, M.,** Albumin, transferrin and the thyroxine-binding prealbumin/retinol-binding protein (TBPA-RBP) complex in assessment of malnutrition, *Clin. Chim. Acta,* 63, 61, 1975.
294. **Scrimshaw, N. S., Taylor, C. E., and Gordon, J. E.,** Interaction of nutrition and infection, Tech. Rep. Ser. World Health Organization No. 57, WHO, Geneva, 1968.
295. **Cunningham-Rundles, S.,** Effects of nutritional status on immunological function, *Am. J. Clin. Nutr.,* 35, 1202, 1982.
296. **Chandra, R. K.,** Immunocompetence, Symp. Lab. Assess. Nutr. Status, Labbé, R. F., Ed., appearing in *Clin. Lab. Med.,* 1, 631, 1981.
297. **Neumann, C. G., Stiehm, R. E., Swendseid, M., Ferguson, A. C., and Lawler, G.,** Cell-mediated immune response in Ghanaian children with protein-calorie malnutrition, in *Malnutrition and the Immune Response,* Suskind, R., Ed., Raven Press, New York, 1977, 77.
298. **Chandra, R. K.,** Immunocompetence as a functional index of nutritional status, *Br. Med. Bull.,* 37, 89, 1981.
299. **Chandra, R. K. and Scrimshaw, N. S.,** Immunocompetence in nutritional assessment, *Am. J. Clin. Nutr.,* 33, 2694, 1980.
300. **Jensen, T. G., Englert, D. M., Dudrick, S. J., and Johnston, D. A.,** Delayed hypersensitivity skin testing: response rates in a surgical population, *J. Am. Dietet. Assoc.,* 82, 17, 1983.
301. **WHO Scientific Group,** Cell-Mediated Immunity and Resistance to Infection, Tech. Rep. Ser. World Health Organization No. 519, WHO, Geneva, 1973.
302. **Sokal, J. E.,** Measurement of delayed-skin-test responses, *N. Engl. J. Med.,* 293, 501, 1975.
303. **Twomey, P., Ziegler, D., and Rombeau, J.,** Utility of skin testing in nutritional assessment: a critical review, *J. Parenteral Enteral Nutr.,* 6, 50, 1982.
304. **Krantzler, N. J., Mullen, B. J., Comstock, E. M., Holden, C. A., Schutz, H. G., Grivetti, L. E., and Meiselman, H. L.,** Methods of food intake assessment — an annotated bibliography, *J. Nutr. Educ.,* 14, 108, 1982.

305. **Stuff, J. E., Garza, C., Smith, E. O., Nichols, B. L., and Montandon, C. M.,** A comparison of dietary methods in nutritional studies, *Am. J. Clin. Nutr.,* 37, 300, 1983.

306. **Burk, M. C. and Pao, E. M.,** Methodology for large-scale surveys of households and individual diets, Home Econ. Res. Rep. No. 40, Agriculture Research Service, United States Department of Agriculture, Washington, D.C., 1976.

307. **Campbell, C., Roe, D. A., and Eickwort, K.,** Qualitative diet indexes: a descriptive or an assessment tool? *J. Am. Dietet. Assoc.,* 81, 687, 1982.

308. *Assessing Changing Food Consumption Patterns,* Food and Nutrition Board, National Research Council, National Academy Press, Washington, D.C., 1981.

309. **Christakis, G.,** Nutritional assessment in health, *Am. J. Public Health,* 63(Part 2) 1973.

310. **Smiciklas-Wright, H. and Guthrie, H. A.,** Dietary methodologies: their use, analyses, interpretations, and implications, in *Nutrition Assessment: A Comprehensive Guide for Planning Intervention,* Simko, M. D., Cowell, C., and Gilbride, J. A., Aspen Systems Corp., Rockville, Md., 1984, 119.

311. **Kerr, G. R., Lee, E. S., Lam, M. K., Lorimor, R. J., Randall, E., Forthofer, R. N., Davis, M. A., and Magnetti, S. M.,** Relationship between dietary and biochemical measures of nutritional status in HANES I data, *Am. J. Clin. Nutr.,* 35, 294, 1982.

312. **Marr, J. W.,** Individual dietary surveys: purpose and methods, *World Rev. Nutr. Dietet.,* 13, 105, 1971.

314. **Hertzler, A. A. and Hoover, L.,** Development of food tables and use with computers, *J. Am. Dietet. Assoc.,* 70, 20, 1977.

315. **Mullen, B. J., Krantzler, N. J., Grivetti, L. E., Shutz, H. G., and Meiselman, H. L.,** Validity of a food frequency questionnaire for the determination of individual food intake, *Am. J. Clin. Nutr.,* 39, 136, 1984.

Section IV
Nutrient Metabolism, Requirements, Nutritional
Status, Imbalance, and Deficiencies

Chapter 7

ENERGY

Loren G. Lipson and George A. Bray

TABLE OF CONTENTS

I. INTRODUCTION

The energy needs of an individual vary according to the person's metabolic rate, size, sex, activity, physical environment, hormonal balance, and age. With increasing age there are marked changes in man's energy consumption and requirements. Age-associated factors in the elderly influence energy intake and lead to alterations in body composition and general energy needs. It is the purpose of this chapter to examine energy requirements of the elderly population, to investigate some special problems in energetics and metabolism involving the aging process, and to formulate a rational approach to meet the energy requirements in our senior citizens.

II. ENERGY HOMEOSTASIS: GENERAL CONSIDERATIONS

Energy homeostasis requires a dynamic balance between nutrient intake and energy expenditure. This complex series of processes begins with the intake of nutrients which are physically and chemically acted upon and transported into various body compartments and then into the cells of these various compartments under the control and regulation of hormonal and neural stimuli. These stimuli are governed in part by the level of nutrient intake. When intake is low, these stimuli act to mobilize the energy which is stored in the body in the form of fat, protein, and carbohydrate. These fuels are metabolized to produce the energy for mechanical work and to maintain the basal metabolic rate (functions such as maintaining constant internal temperature, intracellular environment, and the mechanical processes such as respiration and cardiac function which sustain the body at rest).[1]

Adenosine triphosphate (ATP) is the principal intermediate for energy transfer in the body. Its energy is contained in the high-energy terminal phosphate bond, which can be coupled to biochemical or physical processes and results in the formation of adenosine diphosphate (ADP) or adenosine monophosphate (AMP). These latter compounds may be regenerated to ATP by oxidative phosphorylation. The net result of this process is the consumption of oxygen and fuel and the production of ATP, water, and carbon dioxide. By the process of fuel combustion, two thirds of the energy is released as heat and the other one third is trapped in the terminal phosphate bond of ATP.[1] (See Figure 1.)

In normal man, there is a large endogenous fuel reservoir made up primarily of fat, which accounts for about 85% of the energy reserve (135,000 kcal).[2] Protein stores account for 24,000 kcal (primarily in muscle), while glycogen stores account for only 1200 kcal. These stored fuels can be utilized when nutrient intake is low in order to maintain the basal metabolic rate and to regenerate ATP stores. Triglycerides from the diet or stored in adipose tissue yield fatty acids and glycerol. The latter may be converted back to glucose in the liver or muscle and fatty acids can be metabolized to or, in conditions of insulin deficiency, converted by the liver to ketone bodies.[1] Protein from muscle or other body sources or from food sources is broken down to amino acids; stored glycogen and ingested polysaccharides may be converted to glucose. These key metabolites are further metabolized to produce ATP, heat, water, and carbon dioxide. Some amino acids may enter the Krebs cycle directly and the resulting released ammonia is converted to urea in the liver; other amino acids, glucose, and lactate are first converted into pyruvate and then to acetyl coenzyme A, which is also the product of metabolism of fatty acids and ketone bodies.[1] Thus, changes in the body composition and organ function may have profound changes on energy homeostasis. Regulatory processes by which food intake is coupled to energy needs are not fully understood, but changes in food availability, eating behavior, emotional outlook, and taste perception may influence energy homeostasis.

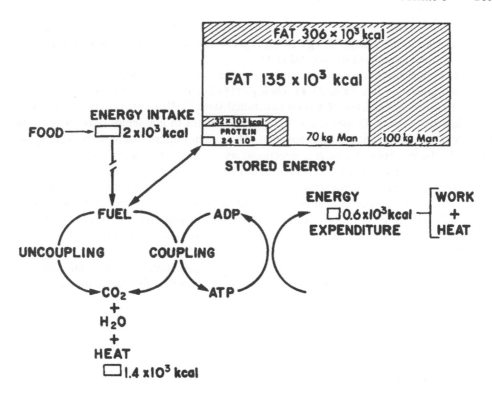

FIGURE 1.   Fuel intake, storage, and utilization. The relationship between the intake of food, the size of the storage depots in both the normal 70-kg man and the 100-kg man and the pathway for the generation of ATP is illustrated. Only about one third of the energy ingested is stored as the high-energy intermediate ATP, the rest being dissipated as heat.

III. ENERGY HOMEOSTASIS IN THE ELDERLY

A. Factors Influencing Energy Intake

Energy intake is largely governed by energy requirements. In younger individuals, the physical and psychological states and the environment may modify nutrient intake, but in the elderly person these factors exert a profound influence on energy intake.

With increasing age, the ability to identify foods in controlled studies of taste is decreased, presumably because of physiological changes in both senses of taste and smell,[3] and this problem is compounded in cases of poor oral hygiene. Changes include loss of taste buds, predominantly involving appreciation of sweet and salty substances, leading to a greater sensitivity to acid and bitter tastes. This may directly lead to decreased food intake or intake of different nutrients.[3]

In addition to alterations in smell and taste, other age-related changes occur either as part of the normal aging process or as a result of chronic disease which may further diminish nutrient intake and/or alter what is eaten. Thus, with increasing age there may be changes in the bite pattern from partial or complete extraction of the teeth. These changes may alter normal mastication. Where complete dentures are required, there may be difficulties in adequately chewing food, and irritation of the buccal mucosa from poorly fitting dentures is frequent. Salivary secretions often decrease with age and tend to make the bollus of food more difficult to swallow.[4]

Chronic changes, such as decreased vision, development of a tremor, and arthritic involvement of finger and hand joints, diminish the older person's ability to prepare his or her own food and make the individual more dependent upon others or upon prepackaged foods, some of which have less-than-ideal nutritive value. There are alterations in the digestive tract which change energy absorption such as delayed esophageal emptying, malabsorption, and constipation[4] (see Table 1).

Psychological and environmental factors also play an important role in controlling energy intake and utilization (see Table 2). Psychological problems such as loneliness and depression are well-known antecedents of a poor nutritional state.[5-7] This situation is worsened if the individual lives alone. Dementias and confusional states in our senior citizens make them dependent upon others to get their nutrition. They often require much aid in feeding themselves. In the presence of cancers, infections, and various chronic diseases, there may be an inability of the older person to take in adequate nutrients and their needs for various nutrients will increase in these stress states, leading to negative protein, vitamin, and energy balance. Lack of adequate education about what is required in a proper diet may lead some individuals into eating fad foods or a diet which is inadequate for their needs. Low income is a reality for many of the aged. This lack of funds will dictate the type and composition of the diet and may lead these economically compromised individuals into states of nutritional imbalance. Excessive alcohol intake in the old person may increase nutritional deficiencies, especially when it is substituted for other forms of nutrients. Similarly, medications may adversely influence nutritional balance.

As man increases in age, there is a diminution in physical activity, and hence, a decreased need for energy intake. Likewise, body composition also changes with increased age. The muscle mass decreases (see below), altering body metabolism and energy needs,[8] and the basal metabolic rate declines. At the same time, the elderly may have increased basal and stress state nutritional requirements which must be met if one is to stay in nutritional and energy balance.[9]

Several studies have shown that there is a progressive decline in energy intake during adult life.[10,11] This decrease in energy requirement represents the combination of physical, psychological, and environmental factors discussed above. The energy requirement decreases after the age of 20 years at a fairly constant rate of 2 to 5%/decade.[1] The energy requirement is greater in men than in women (about 40 kcal/kg for a 22-year-old, 70-kg standard man, and about 36 kcal/kg for a 22-year-old, 56-kg woman[1]). Until the age of 45, men consume approximately 1000 kcal/day more than women, and into the 70s, there is still a 500 kcal/day, greater intake in men. Using the standards given above, men 20 to 24 years of age consume 2800 kcal/day, while women of the same age take in 2016 kcal/day. The relationship between age and sex to caloric intake was closely investigated by the National Center for Health Statistics.[10] There was a steady decline in the energy intake by both men and women from the age of 20 to the age of 74. Men consumed 2888 kcal/day between ages 20 to 24, while at the age range 65 to 74 they took in only about 1805 kcal/day — a decrease of 37.5% in energy intake. Women between the ages of 20 to 24 used 1691 kcal/day, while requiring only 1307 kcal/day between the ages of 65 to 74 (a drop of 23% in energy intake; see Table 3).

In a study of 252 well-educated, financially secure men 23 to 99 years of age who were professionals or retired professionals, there was a decline in energy intake related to physical activity.[11] Men 23 to 34 years old consumed an average of 2688 kcal/day while men 75 to 99 years of age took in 2093 kcal/day (a 22% decline with age). These findings are borne out in several other studies of energy intake and age.[1,12]

B. Changes in Body Composition with Age

Beyond changes in the elderly which influence energy intake, there are significant age-related changes in body composition which may alter nutrient metabolism. There is a steady decline in the lean body mass with increasing age. This process has been shown to occur between the ages of 25 and 65 years,[8] and even longer with some acceleration in later life.[9]

Table 1
AGE-RELATED CHANGES AFFECTING
ENERGY INTAKE OR UTILIZATION

I.   Changes in sensation
    A. Changes in smell
    B. Changes in taste
    C. Decreased vision
II. Mechanical difficulties in preparing food
    A. Altered mental status
    B. Tremor
    C. Arthritis
III. Mechanical difficulties in the intake of food
    A. Poor dentition and dentures
    B. Decreased salivary gland function
IV. Difficulties in digesting and absorbing nutrient energy:
    alterations in gastrointestinal function
V.   Altered body composition
VI. Diminished exercise

Table 2
SOME FACTORS LEADING TO POOR
NUTRITION IN THE AGED[5-7]

I.   Psychological problems
    A. Loneliness
    B. Depression
II.   Mental deterioration
III.   Physical problems
    A. Immobility
    B. Chronic disease
    C. Infection
    D. Cancer
IV.   Lack of knowledge of proper dietary principles
    A. Lack of education
    B. Poor dietary habits
    C. Following food fads
V.   Living alone
VI.   Poverty
VII.   Alcoholism and medications
VIII. Increased dietary requirements

The principal source of the loss in lean body mass is the loss of skeletal muscle. In the young male adult, the skeletal muscle mass may account for about 45% of the body weight, while at the age of 70, a man of the same weight would have only 27% of his body weight as skeletal muscle.[5] Although there is a decline in the weights of visceral organs with increasing age (about 10% decrease in weight at the age of 70 years), the major contributor to the loss of lean body mass is the atrophy of skeletal muscle.

Despite the loss of muscle mass, there is only a slight increase in total body weight with increasing age.[13] The relative stability of the body weight in aging individuals is due to an increase in body fat. This increased adiposity occurs in both sexes and continues into late life. The changes in body weight and composition with age may be partially caused by an imbalance of energy intake and expenditure.[1]

The overall metabolic rate has been shown to decrease with increasing age. This decrease is thought to represent changes in energy intake, body composition, and various metabolic functions and physical activity.[1,11] The basal metabolic rate also has been reported to decrease with increasing age. The basal rate has been reported to fall by 25% with increasing age[1]

Table 3
RELATIONSHIP OF
AGE AND SEX TO
CALORIC INTAKE

Age range	Caloric intake (kcal/day)	
	Men	Women
20—24	2888	1691
25—34	2739	1638
35—44	2554	1558
45—54	2301	1533
55—64	2076	1382
65—74	1805	1307

Adapted from U.S. National Center for Health Statistics, Dietary Intake Findings United States, DHEW (MRA) 77-1647, 1971—1974.

when the data are expressed per unit body surface area. When one set of data were recalculated per unit of lean body mass, the age-related differences in the basal rate were no longer present.[9] Others[15] also have questioned this reported decrease in basal rate and have shown it to be smaller than previously thought. In another study of a large number of men between 23 and 99 years, the basal rate fell only 12% between the ages of 30 and 80.[11] Thus, the decrease in basal metabolic rate is small, but is consistently seen. These decreases in overall metabolic rate and in basal metabolic rate may be reflections of decreases in many of the metabolic components involved in energy homeostasis. Such findings as a decrease in whole-body protein turnover with increasing age may contribute significantly to the age-related decrease in metabolic rate. Hyperglycemia occurs frequently in the elderly and may represent slowing of metabolic processes occurring with age (see below). Thus, it is possible that changes secondary to decreased metabolite turnover may lead to the changes in body composition seen in the elderly.[1]

C. Changes in Energy Utilization with Age

There is a decreased need for energy as people age. This finding has been shown in a number of studies[1,9,12] and has been illustrated in Section III.A. In an important study of energy consumption, basal metabolic rate and energy expenditure of 252 men ranging in age from their early 20s to 99 years of age, this decreased energy requirement was well documented.[11] This group of men were selected on the basis of health, education, and economic success; hence, they had no financial constraints on dietary intake. These older men showed a 22% fall in total energy consumption in comparison of the 20- to 34-year-old group with the 75- to 99-year-old group (a decrease of 600 kcal/day). The overall decrement in dietary calories was 12.4 cal/day/year. From calculations of the basal metabolic rate in these men, it was found that about one third of the decrease in caloric intake represented decreased basal metabolic requirements (a decrease of 200 kcal/day). The remaining decrease of caloric consumption was largely accounted for by changes in physical activity between these two age groups (about 400 kcal/day).[11] (See Table 4.)

Thus, the overall energetics of aging in man are as follows: there is decreased energy consumption with increasing age on the basis of altered body composition and cellular biochemistry which probably leads to a decreased metabolic rate; there is also a decreased energy need because of diminished physical activity of people as they age.

Table 4
THE RELATIONSHIP OF
ENERGY EXPENDITURE
TO INCREASING AGE IN
A GROUP OF ACTIVE,
HEALTHY MEN[11]

Age (years)	Energy expenditure (kcal/day)[a]
20—34	1175
35—44	1166
45—54	982
55—64	950
65—74	928
75—99	640[b]

[a] Total calories consumed minus ca-
    lories required for basal metabolism.
[b] 46% decline in energy expenditure
    compared with the group 20 to 34
    years old.

IV. OBESITY AND AGING

There is an increase in adiposity with increasing age. Despite a gain in weight between the ages of 21 and 30 years, body weight is relatively stable through the age of 70 years, after which time it begins to decline slowly. In spite of this relative stability in body weight the percentage of fat in the total body weight rises progressively with increasing years.[8,13] Thus, the ideal body weight decreases with age.

Older individuals with increased adiposity, and especially those with frank obesity (greater than 20% above ideal body weight), are at increased risk for morbidity and mortality from such age-associated diseases as coronary artery disease, atherosclerosis, hypertension, renal disease, and diabetes mellitus.[7,16-18] The effect of obesity in the elderly is thus of concern in maintaining normal longevity.

It has been frequently postulated that leanness and caloric restriction leads to increased longevity.[16] Thus, it has been advocated that weight reduction is of prime importance in the obese individual. This concept has been challenged recently in an intensive review of mortality as related to obesity.[19] In this study it is concluded that a slight degree of adiposity is beneficial for increased longevity in those elderly who are healthy and not hypertensive or diabetic.[19] However, most of the studies cited which support this hypothesis have lumped smokers and nonsmokers together. When the detrimental effects of smoking are taken into account there is no beneficial effect of extra weight.[20]

As an example of an obesity and age-related problem, weight gain frequently precedes or is associated with the development of diabetes mellitus. There is a 4.3 times higher incidence of diabetes mellitus when relative body weight is 10% above normal. In addition, in some American Indian populations 50% of those individuals who are greater than 20% above ideal body weight have an abnormal glucose tolerance test.[21] It is clear that as weight levels increase, the prevalence of diabetes increases in each decade examined. In addition, for any given body weight, the prevalence of diabetes increases with age.[22] These findings are illustrated in Figure 2 and represent findings from several thousand women. We cannot prevent people from getting older, but if we can prevent them from becoming more obese, we may expect to reduce the prevalence of many of the obesity-associated diseases which occur with increasing age.

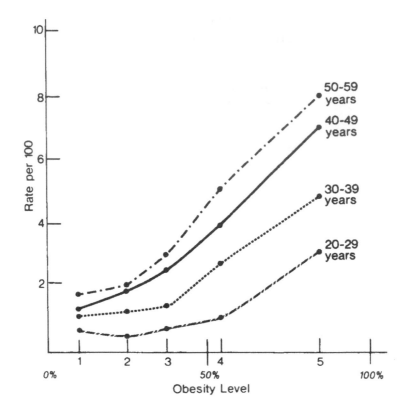

FIGURE 2. Obesity and diabetes mellitus. The relationship of age and degree of obesity in women with rate of occurrence of diabetes mellitus. The percentage of women in various age groups with diabetes mellitus increases with body weight. The weight categories divide the patients into groups according to the percentage above ideal weight.[21,22]

V. THE HYPERGLYCEMIA OF AGING

Associated with a decreased metabolic rate with increasing age is the appearance of hyperglycemia in the elderly population. Whether the decreases in metabolic rate are partially a manifestation of alterations in glucose homeostasis is unknown; however, age-related changes in glucose metabolism may offer insight into how many metabolic pathways are altered in the elderly and may turn physiology into pathology.

In older human populations there is an increase in plasma glucose levels with increasing age.[23-27] This age-related phenomenon is seen in both sexes and is manifest by small increases in the fasting plasma glucose and more marked increases in the post-prandial plasma glucose with increasing age. Table 5 illustrates this finding in a group of aging men and women. There is an increase in the fasting plasma glucose of about 2 mg/dℓ/decade over the age of 30 to 40 and a more marked increase in the 1- to 2-hr post-prandial plasma glucose of 8 to 15 mg/dℓ/decade over the age of 30 to 40. Hence, a 30-year-old man destined to have this form of hyperglycemia may initially have a fasting plasma glucose of 80 mg/dℓ and a 2-hr post-prandial glucose of 120 mg/dℓ, but by the age of 70 years might have a fasting plasma glucose of 90 mg/dℓ (not a striking increase) and a 2-hr post-prandial plasma glucose of 170 to 190 mg/dℓ. The proportion of the population, having this problem after excluding diabetic patients may vary from 10 to 30% of the elderly members of the community.[25] The high prevalence of this problem makes one concerned as to whether the hyperglycemia is a normal physiological consequence of aging about which one need not worry or is it a

Table 5

THE ROLE OF AGE ON THE
GLUCOSE TOLERANCE
TEST[26,27]

	Individuals with blood glucose greater than 160 mg/dℓ (%)		
Age range	Men	Women	Total
18—24	1	5	3
25—34	5	7	6
45—54	13	20	16
65—74	27	43	36
75—79	25	58	42

Note: 1 hr post-oral glucose challenge data shown.

Table 6

POSSIBLE CAUSES OF THE
HYPERGLYCEMIA OF AGING

1. Decreased insulin biosynthesis and/or secretion
2. Alterations in insulin action
   A. Receptor
   B. Post-receptor or intracellular
3. Decreased peripheral glucose utilization
4. Changes in body composition
5. Changes in diet
6. Decreased physical activity

pathological process which may need therapeutic intervention? The Framingham Study has shown those individuals with the hyperglycemia of aging have an increased incidence of coronary artery disease,[23] and more recent studies have shown a progressive rise in hemoglobin A_1c levels in the nondiabetic elderly — both of which tend to support a pathological significance for this hyperglycemia.

Table 6 outlines some of the potential causes of this hyperglycemia. Whether insulin secretion is altered with increasing age has been a source of much controversy since clinical studies have been indecisive, with much of the disagreement resulting from differences in patient selection.[28] In vitro studies of the older rat (as a model of aging) have shown there is a profound defect in glucose-stimulated insulin secretion. Further in vitro dynamic studies have shown islets from aging rats have a blunted first phase insulin release and a significant decrease in second phase release — results comparable with those obtained in the fasted state in rats[29,30] and similar to that seen in mild type II diabetes mellitus.[31] Changes in insulin action also occur with increased age. Although recent studies have not demonstrated alteration in avidity or number of insulin receptors with increasing age,[28] there appears to be increased insulin resistance even in those older individuals in whom glucose homeostasis is normal. Thus, a decrease in insulin sensitivity without a change in responsiveness has been found to occur in aging man. Noninsulin-dependent glucose removal is also decreased with increasing age. The area of diabetes mellitus, insulin resistance, and insulin secretion has been covered in depth in Chapter 5 of Volume II. Decreases in energy expenditure, decreased activity, and altered nutrient intake all further complicate glucose homeostasis (as has already been described above). These changes tend to act in concert and lead to the prevalence of glucose homeostasis in the elderly population.

Thus, the hyperglycemia of aging may be but a shade of gray in the spectrum of age-related changes in carbohydrate tolerance from mild insulin resistance to frank type II diabetes mellitus. And this problem is not entirely due to any one change, but to a combination of several events occurring together during the aging process. Hence, in approaching such a person, one should consider what therapies might improve the carbohydrate tolerance and thus improve the quality of life.

VI. THERAPEUTIC INTERVENTIONS

If the elderly have decreased metabolic rate, increased adiposity, and the tendency to develop hyperglycemia, it is wise to develop general dietary guidelines based on the principles of energy homeostasis and the changes in energy homeostasis observed with aging. First, food intake should match energy expenditures. As exercise and general physical activity and the rate of metabolic processes decrease with age, energy intake should also be reduced.[1] Further, excessive nutrient intake with aging usually leads to obesity, which is a risk factor for the development of cardiovascular disease, diabetes mellitus, renal disease, and hypertension. Whether the relative increase in body fat with aging has similar implications to adiposity in general still remains unclear.

Second, essential vitamins, minerals, trace metals, and nutrients must be included in the diet especially when an effort is made to restrict total energy intake. Altered food selection by the elderly based upon the psychological and physiological changes previously described may contribute to the increased incidence of multiple vitamin deficiencies in the elderly population as a whole. Thus, unless caloric restriction is carefully planned, these deficiencies could be worsened.

It has been recommended that there be a gradual caloric decrease in both men and women over the age of 50, but the vitamin supplementation should be at the level of that required by the young adult. Further, it has been recommended that the diet include 0.8 g/kg of mixed proteins. The composition of carbohydrate and fat may be somewhat variable, but it is preferable to stress the need to eat fruits and vegetables, breads, and grains, and to reduce the intake of saturated fat and cholesterol as well as salt and sugar.[1]

Finally, the guidelines for diets for the elderly must consider the physical and psychological aspects of food ingestion. Age-related alterations in taste change the palatability of many foods commonly eaten by younger individuals. Changes in gastrointestinal function may lead to the addition or elimination of specific types of food. Thus, attempts must be made to promote the compliance of the elderly with diets incorporating recommended changes in energy and nutrient intake.

Energy balance may be improved further in those who do not have significant impairment in the cardiovascular or musculoskeletal system by encouraging increased energy expediture. Such expenditures may be accomplished by 10 to 15 min of mild to moderate exercise per day. Exercise such as walking is useful to increase muscle tone and to better utilize ingested energy and promote glucose removal. In those older persons with hyperglycemia of aging, a large number may have improved glucose homeostasis by the above changes in diet and exercise. For those who do not improve, medical treatment may be needed.

As our older population increases in size, the decreasing quality of life with increasing age is becoming more of a problem. While some factors are beyond the control of the individual, improved nutrition is an attainable goal that would be likely to improve the quality of life among the elderly.

REFERENCES

1. **Heber, D. and Bray, G. A.,** Energy requirements, in *Metabolic Nutritional Disorders*, Exton-Smith, A. N. and Caird, F. I., Eds., John Wright & Sons, Bristol, 1980, 1.
2. **Moore, F. D. and Brennan, M. F.,** *Manual of Surgical Nutrition*, Ballinger, W. F., Ed., W. B. Saunders, Philadelphia, 1974, 109.
3. **Schiffman, S. S.,** Food recognition by the elderly, *J. Gerontol.,* 5, 586, 1977.
4. **Bhanthumnavin, R. and Schuster, M. M.,** Aging and gastrointestinal function, in *Handbook of the Biology of Aging*, Finch, C. E. and Hayflick, L., Eds., Van Nostrand Reinhold, New York, 1977, 709.
5. **Young, V. R.,** Nutrition, in *Health and Disease in Old Age*, Rowe, J. W. and Besdine, R. W., Eds., Little, Brown, Boston, 1982, 317.
6. **Watkin, D. M.,** Nutrition for aging and the aged, in *Modern Nutrition in Health and Disease*, Goodhart, R. S. and Shils, M. E., Eds., Lea & Febiger, Philadelphia, 1980, 781.
7. **Worthington-Roberts, B. S. and Hazzard, W. R.,** Nutrition and aging, in *Annual Review of Gerontology and Geriatrics*, Vol. 3, Eisdorfer, C., Ed., Springer-Verlag, New York, 1982, 297.

8. **Forbes, G. B. and Reina, J. C.,** Adult lean body declines with age: some longitudinal observation, *Metabolism,* 19, 653, 1970.

9. **Munro, H. N.,** Nutritional requirements in the elderly, *Hosp. Pract.,* 17, 143, 1982.

10. National Center for Health Statistics, Dietary Intake Findings: United States 1971—1974, DHEW Publ. #(MRA 77-1647, Health Services Administration, U.S. Department of Health, Education and Welfare, Washington, D.C., 1977.

11. **McGandy, R. B., Barrows, C. H., Jr., Spanias, A., Meredith, A., Stone, J. L., and Norris, A. H.,** Nutrient intakes and energy expenditure in men of different ages, *J. Gerontol.,* 21, 581, 1966.

12. **Barrows, C. H., Jr. and Roeder, L. M.,** Nutrition, in *Handbook of the Biology of Aging,* Finch, C. E. and Hayflick, L., Eds., Van Nostrand Reinhold, New York, 1977, 561.

13. Center for Disease Control, Ten-State Nutrition Survey, 1968—1970, DHEW Publ. # (HSM) 72-8134, Health Services and Mental Health Administration, U.S. Department of Health, Education and Welfare, Washington, D.C., 1972.

14. **Munro, H. N.,** Nutrition and aging, *Br. Med. Bull.,* 37, 83, 1981.

15. **Keys, A., Taylor, H. L., and Grande, R.,** Basal metabolism and age of adult man, *Metabolism,* 22, 578, 1973.

16. **Bray, G. A.,** Obesity, *Curr. Concepts,* 1984.

17. **Brunzell, J. D.,** Obesity and risk for cardiovascular disease in *Obesity,* Greenwood, M. R. C., Ed., Churchill Livingstone, New York, 1983.

18. **Jeffay, H.,** Obesity and aging, *Am. J. Clin. Nutr.,* 36, 809, 1982.

19. **Andres, R.,** Effect of obesity on total mortality, *Int. J. Obesity,* 4, 381, 1980.

20. **Stunkard, A. J.,** Nutrition, aging and obesity: a critical review of a complex relationship, *Int. J. Obesity,* 7, 201, 1983.

21. **Bray, G. A.,** Diabetes mellitus and obesity, in *Medical Complications of Obesity,* Mancini, M., Lewis, B., and Contaldo, F., Eds., Academic Press, London, 1979, 27.

22. **Rimm, A. A., Werner, L. H., and Yserloo, B. V.,** Relationship of obesity and disease in 73,532 weight conscious women, *Public Health Rep.,* 90, 44, 1975.

23. **Epstein, F. H.,** Hyperglycemia — a risk factor in coronary artery disease, *Circulation,* 36, 609, 1967.

24. Report of a working party, 1963, Glucose tolerance and glycosuria in the general population, *Br. Med. J.,* 1, 655, 1963.

25. **Davidson, M. B.,** Diabetes in the elderly: diagnosis and treatment, *Hosp. Pract.,* 17, 113, 1982.

26. **Bennett, P. H.,** Report of work group on epidemiology. National Commission on Diabetes, Vol. 3, Part 1. DHEW Publ. # (NIH) 76-1021, Department of Health, Education and Welfare, Washington, D.C., 1976, 65.

27. **Horwitz, D. L.,** Diabetes and aging, *Am. J. Clin. Nutr.,* 36, 803, 1982.

28. **Rowe, J. W., Minaker, K. L., Pallotta, J. A., and Flier, J. S.,** Characterization of the insulin resistance of aging, *J. Clin. Invest.,* 71, 1581, 1983.

29. **Molina, J. M., Premdas, A., and Lipson, L. G.,** Insulin release in aging: dynamic response of isolated islets of Langerhans of the rat to D-glucose and D-glyceraldehyde, *Endocrinology,* in press.

30. **Lipson, L. G., Siegel, E., Wollheim, C. B., and Sharp, G. W. G.,** Insulin release during fasting: studies on adenylate cyclase, phosphodiesterase, protein kinase and phosphoprotein phosphatase in isolated inlets of Langerhans of the rat, *Endocrinology,* 105, 702, 1979.

31. **Lipson, L. G. and Lipson, M.,** The therapeutic approach to the obese maturity-onset diabetic patient, *Arch. Int. Med.,* 144, 135, 1984.

Chapter 8

PROTEIN AND AMINO ACIDS

Malcolm J. McKay and Judith Bond

TABLE OF CONTENTS

I. INTRODUCTION

There is no shortage of literature on the effects of aging on proteins and amino acids in a variety of living systems and there have been several recent review articles and books dealing with age-related changes in protein and amino acid metabolism at the molecular, cellular, tissue, and whole animal levels.[1-5] The picture that emerges is that while there are changes in the protein constituents of living systems with age, the importance of these changes in the senescence and death of the system and the mechanisms responsible for these changes are unknown. Our understanding of the complex process of aging is rudimentary and few generalizations have emerged with regard to the fundamental aspects of protein and amino acid metabolism or how nutrition influences these processes during aging. There is one area, however, in which a consensus appears to be evolving; there is increasing evidence that post-translational modifications of proteins contribute to the accumulation of altered proteins in aging cells. At one time it was believed that altered or defective proteins occurred in aging cells because of an increasing incidence of random errors in protein synthesis. This belief was based on Orgel's hypothesis[6,7] that the increasing incidence of errors will eventually lead to the "error catastrophe" and cell or organism death. The evidence accumulated over the last 10 years, however, does not support the hypothesis that errors in protein synthesis increase with age. Data now indicate that the altered proteins that accumulate result from post-translational reactions such as deamination, oxidation, glycosylation, or limited proteolysis. This article will deal primarily with the evidence for the protein changes observed in aging as studied in a variety of systems. If we can shed light on how protein and amino acid metabolism changes in aging organisms, perhaps the factors (nutritional, environmental, genetic) that regulate the aging process can be uncovered.

II. HUMAN STUDIES

Studies of human beings have established that there are fundamental changes in protein and amino acid metabolism with aging but the specific changes have been difficult to establish definitively.[3,4] There is a loss of body mass, or more specifically, muscle mass in the elderly.[8,9] Creatinine excretion per day, an index of muscle mass,[10] is decreased in elderly man compared to young adults. *N*-Methyl histidine in urine, a product of muscle protein catabolism, also decreases in the elderly.[9] Measurements of protein synthesis and degradation involve extremely difficult techniques, but a significant amount of data has been accumulated. In whole body studies protein synthesis and degradation per unit of creatinine excretion appeared to be *increased* in elderly man compared to young adults. Skeletal muscle proteins turn over slowly under basal conditions relative to several other tissue (e.g., liver, kidney, heart) proteins,[12] and thus the diminished contribution of skeletal muscle proteins could explain these observations. It has been estimated that muscle accounts for 30% of whole body protein degradation in young men and 20% in elderly men.[3] These observations imply that individual tissues age differently with regard to protein metabolism and that the interpretation of results from whole animal studies must consider the complexity of the system. In addition, while some data indicate that skeletal muscle protein synthesis and degradation decrease with age,[3] other studies indicate that these processes both increase in muscle as well as nonmuscle tissue in the elderly.[11] The balance between the rate of synthesis and the rate of degradation will determine the total amount of body protein. While it is clear that muscle protein mass decreases in the elderly, the relative contributions of the processes of synthesis and degradation in muscle to the observed decrease are unknown.

It is established that certain proteins in animal tissues change with age. The best-characterized examples of these include: collagen (increased aggregation and cross-linking) and red blood cell proteins (accumulation of Heinz bodies).[4] These changes appear to be due to post-translational modifications of the proteins. Protein-deficient diets can result in impairment of the cross-linking process in collagen, but the dietary restrictions affect younger animals more than old animals.[13] Dietary restriction of protein is also reported to retard lipofuscin (fluorescent age pigment) accumulation in brain and heart of mice[14] and to extend significantly the life span of rodents.[15,16] Decreased dietary protein intake significantly increases life span in rodents if the diet is initiated before puberty and continued throughout the life span of the animals. These examples show that diet can influence the processes that govern aging.

No consistent picture emerges with regard to amino acid metabolism changes with age in man.[3] In addition, no differences have been detected in the regulation of plasma amino acid concentrations after fasting or the administration of graded amounts of amino acids when young and old men were compared. However, there is some evidence that insulin sensitivity in peripheral tissues is decreased with age, and this may ultimately alter amino acid transport as well as glucose transport.[17]

It is generally agreed that the energy intake required for the elderly is decreased compared to younger people, but a reduced need for dietary protein has not been established.[3] The recommended dietary allowance for protein in older men and women is 0.8 g/kg.[18,19] However, the scientific base for this recommendation is questionable and there is in fact some evidence to indicate that this level of protein intake is insufficient to maintain nitrogen balance in the elderly.[20]

Studies of protein intake of the elderly and their protein status as determined by serum albumin levels indicate that protein is not a problem nutrient among this age group. O'Hanlon and Kohrs[21] reviewed and summarized 28 studies of the dietary intake of older Americans and found the mean intake for protein was adequate in all studies except the Ten-State Survey.[22] More recent reports confirm that the mean intake of protein by the elderly exceeds the Recommended Dietary Allowance (RDA), which is 56 g/day for males and 44 g/day for females. Mean protein intakes of 83 to 88 g/day[23,24] by elderly males and 62 to 67 g/day by females have been reported. The percentage of elderly consuming less than 100% of the RDA for protein range from 2 to 14%.[24,25] Mean serum albumin in a study by Kohrs et al.[23] was 4.5 g/dℓ in subjects over 60 with a range of 3.4 to 5.7 in males and 2.8 to 5.1 in females.[23]

Mean values can mask problems among subgroups in a study. The Ten-State Survey[22] reported protein intakes below 1 g/kg body weight among several subgroups. However, only low-income black females were below 0.8 g/kg in mean protein intake. Lowenstein[26] found in the First National Health and Nutrition Examination Survey (HANES I) that mean protein intake of 65- to 74-year-old subjects was 48 to 78 g/day. The mean intake was greater than 100% of the RDA even among low-income elderly females. However, 55 to 76% of the subjects had intakes below the RDA. Mean serum albumin levels were normal, but 3.4% had chemical signs of protein malnutrition (hepatomegaly and enlarged parotid glands). Bowman and Rosenberg[27] have noted that 19 to 27% of low-income elderly males and 19 to 22% of low-income elderly females in HANES I had protein intakes below two thirds the RDA; also, 2.6% of white subjects had serum albumin levels less than standard (3.5 g/dℓ) and 3.3% of black subjects were below standard.

Jansen and Harrill[28] studied the protein intake and serum albumin level of elderly women. Protein intake was 55 g/day in the 62- to 75-year group, 46 g/day in the 76- to 85-year group, and 47.4 g/day in the 86- to 99-year group. Mean plasma albumin levels were 3.9, 3.8, and 3.7 g/dℓ for the three groups, respectively. The mean values for protein intake exceeded 100% of the RDA, but 24 to 29% of the women consumed less than 0.8 of protein per kilogram of body weight. Thus, reports indicate that protein insufficiency may be a problem for a small proportion of elderly, particularly females and low-income individuals.

An understanding of protein nutrition in aging humans will be facilitated with a knowledge of the aging process itself. The dynamics of protein turnover (protein synthesis and degradation) in aged subjects is of particular relevance as this process governs the protein content and enzymic activity of all body tissues. Protein synthesis will be affected by the composition of amino acid pools of the cells, which are linked to the degradation and uptake of ingested proteins and components derived from the turnover of the subject's own tissue proteins. Limited research has been conducted in the biochemistry of aging humans; however, a variety of other systems are in use today which serve as models for aging. These models can provide conjectures as to the causes of senescence in humans.

III. MODEL SYSTEMS FOR AGING RESEARCH

A. Mammalian Cells in Culture

Mammalian cells in culture offer one model system for aging, although the interpretation and application of the data obtained from cells in vitro are the subject of debate.[2,29] Human embryonic cell cultures undergo approximately 50 population doublings in phases I and II before they enter phase III, where their rate of cell division declines, eventually stops, and the cells die.[30,31] The life span of cells in culture is not a function of time, but the number of population doublings. An interesting feature of this system is that the doubling potential of cells in culture obtained from donors of different age groups reflects the age of the donor.[32,33]

Although many believe that the death of cells in culture is an artifact (see review by Rothstein[2]), Hayflick[29] points out that the finite life span of cultured cells might well be an expression of aging at the cellular level in vivo and that the decrements observed in vitro actually occur in vivo, but need not result in the death of the cells. Indeed, nondividing phase III cells can be maintained in culture for considerable lengths of time,[34] although no one has been able to increase or restore population doubling potential to phase III cells.[29]

Cells in culture have also been used for the analysis of cellular and molecular changes that occur in the tissues of patients with premature aging diseases such as progeria and Werner's syndrome.[5] Many of the ultrastructural and biochemical changes associated with aging normal fibroblasts are seen prematurely in aging diseases.[35] On a broader basis, Hayflick[36] reported that the doubling potential of cell cultures was related to the life span of the animal or species from which the cells were obtained, although these preliminary observations contradicted an earlier report that found no such relationship.[37]

The accumulation of faulty enzyme proteins has been observed primarily in late phase II and phase III cells, which are dividing very slowly or not at all. Hayflick[29] has compiled a glossary of enzymic functions (as well as other biochemical parameters) that decline in aging human diploid cells in phase III. This list is also accompanied by twice as many parameters that either increase or remain unchanged in phase III.

The best examples of enzyme activities that decline in phase III cells are glucose-6-phosphate dehydrogenase (G6PDH) and triose phosphate isomerase (TPI). An increase in the heat lability of G6PDH was first observed in aging cell cultures,[38] and this increase in heat lability was later found to be the result of post-synthetic modification(s).[39,40] TPI also becomes unstable in aging cell cultures, with a concomitant change in the isozyme pattern due to a series of deamination reactions.[5] These observations support the hypothesis that altered enzymes accumulate as a result of post-synthetic modifications and argue against the "error catastrophe" hypothesis. The reports of Buchanan et al.[41] and Wojtyk and Goldstein,[42] who measured the levels of leucine misincorporation in cell-free extracts of early- and late-passage cells, also detract from Orgel's "error catastrophe" hypothesis. The accumulation of such altered enzymes in late-passage cell cultures from normal patients is reported to occur prematurely in cell cultures derived from patients with premature aging diseases.[43,44]

One explanation for the observation that faulty enzymes accumulate in aging systems may be that old cells lack the ability to remove these proteins through degradation. If so, one might expect to find decreased proteolytic activity in aging cell cultures, as is the case in the free-living nematode (see Section III.E). In cell cultures, however, the opposite apparently occurs. With the exception of a report of a neutral proteinase activity that declines in phase II and III cell cultures,[45] it is well documented that there is an increase in the number of lysosomes and autophagic vacuoles in late-passage WI-38 cell cultures.[35,46,47]

Assessments of protein turnover in cell cultures and its disposition with age are complicated by technical problems with essential agents in the culture medium that affect the relative rates of protein synthesis and degradation. Little work has been performed in this area. Bradley et al.,[48] working with WI-38 cell cultures, reported that short-lived proteins were degraded more rapidly in early phase III than in phase II and that in late phase III, long-lived proteins were degraded more rapidly than in phase II. In the same year, Goldstein et al.[49] reported that there was no change in the relative susceptibility of long-lived proteins to proteolysis in aging cultures of human skin fibroblasts and that the degradation of short-lived proteins actually declined in phase III. Thus, the effects of aging on protein degradation in this model system was not well resolved. Research with this model, although facilitated by the availability of cells from patients with premature aging diseases, has produced no clear pattern to date of the factors that are responsible for the accumulation of the relatively few altered enzymes and why they only appear in late-passage cells. This model system, however, does supply evidence for the involvement of post-synthetic modifications in the generation of altered enzymes in old cells.

B. Mammalian Eye Lens

The mammalian lens offers a useful model for studies on the fate of both structural and catalytic proteins in aging. Aging studies may be performed with the whole lens from donors of different age groups or with the constitutive stratified layers of lens tissue which may be separated and analyzed individually. A cross section of a lens reveals a central region of hard material and stratified layers of tissue surrounding this region. The outer layers in the peripheral region are the youngest tissue.[50,51]

In many ways, the lens is similar to mammalian reticulocytes in that old lens cells are anuclear and contain few (if any) organelles. Peripheral cells are active in protein synthesis, whereas tissue at the center has no such activity.[51] The lens has no blood supply and derives the bulk of its energy through glycolysis.[52] By comparing the enzyme activities in peripheral and central lens tissue, Ohrloff and Hockwin[52] compiled a list of 23 enzyme activities that declined in the maturation of lens tissue. These included the bulk of the glycolytic enzymes. Those enzymes that were studied further were also found to have lower substrate affinities and increased heat labilities. Inactive enzyme species were not removed by proteolysis, as immunological assays revealed the presence of large amounts of inactive enzyme molecules in the older tissues.[52] These enzymes undergo changes in conformation due to post-synthetic modifications that render them less stable. The reported decline in many lens enzyme activities may be associated with the decline in energy metabolism in aging lens tissue.[53]

In aging lens tissue, there is an increase in optical density at 440 nm which is due to the accumulation of yellow pigments which are associated with large aggregates of lens proteins.[54,55] In a manner similar to the formation of Heinz bodies in the maturing mammalian erythrocyte,[56] these deposits of yellow pigments are the end product of a serial aggregation of α-crystallin polypeptides, a process that parallels aging.[57] An analysis of the stratified layers of lens tissue from the young periphery to the old central region shows that there is a change in the disposition of α-crystallin.[55] The amount of α-crystallin declines, with a concomitant increase in the amount of β- and γ-crystallins as well as an increase in the formation of high molecular weight aggregates of these latter forms of crystallins, which are believed to be intermediates in the formation of the yellow pigment deposits.[55] These high molecular weight aggregates of crystallins are not soluble in water and have molecular weights in the range of 1×10^6 to 1×10^9. The formation of these high molecular weight aggregates in aging lens tissue can be followed in vivo with laser light scattering techniques.[58] There are also reports of changes in the fluorescence of protein in aging lens tissue; it is believed the UV radiation incident on the in vivo lens generates fluorophors which may be responsible for the generation of cross links in the lens crystallin, leading to the formation of high molecular weight aggregates.[59] An increased incidence of disulfide bonds in and between lens crystallins has also been reported in aging lens tissue; a larger proportion of which are found in high molecular weight aggregates.[55] Such an observation might be explained by the observed reduction in the reductive capacity in the old lens tissue as a result of the associated decrease in the activities of glutathione reductase and peroxidases with age.[55] There is an age-dependent increase in crystallin polypeptide deamination, an increase in aspartic acid residue racemization, and the partial cleavage of lens α-crystallin subunits.[55] Any or all of these phenomena may lead to changes in the structural conformation of the crystallins and account for the aggregation of proteins observed in aging lens tissues. Clearly, the mammalian lens model has potential for aging studies.

C. Reticulocytes

The mammalian erythroid cell line is another choice as a model for studying the factors that regulate the development, maturation, and aging of mammalian cells. Reticulocytes are readily generated by inducing anemia in laboratory animals[60] and easily fractionated into discrete cell-age groups with buoyant density gradient centrifugation.[61] The reticulocyte is a relatively simple cell, which has extruded its nucleus prior to entry into the peripheral bloodstream and thus lacks the ability to divide.[62] The young circulating reticulocyte has the ability to synthesize protein (almost exclusively hemoglobin[63]) and the ability to degrade rapidly and selectively abnormal proteins (i.e., those that differ from the normal gene products).[64,65] During the maturation of the reticulocyte to an erythrocyte, a number of biochemical changes are known to occur. These include the loss of its ability to synthesize proteins[61] and a decline in the ability to degrade abnormal proteins;[64] both of these functions are absent in the erythrocyte.[64] The loss of a number of other cytosolic enzyme activities parallels these declining functions and their activities serve as biochemical markers of reticulocyte cell age.[66,67]

Age-fractionated rabbit reticulocytes are easily manipulated to synthesize large amounts of radiolabeled abnormal proteins in vitro[64] by incubating them in the presence of either S-2-aminoethylcysteine (a lysine analogue) or puromycin dihydrochloride, an inhibitor of protein synthesis that terminates peptide chain elongation at the ribosome. Under such conditions, reticulocytes synthesize proteins (predominantly globin chains) of either abnormal amino acid composition with normal chain length or normal amino acid composition with abnormal (shortened) chain lengths, respectively.[64,68] Both types of abnormal proteins are rapidly degraded in whole cell and cell-free preparations.[69]

Figure 1 represents some typical results and illustrates the age-related decline in the ability of the reticulocyte to degrade the two forms of abnormal proteins described above. These abnormal proteins form at least two types of high molecular weight aggregates prior to their degradation. One form involves an enzyme-catalyzed covalent conjugation of ubiquitin with the substrate proteins,[65] and a second form involves noncovalent associations between substrate proteins.[69] When cell-free extracts of rabbit reticulocytes were prepared, a proportion of the radiolabeled aggregates sedimented with the debris/membrane fraction following cell lysis and centrifugation. This material may represent membrane-associated abnormal protein aggregates of very high molecular weight. In the residual cell-free extracts, there were at least two classes of abnormal protein aggregates which were discernible by Sephadex® G-100 gel filtration; one eluted in the void volume of the column with a molecular weight of

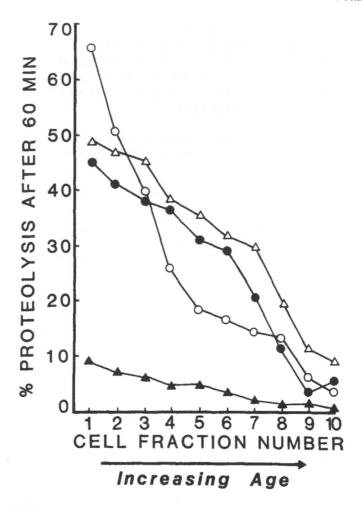

FIGURE 1.   The degradation of abnormal proteins in age-fractionated rabbit reticulocytes. Rabbit reticulocytes were fractionated into 10 cell-age groups by buoyant density gradient centrifugation and each cell age-group individually pulse-labeled with [^{14}C]-leucine in the presence of either (1) 5 m*M* *S*-2-aminoethyl-cysteine (△———△), (2) 5 µg/mℓ (●———●) or 25 µg/mℓ (○———○) puromycin dihydrochloride, or (3) in the presence of a full amino acid compliment (control conditions) (▲———▲). The extent of proteolysis was then determined[64] in each cell age-group over a period of 60 min.

>10^6 and another class eluted with the hemoglobin peak.[68] Examination of the same cells following incubation for 60 min revealed the amount of soluble high molecular weight aggregates decreased, with a concomitant increase in the amount of the degradation products. It has been shown that the extent of abnormal protein degradation in rabbit reticulocytes was directly proportional to the amount of radiolabeled abnormal protein associated with this fraction.[68]

The subsequent catabolism of these high molecular weight aggregates is understood to be initiated by an ATP-dependent proteinase to generate peptide fragments that are then degraded to their constitutive amino acids by ATP-independent proteinases and peptidases.[65] Initial studies with the rabbit reticulocyte have led to the discovery of similar proteinases in other tissues including heart, skeletal muscle, kidney, and liver.[65]

Table 1 shows the disposition of radiolabeled abnormal proteins in age-fractionated rabbit reticulocytes. It can be seen that immediately after the synthesis of abnormal proteins, there

Table 1
THE RELATIVE DISTRIBUTION OF RADIOLABELED RETICULOCYTE ABNORMAL PROTEINS IN THE DIFFERENT SUBCELLULAR FRACTIONS OF AGE-FRACTIONATED RABBIT RETICULOCYTES

% Relative distribution of radiolabeled abnormal proteins in respective rabbit reticulocyte cell fractions

| | Sedimented debris fraction | | Sephadex® G-100 gel filtration fractions | | | |
| | | | Void vol. eluate | | Hemoglobin peak | |
Cell age group	t = 0	t = 60	t = 0	t = 60	t = 0	t = 60
1 (Youngest)	8.5	1.45	43.7	2.05	47.8	41.2
2	9.4	1.95	39.4	6.53	51.2	45.0
3	7.5	2.29	36.4	7.29	56.1	51.1
4	12.4	11.0	32.5	9.02	55.1	55.0
5	11.6	17.8	32.8	15.3	55.6	55.8
6	19.7	21.4	24.1	12.7	56.2	51.0
7	24.7	20.7	20.9	13.6	54.5	40.6
8	23.8	27.1	19.5	19.0	56.7	75.1
9	29.0	31.9	20.5	19.9	50.5	48.9
10 (Oldest)	37.5	39.1	9.9	9.2	52.6	51.4

Note: Age-fractionated rabbit reticulocytes were pulse-labeled with ^{14}C-leucine in the presence of 25 μg/mℓ puromycin.[64] Immediately after pulse labeling, samples of each cell-age group were individually lysed and centrifuged (2000 × g/10 min). The supernatant fractions were subjected to Sephadex® G-100 gel filtration. The remaining cells were incubated for 60 min, the extent of protein degradation determined in each cell-age group, and the individual cell-age groups subjected to subcellular fractionation as above. The relative distribution of radiolabeled abnormal proteins was determined in each subcellular fraction of the age-fractionated reticulocytes before (t = 0) and after a 60-min period of incubation (t = 60).[68]

is a differential distribution of radiolabel between young and old cells. Older cells contain a greater proportion of very high molecular weight abnormal proteins which have sedimented with the debris fraction. Of the soluble abnormal proteins eluting in the void volume of the Sephadex® column, a decreasingly smaller proportion of this material is degraded during the 60-min period of proteolysis.

The age-related loss of proteolytic ability and differential distribution of induced abnormal proteins of the rabbit reticulocytes described above has an interesting parallel in the reticulocytes of humans afflicted with the disease β-thalassemia. In β-thalassemia, the imbalanced rates of α- and β-globin chain synthesis are not evident until the reticulocyte matures to become an erythrocyte,[70,71] at which point the excess α-globin chains precipitate to form high molecular weight aggregates termed "Heinz bodies".[72,73] Of particular interest is that early and intermediate stages of Heinz body formation have been observed in the relatively mature reticulocyte, but not in younger cells; larger condensed forms are not observed in any but the mature erythrocyte.[56] It is suggested that as the human reticulocyte matures, it has a decreasing ability to degrade the excess α-globin chains.[74]

With numerous reports in the literature of inactive enzymes accumulating in the cells of many aging organisms and the reports of decreased protein degradative activity in the same systems, the observations in the maturing rabbit reticulocyte may be relevant to cellular senescence as it occurs in vivo.

D. Erythrocytes

The use of the mammalian erythrocyte as a model for aging is a natural extension of the reticulocyte model. The circulating erythrocyte has a life span of approximately 120 days[62] and, like the reticulocyte, is easily fractionated into discrete cell-age groups by buoyant density gradient centrifugation.[75]

As the erythrocyte has no protein synthetic capacity and relatively little proteolytic activity, analyses with this model are essentially limited to the fate of "nonrenewable" enzyme proteins aging in vivo. Many soluble and membrane-associated enzyme activities decline and there are changes in the structural protein components of the erythrocyte membrane with increasing cell age. Seaman et al.[76] reported that although the activities of all 11 glycolytic enzymes assayed declined with erythrocyte cell age, the half-lives of the bulk of these glycolytic enzymes exceeded the 120-day life span of the erythrocyte in vivo; thus, the oldest erythrocytes had retained more than half of their original maximal enzyme activities. There were three exceptions: aldolase, hexokinase, and pyruvate kinase, which had estimated half-lives that were significantly shorter than that of the erythrocyte. It was suggested that the loss of these three glycolytic enzymes may account for a decline in the metabolic rate of the erythrocyte that ultimately threatens the structural integrity of older cells, leading to their death.[76] Declines in enzyme activity with erythrocyte age were also reported for galactokinase,[77] 6-phosphogluconate, glucose-6-phosphate dehydrogenases,[78] and nucleoside monophosphate kinases.[79] In red blood cell membranes, Gambert and Duthie[80] reported an age-related change in the activity of Na^+-K^+ ATPase. Butterfield et al.[81] performed spin label studies with human erythrocyte membranes and reported differences in the physical composition of aging cells that were not due to changes in lipid membrane fluidity, but altered physical states of the protein components due to altered protein conformation or organization. Bartosz et al.[82] reported an increased content of high molecular weight aggregates in red cell aging as a result of enhanced binding of nonhemoglobin proteins to the membranes of old cells.

Harman[83] first proposed the oxidative theory of aging, and the erythrocyte would seem to be a suitable model with which to examine such a possibility. Superoxide agents may be particularly relevant in the erythrocyte model as a result of the potentially high levels of activated oxygen species that could be generated by the auto-oxidation of oxyhemoglobin.[72] Erythrocytes lose superoxide dismutase activity as they mature;[84] loss of this function would impair the ability of the most mature erythrocyte to counter oxidative stress and could account for the decline of some of the glycolytic enzyme activities described above. Preliminary data with cell-free erythrocyte extracts and superoxide radicals[85] have shown that these agents could inactivate a number of different glycolytic enzymes in vitro but the individual rates of enzyme activity decay did not parallel those previously observed in vivo. Acid phosphatase, adenosine deaminase, and inosine triphosphatase have been shown to be inactivated in aging red cells by as yet uncharacterized reactions that were reversed by thiol reagents.[86] There are at least two age-related post-translational modifications of hemoglobin in aging red cells. Hemoglobin A_{1c} is generated as a result of the glycosylation of the N-terminus of β-globin chains,[87] especially in diabetic patients,[88] and hemoglobin A_{1b} by the deamination of several globin asparagine residues.[89] Many of the enzymes examined in old red cells display different electrophoretic patterns from those in young red cells. Triose phosphate isomerase is understood to be progressively inactivated by deamination in old red cells.[90] Aldolase and glucose-6-phosphate dehydrogenase are also modified to produce altered electrophoretic forms in old red cells by factors that remain obscure.[1]

Melloni et al.[91] have reported the presence of a number of membrane-bound and cytosolic proteinases and peptidases in human and rabbit erythrocytes, all of which (with the exception of dipeptidyl aminopeptidase III) declined exponentially throughout the life span of the aging erythrocyte.

E. Nematodes

The free-living nematode *Turbatrix aceti* is commonly used as a model for aging. Large quantities of synchronized organisms are easily produced.[92] They are bisexual and born with all their cells except the reproductive system. Nematodes have few cell types (mainly gut and muscle), with a very low proportion of somatic cells (thus, one finds only old cells in old organisms) and their life spans range from 15 to 17 days.[2]

Gershon[93] first described the merits of the free-living nematode *T. aceti* as a model for aging studies and went on to report a decline in the activities of α-amylase, malate dehydrogenase, and acetylcholine esterase in aging nematodes.[94] The list of declining enzyme activities has subsequently been increased to include isocitrate lyase, phosphoglycerate kinase, aldolase, and enolase.[2] The specific activity of enolase was reported to decrease approximately 50% with age and this decrease is observed in both crude tissue homogenates and purified enzyme preparations.[95] Enolase from old nematodes was more heat labile (exhibiting a biphasic heat inactivation pattern) and immunoprecipitation studies revealed that more antiserum was required to precipitate the old enzyme. Isozymes of nematode enolase were not detected in old animals, neither were there any age-related changes in the molecular weight, K_m, or charge of the enzyme.[96] Sharma et al.[97] reported that the accumulation of altered forms of enolase in old nematodes was associated with decreased rates in both the synthesis and degradation of the enzyme in old animals.

Aldolase was also reported to decline in specific activity in the aging nematode, with an accumulation of inactive aldolase molecules in old cells.[98] In crude homogenates of old nematodes, the heat stability of old aldolase was reported to be decreased; however, after purification, it was discovered that old aldolase was in fact more stable to heat denaturation.[99] The nature of the differences between young and old nematode aldolases is not clear at present.[2]

The specific activity of triosephosphate isomerase was thought to decline in old nematodes; however, upon purification of the enzyme it was found that there was less of this protein in old cells, and no altered forms were detected.[100]

The specific activity of isocitrate lyase also declines with age in nematodes, and immunoprecipitation experiments determined that altered forms of this enzyme were accumulating in older cells.[101] Old isocitrate lyase had a decreased (biphasic) heat lability and displayed an age-related change in its isozyme pattern.

Phosphoglycerate kinase is another enzyme that has a marked reduction in specific activity in old nematodes. However, the nature of this decline in specific activity remains unclear as phosphoglycerate kinase purified from young and old nematodes has identical molecular weights, K_m values, charge, and thermal stability.[102]

Thus, in the nematode model, there is convincing evidence that the decline in the specific activities of the many enzymes described above are generated by post-synthetic modifications of the enzymes as opposed to errors in their primary sequences. The accumulation of inactive enzymes in old nematodes is believed to be a function of a four- to fivefold decrease in protein degradative capacity observed in old nematodes.[103,104] The exact nature of these altered enzymes in old nematodes is not known. Of all the aging models in use today, the nematode would appear to be the most developed and well defined.

F. Insects

A variety of different insects are used as models for aging as they contain post-mitotic cells in their adult life. These organisms have relatively short life spans — typically 30 days for the house fly *Musca domestica*,[105] 100 days for the fruit fly *Drosophila melanogaster*,[106] and 40 days for the mosquito *Aedes aegypti*.[107] Insects are grown and age rapidly under controlled conditions,[108,109] and the wide variety of available mutants are further assets in genetic studies on the basis of aging.[110]

A noticeable manifestation of senescence in adult insects is the loss of their ability to fly,[111] a result of an age-related functional decrement in their flight muscles.[112] Insect flight muscle has been the source of much research concerning the fate of mitochondria in aging insects;[112] however, relatively little work has been performed on the status of protein turnover in this model. A number of enzyme activities are reported to decline with age.[112] In the mosquito,[107] there is approximately a 75% decline in the activities of three pentose phosphate pathway enzyme activities: glucose-6-phosphate, 6-phosphoglycerate, and isocitrate dehydrogenases. Work with the post-emergent maturation and aging of the house fly revealed a decline in the activities of arginine phosphokinase, α-glycerophosphate dehydrogenase,[113,114] Mg^{2+}-activated ATPase, and actomysin ATPase.[115,116] In the fruit fly, the levels of peroxidase and catalase activities were reported to decline 84 and 38%, respectively, with maturation.[117] In some insects, total superoxide dismutase activity is reported unchanged, but there is a 20% decrease in the activity of mitochondrial superoxide dismutase.[118]

These and other observations of enzyme activities that decline, increase, or remain unchanged in insect maturation and aging are indeed fragmentary.[112] The biochemical causes of the observed changes are unclear and complicated by the unique biology of insects.

G. Bacteria

The bacterium *Escherichia coli* has been especially useful in analyses of Orgel's "error catastrophe" hypothesis. Orgel[6] first proposed that in aging cells, the fidelity of translation in protein synthesis progressively deteriorated with age, and further, that there was an element of positive feedback in that errors in the protein components of the protein synthetic apparatus itself would increase the probability of subsequent mistakes in translation leading to the "error catastrophe" which leads to cellular senescence and death. Orgel[7] expanded his hypothesis and suggested that the error frequency need not increase indefinitely, but approach some elevated resting level. With the construction of mathematical models of cellular error protein propagation,[119-121] it is possible to compare predicted and experimentally observed error propagation in vivo. Work with the *E. coli* model has been directed toward experiments involving the deliberate in vivo modification of the accuracy of transcription and translation, with a view to measuring the stability of the translational apparatus.

Incubating *E. coli* in the presence of agents such as streptomycin or ethanol increases the frequency of translational errors.[119,120] The consequent rate of error protein synthesis is most commonly monitored by measuring the rate of cysteine incorporation into flagellin,[122] the mistranslation of nonsense codons in the genes for alkaline phosphatase,[123] or β-galactosidase.[121] Such experiments have established that the error frequency in *E. coli* increases rapidly and eventually stabilizes after about 5 generations to levels that are as much as 50 times the normal levels in the synthesis of error proteins. Under such conditions, these cells continue to divide slowly and apparently retain their viability. From such observations, one might conclude that the propagation of a large number of errors need not be lethal to the cell and that in the bacterium error propagation does not escalate to catastrophic levels. These observations diminish the likelihood that Orgel's hypothesis, suggesting that translational errors lead to death, is correct. To account for the elevated yet stable rate of error frequency, it has been suggested that there are biological controls that serve to limit error propagation. Such homeostatic controls may include the rapid and selective degradation of abnormal proteins;[65] experiments with bacterial strains lacking in certain cellular proteinases may be particularly illuminating.[124-126] In addition, a method of ribosomal editing has been proposed which involves the dissociation of inappropriate peptidyl-tRNAs if their structures do not correctly compliment the codon of the mRNA.[127]

IV. POST-TRANSLATIONAL MODIFICATIONS OF PROTEINS

A common thread in research into the fate of proteins in old cells is the search for altered proteins. This line of research was initiated by Orgel's "error catastrophe" hypothesis[6] and stimulated by the early observations of the decline of a number of enzyme activities in the aging nematode model.[94] Although the "error catastrophe" hypothesis is questionable, the search for altered enzyme proteins has continued and been expanded with the underlying belief that the accumulation of altered enzyme proteins is responsible (at least in part) for cellular senescence. If the synthesis of error-containing proteins is not accelerated with age, what is the source of the observed altered enzyme proteins and why do they accumulate in old cells? The most attractive hypothesis is that cellular enzymes are subject to post-translational modifications that render them inactive and that a concomitant decline in the proteolytic capacity of senescent cells allows these altered enzyme forms to accumulate. The best support for this hypothesis comes from the aging nematode model, in which there is both an age-related accumulation of some enzymes and a decline in proteolytic activity. The maturing reticulocyte and erythrocyte systems also lend weight to this argument.

A variety of post-translational modifications may be responsible for the inactivation of cellular enzymes associated with old tissues and cells. These modifications may include deamination, glycosylation, partial proteolysis, and oxidation. Although evidence that any one of these post-translational modifications participate in the biochemistry of protein aging is limited, their relative contributions will be examined.

A. Oxidative Reactions

The oxidative effects of superoxide radicals are thought to contribute to the process of enzyme inactivation associated with cellular senescence. The widespread observations of age-related accumulations of aging pigments (the result of lipid peroxidation) suggests that this is an ongoing function of senescent cells. There are a variety of sources of oxidative potential in a cell and some of these may be derived from the diet.[169] Free radical forms include superoxide $[O_2^{\cdot}]$, peroxide $[ROO^{\cdot}]$, singlet oxygen $[^1O_2]$, and the hydroxyl radical $[HO^{\cdot}]$. There are many ways in which superoxide radicals may affect proteins;[128] these include addition reactions resulting in covalent bond formation, and cross-linking reactions. These latter reactions may be catalyzed directly by the free radicals themselves or by agents such as malondialdehyde, which can form Schiff bases with protein amine residues. Such reactions are interesting possibilities in the light of observations of the increased incidence of connective tissue cross-linking in senescent mammals.[129,130] Cellular defenses against these oxidizing radicals include superoxide dismutase, peroxidases, catalase, vitamins C and E, quinones, and glutathione.[128] The contribution of these defenses to protein aging hinges on whether these protective mechanisms deteriorate with age. There is no clear evidence that the protective mechanisms decline with age; it is also unclear whether the administration of antioxidants generally increases the life span of organisms.[2]

The increased oxidizing potential of an aging cell may well induce the formation of intramolecular protein disulfides, altering the conformation of structural proteins and rendering enzymes inactive. There are few reported incidents of disulfide bond formation in aging proteins: one of them is the formation of disulfide bonds in protein high molecular weight aggregates of aging lens tissue;[131] another is the increased incidence of disulfide bond formation in chromatin in older mice and rats.[132] By contrast, there is evidence that some altered enzymes do not have oxidized sulfhydryl groups. For example, enolase in nematodes has four cysteine residues and although the old enzyme has lowered specific activity, all four of its cysteine residues are in the reduced forms.[2] However, in the purification of some enzymes from crude extracts, oxidized protein sulfhydryls may be rapidly reduced by cellular metabolites (e.g., glutathione). When rabbit muscle aldolase is inactivated in vitro by oxidizing agents such as oxidized glutathione, cysteine, cystamine, or sodium periodate, the enzyme is rapidly inactivated, and rendered thermodynamically unstable and highly susceptible to complete proteolysis in vitro.[133] The oxidized forms of other enzymes may be highly susceptible to rapid proteolysis and thus may not accumulate in old cells.

Oxidation of amino acid residues other than sulfhydryl groups (e.g., histidine, tyrosine) has been proposed as the initial event in the degradation of bacterial enzymes such as glutamine synthetase[134] and glutamine phosphoribosylpyrophosphate amidotransferase.[135] Again, these oxidative reactions result in unstable forms of the enzymes which are rapidly degraded by proteases and would not likely accumulate as altered enzymes in cells.

B. Amino Acid Racemization

Reports in the literature concerning the racemization of amino acids in proteins essentially consider such events as a dating technique rather than a phenomenon that participates in the aging process. Masters et al.[136] reported that amino acid racemization was not detectable in hemoglobin and pointed out that racemization may well be limited to metabolically inert proteins such as proteolipid proteins in the myelin sheath, certain collagen components, and β-crystallins in lens tissue. The racemization of aspartic acid has been examined in tooth proteins and found to occur in enamel and the acidic phosphoprotein fraction of dentin.[137] Racemization of aspartic acid in β-crystallin has also been reported in the mammalian lens nucleus.[136] The rates of racemization are very slow, approximately 1% every 10 years in calcified tissue[138] and approximately 1.5% every 10 years in lens β-crystallin.[136] The racemization process has not been examined in soft tissue proteins. Although amino acid racemization is likely to perturb protein conformation and possibly inactivate an enzyme and some mechanisms have been proposed that could accelerate this process in vivo,[136,139] it is doubtful that racemization contributes to the aging process.

C. Glycosylation

The glycosylation of proteins in aging tissues and cells involves the nonenzymic condensation of glucose to the N-terminus of the polypeptide chain by a keto amine linkage. In old erythrocytes, hemoglobin is found in the glycosylated form; glucose is condensed to the N-terminus of β-globin and to some extent the N-terminus of α-globin and ε-lysine amino groups at other sites.[140] This reaction results in a change of charge for the protein molecule. Glycosylation has also been noted at lysine residues of aging bovine and rat lens crystallins[141,142] as well as collagen.[143] Glycosylation may be more relevant as an effector of protein aging in diabetic patients.[144] The observed increased incidence of glycosylation of old proteins is limited to proteins with long half-lives in vivo, but its relevance to other cellular proteins and the aging process is unknown.

D. Deamination

Both asparagine and glutamine are susceptible to deamination in vitro and in vivo and there are numerous reports in the literature concerning the accumulation of deaminated inactive enzymes or deaminated isoenzymes.[90] Deamination of one amino acid residue in cytochrome c inactivates the enzyme 40%, deamination of a second residue inactivates the enzyme by 80%.[145] The deamination of an asparagine residue, four amino acids from the C-terminus of rabbit muscle aldolase, has also been reported; however, there was no change in the activity of this enzyme.[146] The half-time for the deamination reaction was estimated to be approximately 8 days, and closely matched the in vivo half-life of the enzyme estimated at that time.[147] It was suggested that this deamination reaction may represent an initial event in the proteolysis of this enzyme, but more recent estimates of the half-life of aldolase are considerably shorter (1 to 2 days), making this proposal unlikely.[148,149] Triosephosphate isomerase is another enzyme that has been found to be deaminated in old cells. Analyses of the enzyme isolated from both old red cells[90] and old lens tissues[150] has revealed that progressive deamination results in the accumulation of acidic, heat-labile isoenzymes. Indeed, a number of enzymes isolated from old cells display similar changes in charge, believed to be the result of progressive deamination. Isocitrate lyase in nematodes undergoes a change in isoenzyme pattern with a switch from isozymes I and III, which predominate in young nematodes, to isoenzymes I and II — the latter isozyme being more acidic and heat labile.[101] The relative merits of deamination reactions as a possible biological timer that determines the rate of turnover for different enzymes have been discussed.[151] Gracy[90] describes the widespread nature of these reactions in a number of different aging models. Although a ubiquitous process, affecting many proteins, its rate is difficult to assess as an equal number of inactive enzymes in old cells have no apparent changes in their electrophoretic mobility.

E. Partial Proteolysis

The case for the role of partial proteolysis in the accumulation of altered enzymes in aging tissues is limited as the bulk of inactive enzymes purified from homogenates of old tissues retain their original molecular weights. When analyzing an enzyme purified from crude tissue homogenates one must consider that the protein may be vulnerable to proteolysis at any stage in the purification process and may have come in contact with proteinases following the disruption of the cell. In the light of reports concerning the increased fragility of lysosomes in older tissues,[2] enzymes in old cells may be at greater risk of artifactual partial proteolysis.

The only consistent report of an age-related partial proteolysis of proteins is that of α-crystallin in lens tissue. During the maturation of lens tissue there occurs a progressive shortening of the α-crystallin polypeptide from the C-terminal end.[152] These shortened peptides are also found in young peripheral lens tissues but predominate in the high molecular weight material found in the central region of the lens. That these peptides are cleaved at well-defined points suggests that the hydrolysis of the peptide bonds may be the result of nonenzymic chemical cleavage or the action of a very specific proteinase(s).

Rabbit muscle aldolase has been the subject of a number of studies and initially it was thought to be an example of an enzyme altered by partial proteolysis in aging cells. The aldolase molecule is very susceptible to partial proteolysis in the C-terminus (up to 20 amino acid residues may be removed) and this alteration inactivates the enzyme toward the substrate fructose-1,6-bisphosphate.[133] The bulk of the enzyme is resistant to proteolysis. Homogenized tissues from senescent animals appear to accumulate inactive forms of aldolase;[153,154] however, if inhibitors of proteolytic enzymes are added to homogenates, the increase in the "altered" forms of aldolase in aging cells was not observed.[155] Thus, the "defective" molecules appeared to be produced during the homogenization and storage of this tissue rather than in vivo.

V. PROTEIN TURNOVER

Cellular proteins are continually being synthesized and degraded at rates characteristic for each protein. In mammalian tissues proteins are in a very dynamic state; the half-life for an average rat liver protein is 3 days, for a soluble heart protein 4 days, and for a soluble skeletal muscle protein 10 days.[12] The initial events and the proteolytic systems responsible for the extensive degradation of intracellular proteins are unknown. Lysosomal and extra-lysosomal proteinases appear to be involved in the degradation of intracellular proteins as well as the degradation of polypeptides taken into the cell by endocytosis.[65,156-159] The discovery of the ATP-stimulated proteinases[65,160] has evoked much excitement; however, the role of these proteinases in degrading the bulk of cellular proteins is unknown and there is much controversy about the quantitative importance of these proteinases compared to lysosomal and other ATP-independent proteinases, such as Ca^{2+}-dependent muscle proteinases, tyrase (a serine proteinase in liver), and metalloproteinases in kidney.[161-163] Discovering what proteinases exist in cells is essential to gaining insights into how the process of degradation works and is regulated. However, as things stand, none of the proteolytic systems or individual proteinases known can in themselves explain the heterogeneity of protein half-lives or the regulation of accelerated or suppressed protein catabolism observed in tissues under various metabolic conditions. Microautophagy and macroautophagy (pinching off of cellular contents and digestion by lysosomal enzymes) cannot explain why some proteins are degraded faster than others unless structural determinants of proteins direct proteins to autophagic vesicles. Likewise, the ATP-stimulated proteinases themselves do not explain heterogeneity in protein half-lives. These proteinases appear to preferentially degrade "abnormal" or denatured proteins, once again suggesting that structural properties of proteins are determinative in protein catabolism. The evidence indicating that the physicochemical properties of proteins are important in determining rates of degradation is strong.[164-167] Based on previous work, one hypothesis is that factors or ligands that alter the conformation of a protein are important elements in regulating the degradative process in vivo. Thus, the elucidation of factors that alter protein conformation is critical to our understanding of the degradative process; it may also be critical to our understanding of the accumulation of altered enzymes in the aging process. The post-translational modifications described above may be features of the initial events in the degradation of cellular proteins; thus, if the aging of proteins in old cells is a corollary of the initial events in the degradation of cellular proteins, why are inactive enzymes accumulating in old cells? With the exception of the nematode and reticulocyte, there is no concrete evidence that protein degradative capacity is significantly diminished in old tissues and cells.[2] Normal cells also have the ability to rapidly and selectively degrade abnormal proteins, a system that is possibly discrete from the pathway of normal protein degradation.[65]

It is conceivable that a typical cellular enzyme is subject to a number of modification reactions before it is degraded in the cytoplasm or sequestered into the lysosome. If any one of these modification reactions is particularly prevalent in old cells, an enzyme may become inactive but not sufficiently modified for the enzyme to be recognized as a substrate for protein degradation. If protein degradation is a first-order reaction, these partially modified enzymes might have to await further modifications before they can be degraded. This effect would be accentuated in enzymes with relatively long half-lives. Altered enzymes may thus accumulate in old tissues and cells that lack the ability to compensate for a decline in the specific activity of an intact enzyme protein pool. Altered enzymes in old cells are not apparently recognized as abnormal proteins and removed by the appropriate mechanisms.

An enzyme pool with a lowered specific activity may suffice for the metabolism of the cell under resting conditions but may be insufficient for the needs of a cell during periods of stress. Such events may prove to be deleterious to a cell that cannot divide to dilute these altered enzymes and lead to senescence as a result of impaired function. That so many enzymes are not altered with age suggests that diminishing the specific activities of the pools of a few critical enzymes could have far-reaching effects. Studies with the formation and accumulation of altered enzymes in old tissues and cells have lead to few generalizations to date. Examining detailed accounts of altered enzymes in many different aging tissues and cells, as reviewed by Wilson,[168] Rothstein,[2] and Dreyfus et al.,[1] one notes that the activities of relatively few enzymes decline with age and many are unaffected or increase in specific activity with age. Of those enzymes that do decline with age, many only decline approximately 25%; thus, their significance is questionable. Comparing the accumulation of altered enzymes in different tissues also raises questions as to whether there may be differences in the types of altered enzymes seen to accumulate in red cells which have little or no protein synthetic or degradative capacity when compared to the same enzymes in other aging models which are active in both protein synthesis and degradation. In analysis, the molecular biology of proteins in aging is still descriptive, the data is fragmentary, and there is much to gain in insight in the future in this subfield of gerontology.

Rationale for Getting a Haircut (and taking a bath)

We say we are alive, yet we are always dying on the outside.

Strands of hair, nails, external scurf, all the dead parts, constantly increase.

The living parts insist on making more of them, trying to tell us something.

Unchecked, the zone between life and death could slip dangerously deep, disturbing equilibrium.

Negative feedback would become a possibility.

Production of new tissue might stop at the center, precluding turnover, leaving us mired in necrosis.

If we rid ourselves of these vestures that capture us like snakeskins, we will continue to keep death in its proper perspective.

We will survive centrifugally.

William H. Blackwell[170]

VI. SUMMARY

It is recognized that there are a number of age-related changes in the status of proteins in aging tissues. There are changes in the turnover of body proteins and the relative contributions of different tissues to this process. There are also subtle age-related changes in the nature of some constitutive body proteins in humans and other mammals. How diet contributes to these processes is not known, although it appears that controlled dietary regimes over long periods of time may affect the aging process.

Investigations at the molecular level in this field have concentrated on model systems for aging and have generated information that is largely descriptive. The accumulation of altered enzymes is a common feature of all present model systems for aging as evidenced with cells in culture, erythroid cells, eye lens, nematodes, and insects. The general consensus is that altered enzyme molecules accumulate in old cells as a result of post-synthetic modifications of their structures which inactivate them. The factors responsible for these events were briefly assessed and included oxidation, amino acid racemization, glycosylation, deamination, and partial proteolysis. This accumulation of altered enzyme molecules reported in old cells and tissues may be the result of a decline in the ability of the old cells to degrade cellular proteins. This decline may be due to the loss of the proteinases of the cells themselves or a degeneration in the mechanisms that initiate intracellular proteolysis.

Further elucidation of the fundamentals of cellular protein turnover will facilitate our understanding of the changes that occur in senescent cells.

REFERENCES

1. **Dreyfus, J., Kahn, A., and Schapira, F.,** Post-translational modifications of enzymes, *Curr. Topics Cell. Regul.*, 14, 243, 1978.
2. **Rothstein, M., Ed.,** *Biochemical Approaches to Aging,* Academic Press, New York, 1982.
3. **Young, V. R., Gersovitz, M., and Munro, H. N.,** Human aging: protein and amino acid metabolism and implications for protein and amino acid requirements, in *Nutritional Approaches to Aging Research,* Moment, G. B., Ed., CRC Press, Boca Raton, Fla., 1982, 47.
4. **Kao, K.-Y. T. and Lakshmanan, F. L.,** Protein nutrition and aging, *Handbook of Geriatric Nutrition,* Hsu, J. and Davis, R., Eds., Noyes Publications, Park Ridge, N.J., 1981, 56.

5. **Tollefsbol, T. O. and Gracy, R. W.,** Premature aging disease: cellular and molecular changes, *BioScience,* 33, 634, 1983.
6. **Orgel, L. E.,** The maintenance of the accuracy of protein synthesis and its relevance to aging, *Proc. Natl. Acad. Sci. U.S.A.,* 49, 517, 1963.
7. **Orgel, L. E.,** Ageing of clones of mammalian cells, *Nature (London),* 243, 441, 1973.
8. **Winterer, J. C., Steffee, W. P., Davy, W., Perera, A., Uauy, R., Scrimshaw, N. S., and Young, V. R.,** Whole body protein turnover in aging man, *Exp. Gerontol.,* 11, 79, 1976.
9. **Uauy, R., Winterer, J. C., Bilmazes, C., Haverberg, L. N., Scrimshaw, N. S., Munro, H. N., and Young, V. R.,** The changing pattern of whole body protein metabolism in aging humans, *J. Gerontol.,* 33, 663, 1978.
10. **Chinn, K. S. K.,** Prediction of muscle and remaining tissue protein in man, *J. Appl. Physiol.,* 23, 713, 1967.
11. **Lundholm, K. and Schersten, R.,** Leucine incorporation into proteins and cathepsin D activity in human skeletal muscles. The influence of the age of the subject, *Exp. Gerontol.,* 10, 155, 1975.
12. **Waterlow, J. C., Garlick, P. J., and Millward, D. J.,** *Protein Turnover in Mammalian Tissues and in the Whole Body,* North-Holland, New York, 1978.
13. **Rao, J. S., Rao, V. H., and Bose, S. M.,** Effect of protein malnutrition on the cross-linking of dermal collagen in aging, *Exp. Gerontol.,* 17, 227, 1982.
14. **Enesco, H. E. and Kruk, P.,** Dietary restriction reduces fluorescent age pigment accumulation in mice, *Exp. Gerontol.,* 16, 357, 1981.
15. **Barrows, C. H. and Kokkonen, G. C.,** *Advances in Nutrition Research,* Vol. 1, Draper, H. H., Ed., Plenum Press, New York, 1977.
16. **Goodrick, C. L.,** Body weight increment and length of life: the effect of genetic constitution and dietary protein, *J. Gerontol.,* 33, 184, 1978.
17. **DeFronzo, R. A.,** Glucose intolerance and aging. Evidence for tissue insensitivity to insulin, *Diabetes,* 28, 1095, 1979.
18. Food and Nutrition Board, *Recommended Dietary Allowances,* 4th rev. ed., National Research Council, National Academy of Sciences, Washington, D.C., 1980.
19. FAO/WHO, Joint FAO/WHO ad hoc Expert Committee on Energy and Protein Requirements, *WHO Tech. Rep. Ser.,* 552, 1973.
20. **Gersovitz, M., Motil, K., Munro, H. N., Scrimshaw, N. S., and Young, V. R.,** Human protein requirements: assessments of the adequacy of the current recommended dietary allowance for dietary protein in elderly men and women, *Am. J. Clin. Nutr.,* 35, 6, 1982.
21. **O'Hanlon, P. and Kohrs, M. B.,** Dietary studies of older Americans, *Am. J. Clin. Nutr.,* 31, 1257, 1978.
22. U.S. Department of Health, Education and Welfare, Ten-State Nutrition Survey, 1968—70: V. Dietary, Center for Disease Control, DHEW Publ. No. (HSM) 72-8133, DHEW, Washington, D.C., 1972.
23. **Kohrs, M. B., O'Neal, R., Preston, A., Eklund, D., and Abrahams, O.,** Nutritional status of elderly residents in Missouri, *Am. J. Clin. Nutr.,* 32, 2186, 1978.
24. **Garry, P. J., Goodwin, J. S., Hunt, W. C., Hooper, E. M., and Leonard, A. G.,** Nutritional status of a healthy elderly population: dietary and supplemental intakes, *Am. J. Clin. Nutr.,* 36, 319, 1982.
25. **Yearick, E. S., Wang, M. S., and Pisias, S. J.,** Nutritional status of the elderly: dietary and biochemical findings, *J. Gerontol.,* 35, 663, 1980.
26. **Lowenstein, F. W.,** Nutritional status of the elderly in the United States of America, 1971—1974, *J. Am. College Nutr.,* 1, 165, 1982.
27. **Bowman, B. B. and Rosenberg, I. H.,** Assessment of the nutritional status of the elderly, *Am. J. Clin. Nutr.,* 35, 1142, 1982.
28. **Jansen, C. and Harrill, I.,** Intakes and serum levels of protein and iron from 70 elderly women, *Am. J. Clin. Nutr.,* 30, 1414, 1977.
29. **Hayflick, L.,** Recent advances in the cell biology of aging, *Mech. Ageing Dev.,* 14, 59, 1980.
30. **Hayflick, L. and Moorhead, P. S.,** The serial cultivation of human diploid cell strains, *Exp. Cell Res.,* 25, 585, 1961.
31. **Hayflick, L.,** The limited *in vitro* lifetime of human diploid cell strains, *Exp. Cell Res.,* 37, 614, 1965.
32. **Martin, G. M., Sprague, C. A., and Epstein, C. J.,** Replicative life span of cultivated human cells; effects of donor's age, tissue and genotype, *Lab. Invest.,* 23, 86, 1970.
33. **Goldstein, S., Moerman, E. J., Soeldner, J. S., Gleason, R. E., and Barnett, D. M.,** Chronologic and physiologic age affect replicative life span of fibroblasts from diabetic, prediabetic and normal donors, *Science,* 199, 781, 1978.
34. **Matsumura, T., Pfendt, E. A., and Hayflick, L.,** DNA synthesis in the human diploid cell strain WI-38 during *in vitro* aging: an autoradiography study, *J. Gerontol.,* 34, 323, 1979.
35. **Robbins, E., Levine, E. M., and Eagle, M.,** Morphologic changes accompanying senescence of cultured human diploid cells, *J. Exp. Med.,* 131, 1211, 1970.

36. **Hayflick, L.,** in *Biology of Aging,* Finch, C. E. and Hayflick, L., Eds., Van Nostrand Reinhold, Princeton, N.J., 1977, 159.

37. **Stanley, J. F., Pye, D., and McGregor, A.,** Comparison of doubling numbers attained by cultured animal cells with life span of species, *Nature (London),* 255, 158, 1975.

38. **Holliday, R. and Tarrant. G. M.,** Altered enzymes in ageing human fibroblasts, *Nature (London),* 238, 26, 1972.

39. **Kahn, A., Guillouzo, A., Leibovitch, M.-P., Cottreau, D., Bourel, M., and Dreyfus, J.-C.,** Heat lability of glucose-6-phosphate dehydrogenase in some senescent human cultured cells. Evidence for its postsynthetic nature. *Biochem. Biophys. Res. Commun.,* 77, 760, 1977.

40. **Alekseeva, O. M. and Ritov, V. B.,** Two forms of Ca^{2+}-dependent ATPase of the sarcoplasmic reticulum, *Biokhimiya,* 44, 1245, 1979.

41. **Buchanan, J. H., Bonn, C. L., Lappin, R. I., and Stevens, A.,** Accuracy of *in vitro* protein synthesis: translation of polyuridylic acid by cell-free extracts of human fibroblasts, *Mech. Ageing Dev.,* 12, 339, 1980.

42. **Wojtyk, R. I. and Goldstein, S.,** Fidelity of protein synthesis does not decline during aging of cultured human fibroblasts, *J. Cell. Physiol.,* 103, 299, 1980.

43. **Houben, A., Houbion, A., and Remacle, J.,** Lysosomal and mitochondrial heat labile enzymes in Werner's syndrome fibroblasts, *Exp. Gerontol.,* 15, 629, 1980.

44. **Tollefsbol, T. O., Zaun, M. R., and Gracy, R. W.,** Increased lability of triosephosphate isomerase in progeria and Werner's syndrome fibroblasts, *Mech. Ageing Dev.,* 20, 93, 1982.

45. **Bosmann, H. B., Gutheil, R. L., and Case, K. R.,** Loss of a critical neutral protease in aging WI-38 cells, *Nature (London),* 261, 499, 1976.

46. **Leipetz, J. and Cristofalo, V. J.,** Ultrastructural changes accompanying the aging of human diploid cells in culture, *J. Ultrastruct. Res.,* 39, 43, 1972.

47. **LeGall, J. Y., Khoi, T. D., Glaise, D., LeTreut, A., Brissot, P., and Guillouzo, A.,** Lysosomal enzyme activities during ageing of adult human liver cell lines, *Mech. Ageing Dev.,* 11, 287, 1979.

48. **Bradley, M. O., Hayflick, L., and Schimke, R. T.,** Protein degradation in human fibroblasts (WI-38); effects of aging, viral transformation, and amino acid analogs, *J. Biol. Chem.,* 251, 3521, 1976.

49. **Goldstein, S., Stotland, D., and Cordeiro, R. A. J.,** Decreased proteolysis and increased amino acid efflux in ageing human fibroblasts, *Mech. Ageing Dev.,* 5, 221, 1976.

50. **Bloemendal, H.,** The vertebrate eye lens, *Science,* 197, 127, 1977.

51. **Bloemendal, H.,** Biosynthesis of eye lens protein, *Biochemistry of Cell Differentiation II,* Vol. 15, Paul, J., Ed., University Park Press, Baltimore, 1977.

52. **Ohrloff, C. and Hockwin, O.,** Lens metabolism and aging: enzyme activities and enzyme alterations in lenses of different species during the process of aging, *J. Gerontol.,* 38, 271, 1983.

53. **Hockwin, O., Fink, H., and Glasmacher, M.,** Carbohydrate metabolism of the lens depending on age. III. Factor analysis on changes of different enzyme activities and on changes in the concentration of glycolytic metabolites in bovine lens, *Z. Alternsforsch.,* 31, 521, 1976.

54. **Zigman, S., Groff, J., Yuzo, R., and Greiss, G.,** Light extinction and protein in lens, *Exp. Eye Res.,* 23, 555, 1976.

55. **Hoenders, H. J. and Bloemendal, H.,** Lens protein and aging, *J. Gerontol.,* 38, 278, 1983.

56. **Rifkind, R. A. and Danon, D.,** Heinz body anemia — an ultrastructural study. I. Heinz body formation, *Blood,* 25, 885, 1965.

57. **Siezen, R. J., Bindels, J. G., and Hoenders, H. J.,** The interrelationship between monomeric, oligomeric and polymeric α-crystallin in the calf lens nucleus, *Exp. Eye Res.,* 28, 551, 1979.

58. **Jedziniak, J. A., Nicoli, D. F., Baram, M., and Benedek, G. B.,** Quantitative verification of the existence of high molecular weight protein aggregates in the intact normal human lens by light-scattering spectroscopy, *Invest. Ophthalmol.,* 17, 51, 1978.

59. **Spector, A., Roy, D., and Stauffer, J.,** Isolation and characterization of an age-dependent polypeptide from human lens with non-tryptophan fluorescence, *Exp. Eye Res.,* 21, 9, 1975.

60. **Denton, M. J. and Arnstein, H. R. V.,** Characterization of developing adult mammalian erythroid cells separated by velocity sedimentation, *Br. J. Haematol.,* 24, 7, 1973.

61. **Glowacki, E. and Millette, R.,** Polyribosomes and the loss of hemoglobin synthesis in the maturing reticulocyte, *J. Mol. Biol.,* 11, 116, 1965.

62. **Bessis, M.,** *Living Blood Cells and Their Ultrastructure,* Springer-Verlag, New York, 1973.

63. **Woodward, W. R., Adamson, S. D., McQueen, H. M., Larson, J. W., Estvanik, S. M., Wilairat, P., and Herbert, E.,** Globin synthesis on reticulocyte membrane-bound ribosomes, *J. Biol. Chem.,* 248, 1556, 1973.

64. **McKay, M. J., Daniels, R. S., and Hipkiss, A. R.,** Breakdown of aberrant proteins in rabbit reticulocytes decreases with cell age, *Biochem. J.,* 188, 279, 1980.

65. **Hershko, A. and Ciechanover, A.,** Mechanisms of intracellular protein breakdown, *Ann. Rev. Biochem.,* 51, 335, 1982.

66. **Rubinstein, D., Ottolenghi, P., and Denstedt, O. F.,** The metabolism of the erythrocyte. XIII. Enzyme activity in the reticulocyte, *Can. J. Biochem. Physiol.,* 34, 222, 1956.

67. **Chapman, R. G. and Schaumburg, L.,** Glycolysis and glycolytic enzyme activity of ageing red cells in man, *Br. J. Haematol.,* 13, 665, 1967.

68. **Daniels, R. S., Worthington, V. C., Atkinson, E. M., and Hipkiss, A. R.,** Proteolysis of puromycin-peptides in rabbit reticulocytes: detection of a high molecular weight oligopeptide proteolytic substrate, *FEBS Lett.,* 113, 245, 1980.

69. **Hipkiss, A. R., McKay, M. J., Daniels, R. S., Worthington, V. C., and Atkinson, E. M.,** Selective proteolysis of abnormal proteins of shortened chain length in rabbit reticulocytes, *Acta Biol. Med. Germ.,* 40, 1265, 1981.

70. **Clegg, J. B. and Weatherall, D. J.,** Haemoglobin synthesis during erythroid maturation in β-thalassemia, *Nature (London) New Biol.,* 240, 190, 1972.

71. **Wood, W. G. and Stamatogannopoulos, G.,** Globin synthesis in fractionated normoblasts of β-thalassemia heterozygotes, *J. Clin. Invest.,* 55, 567, 1975.

72. **Carrell, R. W., Winterbourn, C. C., and Rachmilewitz, E. A.,** Activated oxygen and hemolysis, *Br. J. Haematol.,* 30, 259, 1975.

73. **Winterbourn, C. C.,** Protection by ascorbate against acetylphenyl-hydrazine-induced Heinz body formation in glucose-6-phosphate dehydrogenase deficient erythrocytes, *Br. J. Haematol.,* 41, 245, 1979.

74. **Hanash, S. M. and Rucknagel, D. L.,** Proteolytic activity in erythrocyte precursors, *Proc. Natl. Acad. Sci. U.S.A.,* 75, 3427, 1978.

75. **Piomelli, S., Lurinsky, G., and Wasserman, L. R.,** The mechanism of red cell aging. 1. Relationship between cell age and specific gravity evaluated by ultracentrifugation in a discontinuous density gradient, *J. Lab. Clin. Med.,* 69, 659, 1967.

76. **Seaman, C., Wyss, S., and Piomelli, S.,** The decline in energetic metabolism with aging of the erythrocyte and its relationship to cell death, *Am. J. Hematol.,* 8, 31, 1980.

77. **Magnani, M., Cucchiarini, L., Dacha, J., and Fornaini, G.,** Galactokinase activity and red blood cell age, *Age,* 3, 39, 1980.

78. **Kahn, A., Boyer, C., Cottreau, D., Marie, J., and Boivin, P.,** Immunologic study of the age-related loss of activity of six enzymes in the red cells from newborn infants and adults — evidence for a fetal type of erythrocyte phosphofructokinase, *Pediatr. Res.,* 11, 271, 1977.

79. **Jamil, T. P., Swallow, D. M., and Povey, S.,** A comparative study of the age-related patterns of decay of some nucleoside monophosphate kinases in human red cells, *Biochem. Genet.,* 16, 1219, 1978.

80. **Gambert, S. R. and Duthie, E. M.,** Effect of age on red cell membrane sodium-potassium dependent adenosine triphosphatase activity in healthy men, *J. Gerontol.,* 38, 23, 1983.

81. **Butterfield, D. A., Ordaz, F. E., and Markesbery, W. R.,** Spin label studies of human erythrocyte membranes in ageing, *J. Gerontol.,* 37, 535, 1982.

82. **Bartosz, G., Soszynski, M., and Wasilewski, A.,** Increased content of high molecular weight protein aggregates with cell aging, *Mech. Ageing Dev.,* 19, 45, 1982.

83. **Harman, D.,** Aging: a theory based on free radical and radiation chemistry, *J. Gerontol.,* 11, 298, 1956.

84. **Bartosz, G., Tannert, C., Fried, R., and Leyko, W.,** Superoxide dismutase activity decreases during erythrocyte aging, *Experientia,* 34, 1464, 1978.

85. **Bartosz, G.,** Aging of the erythrocyte. VII. On the possible causes of inactivation of red cell enzymes, *Mech. Ageing Dev.,* 13, 379, 1980.

86. **Hopkinson, D. A.,** *Isoenzymes,* Vol. 1, Market, C. L., Ed., Academic Press, New York, 1975.

87. **Bookchin, R. M. and Gallop, P. M.,** Structure of hemoglobin A_{Ic}: nature of the N-terminal β-globin blocking group, *Biochem. Biophys. Res. Commun.,* 32, 86, 1968.

88. **Gabbay, K. H.,** Glycosylated hemoglobin and diabetic control, *N. Engl. J. Med.,* 295, 443, 1976.

89. **Krishnamoorthy, R., Gacon, G., and Labie, D.,** Isolation and partial characterization of hemoglobin A_{Ib}, *FEBS Lett.,* 77, 99, 1977.

90. **Gracy, R. W.,** Epigenetic formation of isozymes: the effect of aging, *Isozymes: Current Topics in Biological and Medical Research,* Vol. 7, Alan R. Liss, New York, 1983, 187.

91. **Melloni, E., Salamino, F., Sparatore, B., Michetti, M., Morelli, A., Benatti, U., DeFlora, A., and Pontremoli, S.,** Decay of proteinase and peptidase activities of human and rabbit erythrocytes during cellular aging, *Biochim. Biophys. Acta,* 675, 110, 1981.

92. **Heib, W. F. and Rothstein, M.,** Aging in the free-living nematode *Turbatrix aceti.* Techniques for synchronization and aging of large-scale axenic cultures, *Exp. Gerontol.,* 10, 145, 1975.

93. **Gershon, D.,** Studies on aging in nematodes. I. The nematode as a model organism for aging research, *Exp. Gerontol.,* 5, 7, 1970.

94. **Erlanger, M. and Gershon, D.,** Studies on aging in nematodes. II. Studies of the activities of several enzymes as a function of age, *Exp. Gerontol.,* 5, 13, 1970.

95. **Sharma, H. K., Gupta, S. K., and Rothstein, M.,** Age-related alteration of enolase in the free-living nematode, *Turbatrix aceti, Arch. Biochem. Biophys.,* 174, 324, 1976.

96. **Sharma, H. K. and Rothstein, M.,** Altered enolase in aged *Turbatrix aceti* results from conformational changes in the enzyme, *Proc. Natl. Acad. Sci. U.S.A.,* 77, 5865, 1980.

97. **Sharma, H. K., Prasanna, H. R., Lanes, R. S., and Rothstein, M.,** The effect of age on enolase turnover in the free-living nematode, *Turbatrix aceti, Arch. Biochem. Biophys.,* 194, 275, 1979.

98. **Zeelon, P. E., Gershon, H., and Gershon, D.,** Inactive enzyme molecules in aging organisms. Nematode fructose-1,6-diphosphate aldolase, *Biochemistry,* 12, 1743, 1973.

99. **Reznick, A. Z. and Gershon, D.,** Age-related alterations in purified fructose-1,6-diphosphate aldolase from the nematode *Turbatrix aceti, Mech. Ageing Dev.,* 6, 345, 1977.

100. **Gupta, S. K. and Rothstein, M.,** Triosephosphate isomerase from young and old *Turbatrix aceti, Arch. Biochem. Biophys.,* 174, 333, 1976.

101. **Reiss, U. and Rothstein, M.,** Age-related changes in isocitrate lyase from the free-living nematode *Turbatrix aceti, J. Biol. Chem.,* 250, 826, 1975.

102. **Gupta, S. K. and Rothstein, M.,** Phosphoglycerate kinase from young and old *Turbatrix aceti, Biochim. Biophys. Acta,* 445, 632, 1976.

103. **Reznick, A. Z. and Gershon, D.,** The effect of age on the protein degradation system in the nematode *Turbatrix aceti, Mech. Ageing Dev.,* 11, 403, 1979.

104. **Prasanna, H. R. and Lane, R. S.,** Protein degradation in aged nematodes *(Turbatrix aceti), Biochem. Biophys. Res. Commun.,* 86, 552, 1979.

105. **Sohal, R. S.,** Aging changes in insect flight muscle, *Gerontology,* 22, 317, 1976.

106. **Miquel, J., Lundgren, P. R., and Bensch, K. G.,** Effects of oxygen:nitrogen (1:1) at 760 TORR on the life span and fine structure of *Drosophila melanogaster, Mech. Ageing Dev.,* 4, 41, 1975.

107. **Lang, C. A. and Stephan, J. K.,** Nicotinamide adenine dinucleotide phosphate enzymes in the mosquito during its life span, *Biochem. J.,* 102, 331, 1967.

108. **Massie, H. R. and Williams, T. R.,** Increased longevity of *Drosophila melanogaster* with lactic and gluconic acids, *Exp. Gerontol.,* 14, 109, 1979.

109. **Atlan, H., Miquel, J., Helmle, L. C., and Dolkas, C. B.,** Thermodynamics of aging in *Drosophila melanogaster, Mech. Ageing Dev.,* 5, 371, 1976.

110. **Bozcuk, A. N.,** Genetics of longevity in *Drosophila, Exp. Gerontol.,* 16, 415, 1981.

111. **Ragland, S. S. and Sohal, R. S.,** Mating behaviour, physical activity and aging in the house fly *Musca domestica, Exp. Gerontol.,* 8, 135, 1973.

112. **Baker, G. T.,** Insect flight muscle: maturation and senescence, *Gerontology,* 22, 334, 1976.

113. **Baker, G. T.,** Identical age-related patterns of enzyme activity changes in *Phromia regina* and *Drosophila melanogaster, Exp. Gerontol.,* 23, 58, 1975.

114. **Baker, G. T.,** Age-related activity changes in arginine phosphokinase in the house fly *Musca domestica, J. Gerontol.,* 30, 163, 1975.

115. **Rockstein, M.,** Cellular age changes in insects, *Symp. Soc. Exp. Biol.,* 21, 337, 1967.

116. **Rockstein, M. and Chesky, J.,** Age-related changes in natural actomyocin of the male house fly, *Musca domestica, J. Gerontol.,* 28, 455, 1973.

117. **Armstrong, D., Rinehart, R., Dixon, L., and Reigh, D.,** Changes of peroxidase with age in *Drosophila, Age,* 1, 8, 1978.

118. **Massie, H. R., Aiello, V. R., and Williams, T. R.,** Changes in superoxide dismutase and copper during development and ageing in the fruit fly *Drosophila melanogaster, Mech. Ageing Dev.,* 12, 279, 1980.

119. **Kirkwood, T. B. L.,** Error propagation in intracellular information transfer, *J. Theor. Biol.,* 82, 363, 1980.

120. **Gallant, J. A. and Prothero, J.,** Testing models of error propagation, *J. Theor. Biol.,* 83, 561, 1980.

121. **Rosenberger, R. F., Foskett, G., and Holliday, R.,** Error propagation in *Escherichia coli,* and its relation to cellular ageing, *Mech. Ageing Dev.,* 13, 247, 1980.

122. **Edelmann, P. and Gallant, J. A.,** On the translational theory of aging, *Proc. Natl. Acad. Sci. U.S.A.,* 74, 3396, 1977.

123. **Gallant, J. A. and Palmer, L.,** Error propagation in viable cells, *Mech. Ageing Dev.,* 10, 27, 1979.

124. **Miller, C. G.,** Peptidases and proteases of *Escherichia coli* and *Salmonella typhimurium, Ann. Rev. Microbiol.,* 29, 485, 1975.

125. **Miller, C. G. and Zipser, D.,** Degradation of *Escherichia coli* β-galactosidase fragments in protease-deficient mutants of *Salmonella typhimurium, J. Bacteriol.,* 130, 347, 1977.

126. **Cheng, Y. E., Zipser, D., Cheng, C., and Rolseth, S. J.,** Isolation and characterization of mutations in the structural gene for Protease III, *J. Bacteriol.,* 140, 125, 1979.

127. **Menninger, J. R.,** Ribosome editing and the error catastrophe hypothesis of cellular aging, *Mech. Ageing Dev.,* 6, 131, 1977.

128. **Liebovitz, B. E. and Siegel, B. V.,** Aspects of free radical reactions in biological systems: aging, *J. Gerontol.,* 35, 45, 1980.

129. **Robert, L.,** Aging of connective tissue, *Mech. Ageing Dev.,* 14, 273, 1980.

130. **Nagy, I. Z. and Nagy, K.,** On the role of cross-linking of cellular proteins in aging, *Mech. Ageing Dev.,* 14, 245, 1980.
131. **Truscott, R. J. W. and Augosteyen, R. C.,** Oxidative changes in human lens proteins during senile nuclear cataract formation, *Biochim. Biophys. Acta,* 492, 43, 1977.
132. **Tas, S., Tam, C. F., and Walford, R. L.,** Disulphide bonds and the structure of the chromatin complex in relation to aging, *Mech. Ageing Dev.,* 12, 65, 1980.
133. **Bond, J. S. and Offermann, M. K.,** Initial events in the degradation of soluble cellular enzymes: factors affecting the stability and proteolytic susceptibility of fructose-1,6-bisphosphate aldolase, *Acta Biol. Med. Germ.,* 40, 1365, 1981.
134. **Levine, R. L.,** Oxidative modification of glutamine synthetase, *J. Biol. Chem.,* 258, 11823, 1983.
135. **Bernlohr, D. and Switzer, R. L.,** Reaction of *Bacillus subtilis,* glutamine phosphoribosylpyrophosphate amidotransferase with oxygen: chemistry and regulation by ligands, *Biochemistry,* 20, 5675, 1981.
136. **Masters, P. M., Baba, J. L., and Zigler, J. S.,** Aspartic acid racemization in heavy molecular weight crystallins and water insoluble protein from normal human lenses and cataracts, *Proc. Natl. Acad. Sci. U.S.A.,* 75, 1204, 1978.
137. **Helfman, P. M., Baba, J. L., and Shov, M. Y.,** Consideratioins on the role of aspartic acid racemization in the aging process, *Gerontology,* 23, 419, 1977.
138. **Helfman, P. M. and Baba, J. L.,** Aspartic acid racemization in dentine as a measure of ageing, *Nature (London),* 262, 279, 1976.
139. **Poplin, L. and DeLong, R.,** Accelerated aging due to enzymatic racemization, *Gerontology,* 24, 365, 1978.
140. **Shapiro, R., McManus, M., Garrick, L., McDonald, M. J., and Bunn, H. F.,** Nonenzymatic glycosylation of human hemoglobin at multiple sites, *Metabolism,* 28, 427, 1979.
141. **Stevens, V. J., Rouzer, C. A., and Monnier, V. M.,** Diabetic cataract formation: potential role of glycosylation of lens crystallin, *Proc. Natl. Acad. Sci. U.S.A.,* 75, 2918, 1978.
142. **Cerami, A., Stevens, V. J., and Monnier, V. M.,** Role of nonenzymatic glycosylation in the development of the sequelae of diabetes mellitus, *Metabolism,* 28, 431, 1979.
143. **Rosenberg, H., Modrak, J. B., Hassing, J. M., Al-Turk, W. A., and Stohs, S. J.,** Glycosylated collagen, *Biochem. Biophys. Res. Commun.,* 91, 498, 1979.
144. **Guthrow, C. E., Morris, M. A., Day, J. F., Thorpe, S. R., and Baynes, J. W.,** Enhanced nonenzymatic glycosylation of human serum albumin in diabetes mellitus, *Proc. Natl. Acad. Sci. U.S.A.,* 76, 4258, 1979.
145. **Flatmark, T. and Sletten, K.,** Multiple forms of cytochrome *c* in the rat, *J. Biol. Chem.,* 243, 1623, 1968.
146. **Lai, C. Y., Chen, C., and Horecker, B. L.,** Primary structure of two COOH-terminal hexapeptides from rabbit muscle aldolase: a difference in the structure of αβ subunits, *Biochem. Biophys. Res. Commun.,* 40, 461, 1970.
147. **Midelfort, C. F., and Mehler, A. H.,** Deamination *in vivo* of an asparagine residue of rabbit muscle aldolase, *Proc. Natl. Acad. Sci. U.S.A.,* 69, 1816, 1972.
148. **Dolken, G. and Pette, D.,** Turnover of several glycolytic enzymes in rabbit heart, soleus muscle and liver, *Hoppe-Seyler's Z. Physiol. Chem.,* 355, 289, 1974.
149. **McDonald, M. L., Augustine, S. L., Burk, T. L., and Swick, R. W.,** A comparison of methods for the measurement of protein turnover *in vivo, Biochem. J.,* 184, 473, 1979.
150. **Skala-Rubinson, M., Vibert, M., and Dreyfus, J. C.,** Electrophoretic modifications of three enzymes in extracts of human and bovine lens, *Clin. Chem. Acta,* 70, 385, 1976.
151. **Robinson, A. B.,** Molecular clocks, molecular profiles and optimum diets, three approaches to the problem of ageing, *Mech. Ageing Dev.,* 9, 225, 1979.
152. **Van Kleef, F. S. M., DeJong, W. W., and Moenders, H. J.,** Stepwise degradations and deamination of the eye lens protein α-crystallin in ageing, *Nature (London),* 258, 264, 1965.
153. **Gershon, H. and Gershon, D.,** Inactive enzyme molecules in aging mice: liver aldolase, *Proc. Natl. Acad. Sci. U.S.A.,* 70, 909, 1973.
154. **Anderson, P. J.,** Ageing effects on the liver aldolase of rabbits, *Biochem. J.,* 140, 341, 1974.
155. **Petell, J. K. and Lebherz, H. G.,** Properties and metabolism of fructose diphosphate aldolase in livers of "old" and "young" mice, *J. Biol. Chem.,* 254, 8179, 1979.
156. **Ward, W. F., Cox, J. R., and Mortimore, G. E.,** Lysosomal sequestration of intracellular protein as a regulatory site in hepatic proteolysis, *J. Biol. Chem.,* 252, 6955, 1977.
157. **Goldberg, A. L. and St. John, A. C.,** Intracellular protein degradation, *Ann. Rev. Biochem.,* 51, 335, 1982.
158. **Ballard, F. J.,** Intracellular protein degradation, *Essays Biochem.,* 13, 1, 1977.
159. **Bigelow, S., Hough, R., and Rechsteiner, M.,** The selective degradation of injected proteins occurs principally in the cytosol rather than in lysosomes, *Cell,* 25, 83, 1981.
160. **DeMartino, G. N. and Goldberg, A. L.,** Identification and partial purification of an ATP-stimulated alkaline protease in rat liver, *J. Biol. Chem.,* 254, 3712, 1979.

161. **Dayton, W. R., Goll, D. E., Zeece, J. G., Robson, R. M., and Reville, W. J.**, A Ca^{2+}-activated protease possibly involved in myofibrillar protein turnover. Purification from porcine muscle, *Biochemistry*, 15, 2150, 1976.

162. **Saklavala, J., Bond, J. S., and Barrett, A. J.**, Isolation and characterization of tryase, a serine proteinase from rat liver, *Biochem. J.*, 193, 251, 1981.

163. **Barrett, A. J., Ed.**, *Proteinases in Mammalian Cells and Tissues*, North-Holland, Amsterdam, 1977.

164. **Duncan, W. E., Offermann, M. K., and Bond, J. S.**, Intracellular turnover of stable and labile soluble liver proteins, *Arch. Biochem. Biophys.*, 199, 331, 1980.

165. **Goldberg, A. L. and Dice, J. F.**, Intracellular protein degradation in mammalian and bacterial cells, *Ann. Rev. Biochem.*, 43, 835, 1974.

166. **Momany, F. A., Aguanno, J. J., and Larrabee, A. R.**, Correlation of degradation rates of proteins with a parameter calculated from amino acid composition and subunit size, *Proc. Natl. Acad. Sci. U.S.A.*, 73, 3093, 1976.

167. **Dice, J. F., Mess, E. J., and Goldberg, A. L.**, Studies on the relationship between the degradative rates of proteins *in vivo* and their isoelectric points, *Biochem. J.*, 178, 305, 1979.

168. **Wilson, D. D.**, Enzyme changes in aging mammals, *Gerontologia*, 19, 79, 1973.

169. **Ames, B. N.**, Dietary carcinogens and anticarcinogens: oxidative radicals and degenerative diseases, *Science*, 221, 1256, 1983.

170. **Blackwell, W. H.**, Rationale for getting a haircut (and taking a bath), *J. Irreproducible Results*, 28, 5, 1983.

Chapter 9

FAT-SOLUBLE VITAMINS

Part I

VITAMINS A AND D

Glenville Jones

TABLE OF CONTENTS

I. INTRODUCTION

Vitamins A and D are lipophilic substances required in a continuous supply for specialized but essential functions in the animal body. Vitamin A is required for vision, reproduction, and systemically for differentiation of epithelial cells. Vitamin D is required to maintain blood calcium levels between narrow limits, thereby ensuring adequate calcium ion concentrations within cells for muscle contraction, neurotransmission, and general cellular biochemistry.

Of the two vitamins, only vitamin A must be derived from the diet since vitamin D can be synthesized in the skin from 7-dehydrocholesterol following exposure to adequate amounts of UV light. Both vitamins can be stored in the body, but without a constant supply, storage depots are depleted and eventually used up. At this point one or more of the essential functions listed above is compromised, the health of the animal is affected deleteriously, and death results.

Though vitamins A and D carry out separate functions in the body, the two tend to be associated nutritionally because of their fat-soluble nature and because the two are often combined in foodstuff supplements and pharmaceutical formulations. In fact, there is some evidence that this association of the two vitamins in vitro may even exist in vivo. It has been known for some time that large doses of vitamin A ameliorate the high blood calcium levels induced by excessive amounts of vitamin D. This phenomenon remains unexplained. Recently, it has been shown that $1,25\text{-}(OH)_2D_3$, the hormonal form of vitamin D, plays a key but as yet undefined role in differentiation, as does vitamin A. Thus, we may be entering an era of vitamin research in which the interrelationship of these two fat-soluble vitamins, essential for growth of the skeleton and soft tissues, will be clarified.

Both vitamins A and D require extensive metabolism to generate active metabolites, specific macromolecules to transport these hydrophobic molecules to their metabolizing enzymes and target tissues, and efficient catabolic systems to rid the body of potentially toxic substances generated from them. Thus, the body can regulate the effects of these substances by regulating synthesis, transport, or degradation. There is presently limited evidence to indicate how aging influences the enzymes and proteins of these pathways. However, since both vitamins A and D are required for a number of diverse functions in the body, it is likely that the need for these two substances goes on throughout life. It is also probable that since both are associated with growth of the young animal, nutritional requirements would be greatest during the growing period and as adulthood is reached, daily needs would be expected to fall.

On the other hand, since both vitamins A and D are polyenes, their double bonds are susceptible to attack by molecular oxygen and light. Though both are present in the body in only nanomolar or micromolar concentrations, adequate amounts of antioxidants are required to keep these molecules protected in the membranes and cell organelles of the body. If the popular concept of aging as an inability to control the concentration of free radicals in the body has some truth,[1,2] then vitamins A and D would be molecules susceptible to increased degradation in the aging animal. Thus, one could make a case for increased requirements with aging.

It is clear, however, that much remains to be learned about the effects of aging on the metabolism and nutritional requirements of vitamins A and D and how these effects of aging may contribute to certain disease states.

II. VITAMIN A

A. Functions of Vitamin A*

1. Vitamin A, in the form of 11-cis-retinaldehyde, is required for vision. 11-cis-Retinaldehyde combines with proteins, known as opsins, to form the visual pigments of the rods and cones of the eye. The 11-cis-retinaldehyde of the visual pigment is isomerized by light into all-*trans*- retinaldehyde, and this process generates an electrical impulse. Many electrical impulses from molecules of visual pigment throughout the retina are coordinated and interpreted by the brain in the process of vision.
2. Vitamin A, in the form of retinol, maintains reproduction by promoting proper development of the germinal epithelium. In male animals deprived of vitamin A, spermatogenesis ceases while in females deprived of vitamin A, conception occurs normally but their fetusus are resorbed.
3. Vitamin A, in the form of retinol or retinoic acid, is required for normal growth and differentiation of many types of epithelial cells — skin, intestinal mucosa, mucous epithelium of the trachea, and other organs.

In the absence of vitamin A, animals cease to grow, become susceptible to infections, and eventually die.

B. Metabolism of Vitamin A

The structure of vitamin A and its natural metabolites is shown in Figure 1. Dietary sources of vitamin A can be either β-carotene from plant material or retinyl esters from animal tissue. β-Carotene is split in the mucosa of the intestine to two molecules of retinaldehyde, which are in turn converted to retinol. Similarly, retinyl esters are hydrolyzed in the lumen of the intestine to retinol and the retinol is absorbed into the mucosal cell. Retinol from either source is reesterified in the mucosa and incorporated into lymph chylomicrons for transport to the liver. Retinyl esters present in the chylomicron remnant are taken up almost exclusively by the liver[3] (Figure 2). The retinyl esters are further hydrolyzed and reesterified in the liver and stored as retinyl esters within the hepatocytes. There is some evidence that redistribution of vitamin A occurs within the liver and that stellate cells may subsequently inherit part of these stores.

Vitamin A circulates around the body as retinol bound to a specific plasma protein, retinol-binding protein (RBP). RBP is an α_1-globulin, (20,000 mol wt) with a single binding site for retinol.[4] RBP binds strongly to plasma prealbumin, the macromolecule responsible for the transport of thyroxine. The 1:1 molar complex of RBP:prealbumin helps to reduce the glomerular filtration of RBP and therefore conserve retinol. RBP is extremely specific for retinol so that retinoic acid entering the circulation must be carried on serum albumin. RBP serves many functions in that it transports vitamin A to its target cell, prevents this hydrophobic molecule from entering and labilizing membranes in a random manner, conserves it from loss into the urine, and finally, acts as a storage form of retinol for cells that require it.

Cells that take up retinol from the RBP-retinol complex seem to possess a cell surface recognition receptor for the protein. However, it seems likely that RBP does not accompany retinol in crossing the cell membrane. Retinol inside the cell is transported by a second specific carrier protein known as cellular retinol-binding protein (CRBP), which is both

---

* The term "vitamin A" is used in the general nutritional sense to denote compounds that exhibit qualitatively the biological activity of retinol. The terms "retinol", "retinaldehyde", and "retinoic acid" are used to denote the alcohol, aldehyde, and acid forms of vitamin A. The term "retinoids" is used in a general context to include all the natural and synthetic analogues of vitamin A, with and without the biological activity of retinol.

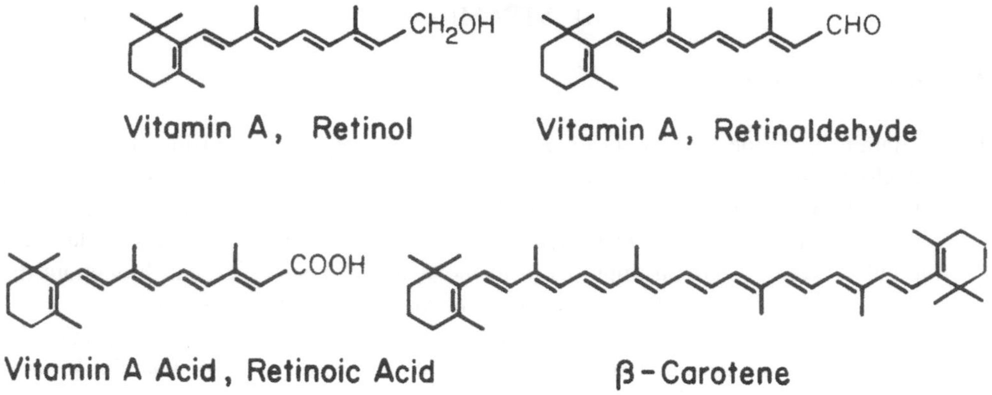

Vitamin A, Retinol

Vitamin A, Retinaldehyde

Vitamin A Acid, Retinoic Acid

β−Carotene

FIGURE 1.   Structures of retinol, retinaldehyde, retinoic acid, and β-carotene.

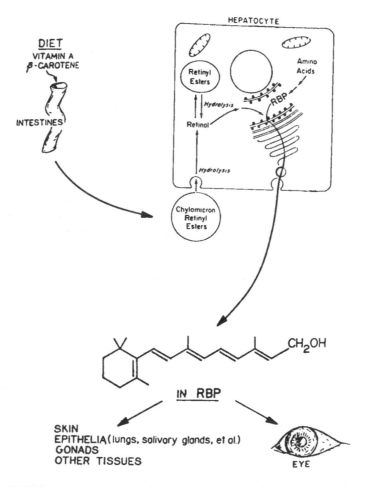

FIGURE 2.   Absorption, hepatic handling, transport, and delivery of vitamin A. (From Goodman, D. S., *Fed. Proc. Fed. Am. Soc. Exp. Biol.*, 38, 2501, 1979; Smith, J. E. and Goodman, D. S., *Fed. Proc. Fed. Am. Soc. Exp. Biol.*, 38, 2504, 1979. With permission.)

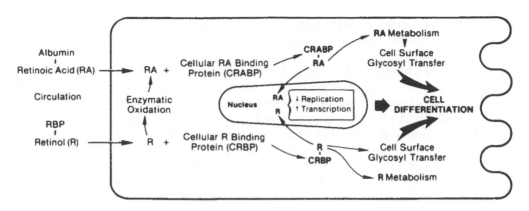

FIGURE 3.   Possible actions of vitamin A in target cell. (Courtesy Dr. M. R. Haussler, University of Arizona and Amersham Corporation, Arlington Heights, Ill.)

structurally and immunologically different from RBP[5] (Figure 3). All-*trans*-retinol within the cells of the retina can be converted to all-*trans*-retinaldehyde and then to 11-cis–retinaldehyde for incorporation into the visual proteins, such as rhodopsin.[6] In the testes and epithelial tissues, retinol may be translocated to the nucleus to initiate new protein synthesis responsible for the functions of spermatogenesis and differentiation ascribed to vitamin A. It is also possible that retinol promotes differentiation of cells into a mucous-secreting phenotype by mediating, through the formation of mannosyl retinyl phosphate, the transfer of glycosyl residues to nascent protein in the rough endoplasmic reticulum.[7] Glycoproteins thus made are incorporated into the cell surface and allow the cell to become a mucous-secreting type. Retinol has been implicated in the synthesis of such glycoproteins as gobletin in the intestine, α_1-macroglobulin in the plasma,[8] and other glycoproteins in the cornea and trachea. It is not yet clear whether this ability for glycosyl transfer is specific to the vitamin A molecule or merely reflects a broad specificity for the enzymes which normally use dolichol phosphate, a similar isoprenoid alcohol which carries out other glycosyl transfers in organisms from bacteria through to mammals.

There is considerable evidence, much of it acquired in vitro, that retinoic acid can execute the actions of vitamin A in promotion of cell growth and differentiation.[9] Retinol can be converted via retinaldehyde into retinoic acid in several tissues, including liver, kidney, and intestine, but it is not clear if other target cells are able to synthesize retinoic acid from retinol delivered to them in the form of RBP-retinol complex.[10] Exogenous retinoic acid can be carried to target cells bound to albumin. Here it freely enters the cell and combines with another specific intracellular binding protein known as cellular retinoic acid-binding protein (CRABP)[5] (Figure 3). Thus, it may be the retinoic acid-CRABP complex and not retinol-CRBP which translocates to the nucleus and initiates new protein synthesis in epithelial cells. Though the biological activity of exogenously administered retinoic acid is less than retinol on a weight basis, this is mainly due to the poor delivery of retinoic acid to the target cell, and biological activity can be improved by more frequent administration of smaller doses. Since retinoic acid cannot be converted by the reverse reaction into retinaldehyde or retinol, retinoic acid alone is incapable of supporting the function of vitamin A in the eye or in reproduction.

Over recent years, in vitro work has shown that analogues of vitamin A with a wide variety of structures (retinoids) are capable of inducing numerous changes in cultured normal and neoplastic cells.[11] For example, the proliferation rate of certain tumor cells is reduced by retinoids.[12] Retinoids can also prevent the development of tumors induced in vivo by chemical carcinogens or even reverse the malignant transformation. The use of retinoids as anticancer drugs has also evoked the nutritional question of whether increased dietary intakes of β-carotene or vitamin A might help prevent or retard growth of tumors in humans by promoting endogenous synthesis of natural retinoids.[13] (See Chapter 4 in Volume II.)

C. Age-Related Changes in Vitamin A Metabolism

More is known about the age-related changes in vitamin A absorption and transport than about the effects of aging on the metabolic enzymes. Since lack of dietary intake represents the major vitamin A-related disease, it is important to note that Fleming and Barrows[14] showed that intestinal absorption of vitamin A remains unchanged with aging in the rat. Thus, only where low dietary intakes due to poor diet are involved would we expect to see a high incidence of night blindness or other vitamin A deficiency symptoms.

Except at very high and low intakes of vitamin A, by far the most important determinant of vitamin A in the circulation, is the rate of synthesis of RBP. The secretion of liver retinol into plasma is governed by the rate of synthesis and release of RBP and not just by the extent of the reserves of retinyl esters. Furthermore, RBP secretion can be influenced by a number of physiological and pathological factors, one of which is age. Normal adult human level RBP is in the range of 40 to 50 μg/mℓ (50 μg retinol per 100 mℓ) — 47 μg/mℓ in males and 43 μg/mℓ in females.[15] Children begin with RBP levels around 25 to 28 μg/mℓ which rise to the adult level around puberty. A value of 20 μg/mℓ RBP (25 μg retinol per 100 mℓ) is low, and some consider it to be vitamin A-deficient.

The rat shows a marked rise of RBP above the adult level during the development of the gonads so that RBP levels start around 25 μg/mℓ and rise to 60 μg/mℓ before falling to 37 μg/mℓ in the male or start around 25 μg/mℓ, rise to 45 μg/mℓ, and fall to 32 μg/mℓ in the female.[16] Where sexual activity is seasonally related, e.g., birds and sheep, there can be two- to threefold changes in the RBP levels.[17]

Serum levels of retinol in elderly humans do not seem to be dramatically different from young adults. One study of Fisher et al.[18] could demonstrate little difference in serum retinol between a population of elderly rural Utahans (mean age 69 years) and normal values from the literature. In their study, serum retinol values in micrograms per 100 milliliters (mean ± SD) were 50 ± 12 for 58 males and 48 ± 11 for 129 females. In another study, Harrill and Cervone[19] found serum retinol levels to be 53 ± 30 μg/100 mℓ in 46 elderly females (aged 62 to 99 years), with the means of three subgroups (62 to 75, 76 to 85, and 86 to 99 years) also at 53 μg/100 mℓ. However, results from this study do show that a small percentage of elderly adults (5 to 17%) have deficient or borderline retinol values indicative of inadequate intakes of vitamin A. Other larger nutritional studies[20,21] indicate that the level of serum retinol increases with age. This may be due to decreased turnover of RBP-retinol complex due to the age-related decline in kidney function.

There have been claims that dietary β-carotene or vitamin A intakes are directly proportional to health and inversely proportional to the incidence of cancer.[13,23] Since the circulating retinol level is determined by the rate of release of RBP and not by dietary vitamin A levels (over the normal range),[24] it is difficult to rationalize whether low dietary carotenoids could influence the incidence of disease.

There are no reports in the literature of age-related changes in cellular retinoid-binding proteins CRBP or CRABP, mannosyl retinyl phosphate, or retinoic acid.

D. Requirements for Vitamin A

Intakes of vitamin A and carotene are today usually expressed in "microgram retinol equivalents". By definition, one retinol equivalent is equal to 1 μg of retinol, 6 μg of β-carotene, or 12 μg of other provitamin A carotenoids (e.g., α-carotene, cryptoxanthin). Older literature still uses the international unit (IU). One retinol equivalent is equal to 3.33 IU in the form of retinol.

The Recommended Daily Allowance (RDA)* for human infants up to 1 year is 400 to 420 μg retinol equivalents per day, 1000 μg retinol equivalents per day for adult males, and 800 μg retinol equivalents per day for adult females. Pregnant and lactating females are encouraged to increase intakes to 1000 and 1200 μg retinol equivalents per day, respectively. No additional recommendations are made for special dietary intakes for the elderly. In North America milk is fortified with vitamin A (450 μg/ℓ).

E. Vitamin A Status of the Elderly
1. Dietary Assessment

In two large national surveys (Health and Nutrition Examination Survey[21] [HANES] and Household Food Consumption Survey[25] [HFCS]) the mean intake of vitamin A was greater than two thirds of the RDA. Mean values can mask problems among subgroups. In the HFCS[25] study, 41% of men between the ages of 65 and 74, 51% of men above the age of 75, and 60% of women above the age of 65 had vitamin A intake below two thirds of the RDA. Studies reported on different elderly population groups by different researchers revealed inadequate intakes of this vitamin to various extents. Guthrie et al.[26] reported that 66% of the elderly in rural Pennsylvania had vitamin A intakes below two thirds of the RDA, and Rawson et al.[27] reported the same intake in 50% of the elderly in rural southwestern Pennsylvania. Pao and Hill,[28] Todhunter and Darby,[29] LeBovit,[30] and Yearick et al.[31] reported vitamin A intake below two thirds of the RDA in 54, 44, 7, and 4% of their male and female subjects in the North-Central and Southern U.S. (Tennessee, Rochester, N.Y., and Corvallis, Ore., respectively). Other studies reported separate figures for inadequate intakes of this vitamin for men and women. Dibble et al.,[32] Steinkamp et al.,[33] and Lyons and Trulson[34] observed vitamin A intake below two thirds of the RDA in 36, 24, and 10% of men and 33, 27, and 4% of women subjects in Syracuse, N.Y., San Mateo, Calif., and Boston, Mass., respectively. Harrill and Cervone[19] and Fry et al.[35] reported vitamin A intake below two thirds of the RDA in 21 and 9% of women subjects in Fort Collins, Colo. and Lincoln, Neb., respectively. It thus appears that in certain regions especially in rural areas, the problem of low vitamin A intake is serious.

2. Biochemical Measurement

The plasma vitamin A level of adequacy set by the Interdepartmental Committee on Nutrition for National Defense (ICNND) is 40 μg/100 mℓ. A plasma level of lower than 10 μg/100 mℓ is considered "deficient", and a level between 10 to 39 μg/100 mℓ is considered "low". All studies have shown that mean plasma vitamin A levels met or exceeded the ICNND level of adequacy. In the Ten-State Nutrition Survey[36] (TSNS), about 98% of individual levels were adequate except Spanish Americans; 14.8% of men and 11.8% of women in this population had inadequate levels. Brin et al.[37] reported that 20% of their subjects studied in Onondago County, N.Y. had low or deficient vitamin A levels. Harrill and Cervone[19] and Yearick et al.,[31] in their studies in Fort Collins, Colo. and Corvallis, Ore., reported that 14 and 10% of the population, respectively, also had low or deficient vitamin A levels. Guillum[38] reported that only 5.6% of subjects in San Mateo, Calif. had serum vitamin A levels below 30 μg/100 mℓ. Dibble et al.[32] reported only 1% of the subjects studied in Syracuse, N.Y. had low or deficient plasma vitamin A values. Fisher[18] reported that no low serum levels of vitamin A or carotene were observed in subjects studied in rural Utah. It appears that the low vitamin A intake reported in many studies was not generally reflected in inadequate plasma levels of vitamin A.

F. Problems of Deficiency and Excess of Vitamin A

Vitamin A deficiency manifests itself differently in different species.[39] In man one rarely

---

* Extracted from *Recommended Dietary Allowances*, 9th ed., Committee on Dietary Allowances Food and Nutrition Board, Division of Biological Sciences, National Academy of Sciences, Washington D.C., 1980.

sees simple vitamin A deficiency uncomplicated by caloric or protein malnutrition. However, mild vitamin A deficiency caused by a fall in the plasma RBP-retinol level is usually associated with impairment of dark adaptation (night blindness). More severe vitamin A deficiency in man can include xerophthalmia and keratomalacia. These are characterized by a metaplasia of the conjuctival membranes and keratinization of the cornea. Left unchecked, pyogenic infections of the eye can arise and result in opacity of the cornea followed by loss of the lens. Other organs affected in man include the skin (follicular hyperkeratosis or toad skin), nervous system (degeneration), and epithelial linings of bronchus, trachea, esophagus, nose, and tongue. Loss of the senses of smell and taste can result.

In the rat, where vitamin A deficiency is allowed to proceed beyond changes to the visual cycle, secondary changes include diarrhea, dental depigmentation, and degeneration of the testes. However, it is the damage to the mucous membranes throughout the body which eventually results in the loss of appetite, cessation of growth, and infections that indicate the animal is entering the later stages of vitamin A deficiency. Death, the eventual consequence of vitamin A deficiency, rarely occurs in rats maintained on a vitamin A-free diet under germ-free conditions.

In other animals, such as chicks, nerve lesions (e.g., increase in cerebrospinal fluid pressure, lack of coordination) are dominant in vitamin A deficiency while in calves optic nerve degeneration caused by "bony-overgrowths" of the skull and/or degeneration of nerve cells is common.

Excessive vitamin A intakes result in increased blood vitamin A concentrations. However, instead of this being in the form of retinol-RBP complex, it is mainly in the form of retinyl esters complexed to lipoproteins of density less than 1.21. Nonspecific and unregulated delivery of vitamin A to tissues in this manner leads to vitamin A toxicity. There appears to be excessive destruction of skeletal tissue in vitro when bone rudiments are cultured in the presence of excess retinol.[40] Growing rats given excessive intakes of vitamin A develop bone fractures, while adult animals show only bone rarefaction. Similarly, humans given excess vitamin A develop thickenings of bone, called hyperostoses, which are characterized by "a shell-like appearance with a zone of diminished density between the subperiosteal thin layer of bone and external surface of the old cortex". Hyperostoses can appear throughout the skeleton in the ulna, tibia, fibula, and skull. Nausea, vomiting, diarrhea, and behavioral changes can also accompany acute poisoning with vitamin A.

III. VITAMIN D

A. Functions of Vitamin D

1. Vitamin D, in the form of 1,25-dihydroxyvitamin D_3 (1,25-$(OH)_2D_3$*), stimulates intestinal calcium absorption and raises blood calcium level.
2. Vitamin D, in the form of 1,25-$(OH)_2D_3$, stimulates intestinal phosphate absorption and raises blood phosphate level.
3. Vitamin D, in the form of 1,25-$(OH)_2D_3$, acts synergistically with parathyroid hormone to stimulate bone ($CaPO_4$) resorption by osteoclasts and thereby raise blood calcium and phosphate levels.
4. Vitamin D acts either directly on bone cells or indirectly through the calcium-raising functions listed above to promote the mineralization of new bone matrix.

---

* The term "vitamin D" is used in the general nutritional sense to denote compounds that qualitatively exhibit the biological activities of vitamins D_2, D_3, and their metabolites. Abbreviations used are 25-hydroxyvitamin D_2 or D_3 = 25-$(OH)D_2$ or D_3; 1,25-dihydroxyvitamin D_2 or D_3 = 1,25-$(OH)_2D_2$ or D_3; 24,25-dihydroxyvitamin D_2 or D_3 = 24,25-$(OH)_2D_2$ or D_3.

In the absence of vitamin D, blood calcium and phosphate levels fall, animals cease to grow, the skeleton becomes devoid of mineral, and rickets or osteomalacia results. If the calcium levels remain unchecked, animals suffer neuromuscular problems such as tetany and convulsions, which can result in death.

B. Metabolism of Vitamin D

Vitamin D can be synthesized in the skin from 7-dehydrocholesterol on irradiation by UV light or can be absorbed in the intestine from the diet.[41,42] Endogenously produced vitamin D is vitamin D_3, whereas dietary vitamin D can be either vitamin D_3 of animal origin or vitamin D_2 from plant sterols. Vitamins D_2 and D_3 differ only in the side chain region of the molecule where D_2 possesses an extra C_{22}–C_{23} double bond and a C_{24} methyl group (Figure 4). The two are similarly metabolized and equally biologically active in most mammals.

Skin synthesis requires light of wavelength 260 to 320 nm, and the initial compound synthesized is pre-vitamin D_3, a steroid-like molecule with a split between carbons 9 and 10. Pre-D_3 is rapidly isomerized by thermal equilibration into vitamin D_3, which is swept away from the skin because of its moderately strong affinity for blood vitamin D-binding globulin (DBP),[43] an α-globulin with molecular weight 51,000.[44]

Alternatively, vitamin D_2 or D_3 in the diet is absorbed in the intestine by an efficient process, enters the plasma through the lymphatics bound to chylomicrons, and is rapidly taken up by the liver. Vitamin D_3, whether bound to chylomicrons or DBP, never achieves a high concentration (i.e., always less than 10 ng/mℓ) in the blood except where pharmacological doses of vitamin D are administered.

The liver acts as the major storage organ for vitamin D, which remains largely unmetabolized, though considerable amounts can also be found in fat and muscle. Vitamin D_3 is converted in the hepatocyte into 25-hydroxyvitamin D_3 (25-(OH)D_3). Similarly, vitamin D_2 is converted into 25-(OH)D_2. The enzyme which carries out 25-hydroxylation is a microsomal enzyme which requires NADPH, molecular oxygen, and Mg^{2+}.[45] This enzyme is inhibited in an end product fashion by 25-(OH)D_3, thereby providing a tight control of liver 25-hydroxylation. However, it is likely that a second 25-hydroxylase present in the liver mitochondria can also utilize vitamin D_3 as its substrate if high plasma concentrations of vitamin D_3 are achieved. Thus, pharmacological doses of vitamin D_3 give rise to high plasma levels of vitamin D_3, and the control of 25-hydroxylation breaks down, resulting in high plasma levels of 25-(OH)D_3 (normal: 5 to 60 ng/mℓ; high: 100 to 500 ng/mℓ).[46]

25-(OH)D_3, made in the liver, is secreted into the bloodstream bound to DBP. Unlike with retinol, however, DBP is not the main determinant of the plasma concentration of the vitamin, since in the normal individual only approximately 2% of DBP present in the bloodstream is complexed to vitamin D. 25-(OH)D_3 is taken up by several tissues such as muscle, kidney, fat, and intestine, though it appears that only the kidney metabolizes it further to any large extent. Uptake by the kidney is poorly understood, but probably involves release of free 25-(OH)D_3 on the outside of the kidney cell followed by passage of the 25-(OH)D_3 into the mitochondrion. Cytochrome P_{450} enzymes, present in kidney mitochondria, metabolize 25-(OH)D_3 into either the hormonal form of vitamin D_3, 1,25-(OH)$_2D_3$,[47] or into a degradation product, 24,25-(OH)$_2D_3$[48] (Figure 5). When blood calcium is low, parathyroid hormone (PTH) is secreted by the parathyroid glands, PTH stimulates renal 1-hydroxylase to synthesize 1,25-(OH)$_2D_3$. Alternatively, low blood phosphate stimulates 1,25-(OH)$_2D_3$ production by an unknown mechanism which does not involve PTH. 1,25-(OH)$_2D_3$, bound to DBP, is then transported to intestine and bone where it acts to raise blood calcium and phosphate by the steps outlined below. 1,25-(OH)$_2D_3$ stimulates intestinal calcium and phosphate absorption and along with PTH stimulates bone resorption (Figure 6). In addition, 1,25-(OH)$_2D_3$ acts to conserve calcium during its reabsorption in the kidney. All of these actions serve to raise serum calcium and phosphate, thereby correcting the original hypo-

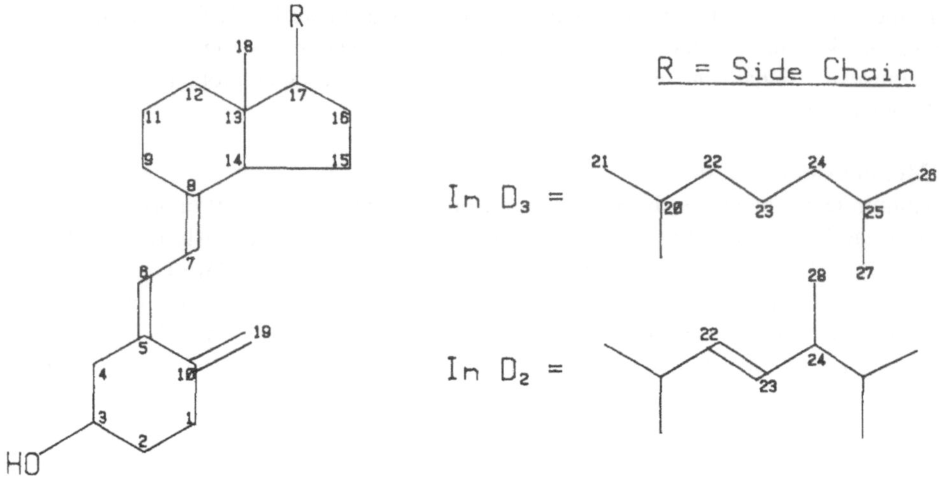

FIGURE 4.   Structure of vitamins D_2/D_3.

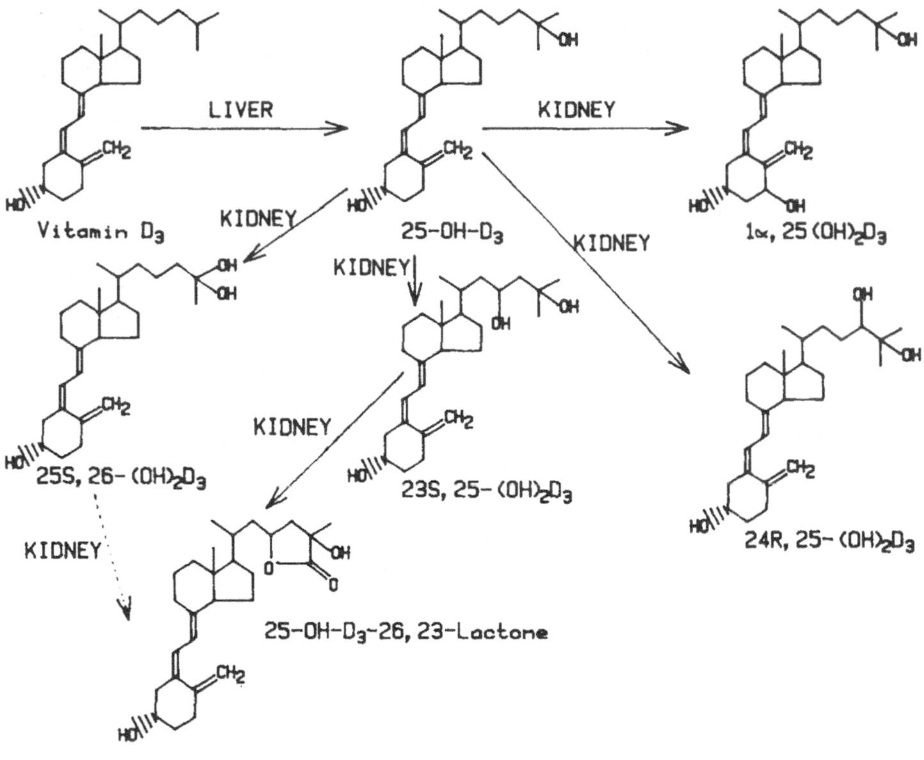

FIGURE 5.   Metabolism of vitamin D_3.

calcemic or hypophosphatemic stimulus. With the restoration of blood calcium to normal, the acute need for $1,25\text{-}(OH)_2D_3$ is past and instead the kidney switches to the synthesis of $24,25\text{-}(OH)_2D_3$, a metabolite which probably represents the first step on an inactivation pathway, as well as to synthesis of other metabolites (see Figure 5).

Plasma $1,25\text{-}(OH)_2D_3$ levels rarely rise above 100 pg/mℓ and in the human average 40 pg/mℓ. Much like that of retinol and retinoic acid, the action of $1,25\text{-}(OH)_2D_3$ on target

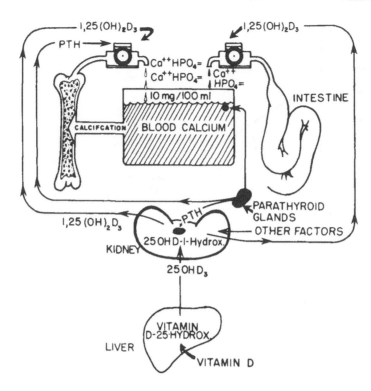

FIGURE 6. Role of vitamin D_3 in calcium homeostasis. (From DeLuca, H. F., *Am. J. Med.*, 58, 39, 1975. With permission.)

cells is mediated by an intracellular binding protein. $1,25\text{-(OH)}_2D_3$ arriving at target cells in intestine, bone, or elsewhere bound to DBP probably crosses the cell membrane in a "free" form to combine inside the cell with the $1,25\text{-(OH)}_2D_3$ receptor protein.[49] There, $1,25\text{-(OH)}_2D_3$ translocates to the nucleus in order to effect increased synthesis of proteins responsible for transporting calcium and phosphate. Its function over, $1,25\text{-(OH)}_2D_3$ is converted to calcitroic acid, a vitamin D molecule with a shortened side chain, for excretion in the bile. Recently it has been demonstrated that $1,25\text{-(OH)}_2D_3$ receptor proteins are present in cells throughout the body (e.g., skin, blood cells, osteoblast, brain, etc.), i.e., in tissues thought not to be target organs. Furthermore, $1,25\text{-(OH)}_2D_3$ appears to play important roles in these cells: in the osteoblast $1,25\text{-(OH)}_2D_3$ stimulates synthesis of osteocalcin, a γ-carboxy glutamic acid containing, vitamin K-dependent protein;[50] in the macrophage it induces fusion and differentiation.[51,52] Thus, $1,25\text{-(OH)}_2D_3$ may yet play additional functions in the body outside of its extracellular calcium regulating functions.

Other vitamin D metabolites ($1,24,25\text{-(OH)}_3D_3$, $25,26\text{-(OH)}_2D_3$, $23,25\text{-(OH)}_2D_3$, and $25\text{-(OH)}D_3\text{-}26,23\text{-lactone}$) are made in the kidney, but their role is unknown.

C. Age-Related Changes in Vitamin D Metabolism

Vitamin D absorption from the intestine is a very efficient process (60 to 80%) and in the rat appears to be unaffected by age.[14] In humans there is some evidence that vitamin D absorption can be less efficient during the neonatal period and in the elderly.[53] This is by no means clearcut since other studies have failed to demonstrate an age-related decline in vitamin D absorption.[54] Skin synthesis of vitamin D_3 from 7-dehydrocholesterol (provitamin D_3) does seem to be affected by aging. Holick et al.[55] have shown that 70-year-old epidermis has only 50% of the provitamin D_3 content of 20-year-old epidermis, suggesting that the ability of skin to synthesize provitamin D_3 on exposure to UV light decreases significantly

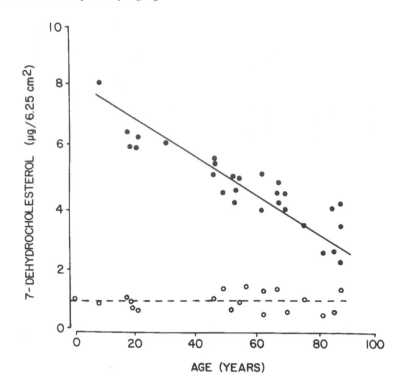

FIGURE 7.   Effects of aging on the skin content of 7-dehydrocholesterol (provitamin D) in 28 healthy human volunteers. Concentration of 7-dehydrocholesterol in epidermis, ●; dermis, ○. (Unpublished results, courtesy J. A. MacLaughlin and M. F. Holick, Massachusetts General Hospital, Boston.)

with age (Figure 7). In addition, a study of the rate of hepatic 25-hydroxylation of vitamin D, assessed by the ability of individuals to metabolize a large intramuscular injection of vitamin D, suggests that the efficiency of this process also diminishes in the elderly.[56,57]

All the above observations are consistent with the frequent finding that the plasma 25-(OH)D level is lower in the elderly population than in young adults.[53,58,59] Inefficient intestinal absorption of vitamin D, decreased skin synthesis of vitamin D, or poor 25-hydroxylation by the liver could all account for the diminished plasma 25-(OH)D level observed in the elderly. However, since plasma 25-(OH)D concentration is a good reflection of dietary intake of vitamin D and sunlight exposure of the individual, we cannot rule out that low levels of 25-(OH)D in the elderly are not also due in part to poor dietary practices or lack of exposure to sunlight brought on by decreased mobility. In fact, some data suggests that increased exposure to UV light can produce perfectly normal 25-(OH)D values in the elderly,[60] despite the decreased intestinal absorption, and lack of efficient skin synthesis and liver hydroxylation alluded to by others.

Though the changes in plasma 25-(OH)D can be marked[59] (in Reference 59, mean 25(OH)D$_3$ levels in normal controls [n = 47] were 29.1 ± 9.7 ng/mℓ [±SD] compared to 15.5 ± 7.2 in the elderly [n = 268] population, p <0.0001), there is no evidence that this difference necessarily causes any deleterious effects provided that the 25-(OH)D level remains above 5 ng/mℓ. Furthermore, the body experiences large fluctuations in plasma 25-(OH)D due to the seasonal variation in UV-exposure without discernable effects on calcium metabolism. This is because the plasma pool of 25-(OH)D merely acts as a reservoir for the kidney to use in synthesis of the small amounts of 1,25-(OH)$_2$D required to maintain normal Ca and PO$_4$ absorption in the intestine and resorption in the bone.

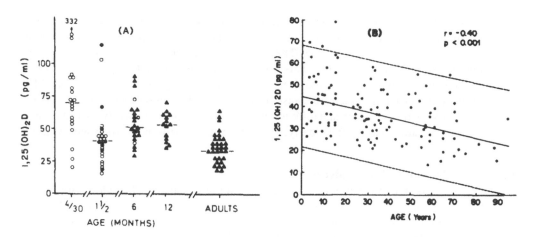

FIGURE 8. Effects of aging on the plasma concentration of 1,25-dihydroxyvitamin D in normal humans. (A) Over the first year of life. Symbols indicate differences in nutrition ○, human milk; ●, formula; ▲, mixed diet. (From Markestad, T., *J. Clin. Endocrinol. Metab.*, 57, 755, 1983. With permission.) (B) Over the range 2 to 94 years. — Regression, ---, 95% confidence limits for a single observation around the line. (From Manolagas, S. C. et al., *J. Clin. Endocrinol. Metab.*, 56, 751, 1983. With permission.)

It is the plasma concentration of 1,25-$(OH)_2$D that more accurately reflects needs for vitamin D. Several workers have demonstrated an age-related decline in plasma 1,25-$(OH)_2$D levels in both rat and man (Figure 8).[61-63] Some workers[62] have shown that the decline in plasma 1,25-$(OH)_2$D correlates well with both the age-related decline in mucosal 1,25-$(OH)_2$D concentration and the age-related decline in intestinal calcium absorption.[64] Whether these changes play any role in the etiology of osteoporosis is debatable (see Chapter 7 in Volume II).

The synthesis of 1,25-$(OH)_2$D by the kidney is stimulated not only by PTH but by a number of hormonal factors, and in the aging adult these hormonal modulators of 1,25-$(OH)_2$D may thus indirectly influence calcium metabolism. Growth hormone may well mediate the stimulatory effect of low blood phosphate on the 25-$(OH)D_3$-1-hydroxylase. Thus, the high 1,25-$(OH)_2$D levels seen in young children[65] may be the result of high plasma growth hormone. Estrogen also stimulates 1,25-$(OH)_2$D synthesis, and lack of this hormone in post-menopausal osteoporotic females may be partially responsible for slightly reduced 1,25-$(OH)_2$D levels.[66] Hormonal factors in the pregnant female result in additional placental synthesis of 1,25-$(OH)_2$D, which causes blood levels to be about twofold higher than in nonpregnant age-matched females.[67,68]

Thus, it appears that calcium homeostatic machinery is closely attuned to the body's needs for calcium in growth, pregnancy, and normal skeletal turnover. The blood 1,25-$(OH)_2$D concentration is an integral part of this homeostasis. Whether the lower blood 1,25-$(OH)_2$D level seen in old age is the reflection of decreased needs for vitamin D or represents a defect in the ability of the aging body to synthesize 1,25-$(OH)_2$D is an important question still to be answered.

D. Requirements for Vitamin D

The traditional unit of vitamin D activity, the international unit (IU), equivalent to 25 ng vitamin D, is based upon a bioassay called the "rat line-test", in which the preparation under test is compared to a standard amount of pure vitamin D for its ability to stimulate mineralization of a vitamin D-deficient rat bone. Such a unit is inadequate for expressing biological activity of newer, more potent metabolites such as 25-$(OH)D_3$ or 1,25-$(OH)_2D_3$.

The Recommended Dietary Allowances* are now expressed in micrograms of vitamin D per day. Suggested intakes for children of 10 μg/day are higher than adults and reflect an increased need over the growing period. The RDA for adolescents (19 to 22) is 7.5 μg/day, and for adults (over 22) it is 5 μg/day. This is greater than the amount (2.5 μg/day) believed adequate to maintain homeostasis. Additional allowances of 5 μg/day are recommended for both pregnant and lactating females because of the increased drain on their calcium metabolism caused by the fetal skeleton and milk for the newborn infant, respectively. No additional recommendations are made for the elderly. In North America milk is fortified with vitamin D (9 μg/ℓ).

E. Vitamin D Status of the Elderly

The studies reported in the literature regarding vitamin D status of the elderly are mainly from European countries. Reports from the U.S. and Canada are limited. Both in the U.S. and Canada, milk is fortified with vitamin D to a similar extent, but the solar exposure is less in Canada. In European countries milk is not fortified with vitamin D. An excellent review by Parfitt et al. on vitamin D and bone health in the elderly is available.[69]

Omadahl et al.[59] reported that in a group of healthy elderly subjects in Albuquerque, N.M., the median dietary intake of vitamin D was 88 IU/day, with 26% of the population taking a median supplement of 400 IU. Among the subjects studied, 48% of males and 54% of females consumed less than the RDA for vitamin D, and one third consumed less than 100 IU/day of vitamin D. The plasma level of 25-(OH)D was 47% lower in the elderly (15.5 ng/mℓ) compared to a younger control (29.1 ng/mℓ) population. Among the elderly population, 14.8% had plasma levels less than 8 ng/mℓ, which indicates borderline vitamin D deficiency. Within the elderly population, the plasma 25-(OH)D level demonstrated a seasonal influence and was consistently higher for males compared to females. People taking vitamin D supplements had higher plasma 25-(OH)D levels regardless of seasonal influence. In an abstract by Lukert et al.,[70] elderly subjects in a rural area, including some nursing home residents, had a mean dietary vitamin D intake of 8.6 μg/day and a mean plasma 25-(OH)D level of 27.2 ng/mℓ. Both are normal levels.

Somerville et al.[54] reported that a group of female elderly subjects in Montreal had a mean vitamin D intake of 169 IU/day, which is well above the minimum recommended Canadian intake of 100 IU/day, and was not significantly different from that of younger subjects. The mean plasma level of 25-(OH)D (23.3 ng/mℓ) was also not significantly different from that of younger subjects. Not only were the mean values normal, but no individual subjects had any evidence of suboptimal vitamin D nutrition.

The reports from European countries showed proportional declines in sun exposure and vitamin D intake with age. A report showed a more serious vitamin D deficiency problem in England than the U.S.[71] Most investigators have found significantly lower mean plasma levels of 25-(OH)D in the elderly compared to young controls.[71-74] Age-related declines in plasma levels of 25-(OH)D have been observed.[74-77] The principal cause of the modest decline of plasma 25-(OH)D level with age is diminished exposure to sunlight. In the elderly, plasma 25-(OH)D level correlates more strongly with sun exposure than with dietary vitamin D intake.[74] However, if dietary intake is high enough to compensate for absence of sun exposure, plasma level of 25-(OH)D may increase.[78]

Thus, inadequate dietary vitamin D intake and inadequate solar exposure appear to contribute to low vitamin D status in the elderly.

F. Problems of Deficiency and Excess of Vitamin D

Simple vitamin D deficiency is rare.[79] It is mainly seen in breast-fed babies born in the autumn, since the vitamin D content of human milk is low and babies born at this time may receive inadequate exposure to sunlight during their first 6 months. In older children and

---

* Extracted from *Recommended Dietary Allowances*, 9th ed., Washington, D.C., 1980.

adults, vitamin D deficiency only occurs where social practices restrict exposure of skin and diets are poor in dairy products and meat.

Vitamin D deficiency is characterized by low plasma levels of 25-(OH)D (less than 5.0 ng/mℓ) which result in inadequate substrate for the kidney and ultimately, a fall in 1,25-(OH)D levels. Inadequate delivery of 1,25-(OH)$_2$D to the intestine results in a decrease in the percent calcium and phosphate absorption from the diet. Plasma Ca and PO$_4$ levels fall below the normal range (Ca: below 8.8 mg/100 mℓ; PO$_4$: below 2.5 mg/100 mℓ). PTH is secreted by the parathyroid gland in response to low blood calcium but fails to restore blood calcium to normal. Inadequate levels of calcium and phosphate produce a number of sequelae. Bone matrix is improperly mineralized and when examined histologically, appears as wide unmineralized "osteoid" (bone matrix) seams on the surface of bone. In children, unmineralized bone is clearly seen at the ends of long bones, where growth occurs. These defects of the growth plates, together with a weakening of the shaft of the bone, causing bowing, are known as "rickets". In adults the growth plates are not involved and the condition is known as osteomalacia. In the elderly, osteomalacia is often accompanied by osteoporosis and it is difficult to make a diagnosis of simple vitamin D deficiency. Restoration of blood calcium and phosphate levels to normal by administration of vitamin D quickly corrects the lack of mineralization of bone in simple vitamin D deficiency.

Left unchecked, plasma Ca can reach a dangerously low level so that inadequate amounts are available for neurotransmission or muscle contraction. Tetany of muscles and convulsions can occur. In vitamin D-deficient babies the costocondral joints of the rib cage become swollen and are known as the "rachitic rosary". In fact, the chest wall can become weakened sufficiently to leave the baby susceptible to pneumonia and this can result in death if severe vitamin D deficiency is left untreated.

Excessive intakes of vitamin D are quite dangerous and can be fatal.[80] The mechanism is still not clear but the levels of several metabolites of vitamin D, including 25-(OH)D, become extremely elevated. 1,25-(OH)$_2$D levels are probably only slightly elevated and some claim that they may be even suppressed. Excessive bone resorption occurs, which results in hypercalcemia, which is difficult to control. Plasma calcium level can rise as high as 15 to 20 mg/100 mℓ. Heart failure can result. Chronic mild hypercalcemia can result in soft tissue calcifications, especially in the kidney. Some renal damage invariably occurs as the result of long-term vitamin D high-dosage therapy. Megavitamin D supplements to the diet are not recommended for these reasons.

REFERENCES

1. **Harman, D.,** Free radical theory of aging: effect of free radical reaction inhibitors on the mortality rate of male LAF mice, *J. Gerontol.,* 23, 476, 1968.
2. **Ames, B. N.,** Dietary carcinogens and anticarcinogens, *Science,* 221, 1256, 1983.
3. **Goodman, D. S.,** Vitamin A and retinoids: recent advances, *Fed. Proc. Fed. Am. Soc. Exp. Biol.,* 38, 2501, 1979.
4. **Smith, J. E. and Goodman, D. S.,** Retinol-binding protein and the regulation of vitamin A transport, *Fed. Proc. Fed. Am. Soc. Exp. Biol.,* 38, 2504, 1979.
5. **Chytil, F. and Ong, D. E.,** Cellular retinol- and retinoic acid-binding proteins in vitamin A action, *Fed. Proc. Fed. Am. Soc. Exp. Biol.,* 38, 2510, 1979.
6. **Wald, G.,** Molecular basis of visual excitation, *Science,* 162, 230, 1968.
7. **DeLuca, L. M., Bhat, P. V., Sasak, W., and Adamo, S.,** Biosynthesis of phosphoryl and glycosyl phosphoryl derivatives of vitamin A in biological membranes, *Fed. Proc. Fed. Am. Soc. Exp. Biol.,* 38, 2535, 1979.
8. **Wolf, G., Kiorpes, T. C., Masushige, S., Schreiber, J. B., Smith, M. J., and Anderson, R. S.,** Recent evidence for the participation of vitamin A in glycoprotein biosynthesis, *Fed. Proc. Fed. Am. Soc. Exp. Biol.,* 38, 2540, 1979.

9. **DeLuca, L. M. and Shapiro, S. S., Eds.,** Modulation of cellular interactions of vitamin A and its derivatives (retinoids), *Ann. N.Y. Acad. Sci.,* 359, 1, 1981.

10. **Pitt, G. A. J.,** Chemical structure and changing concept of vitamin A activity, *Proc. Nutr. Soc.,* 42, 43, 1983.

11. **Orfanos, C. E., Ed.,** *Retinoids,* Springer-Verlag, Berlin, 1981.

12. **Lotan, R.,** Effects of vitamin A and its analogs (retinoids) on normal and neoplastic cells, *Biochim. Biophys. Acta,* 605, 33, 1980.

13. **Peto, R., Doll, R., Buckley, J. D., and Sporn, M. B.,** Can dietary beta-carotene materially reduce human cancer rates? *Nature (London),* 290, 201, 1981.

14. **Fleming, B. B. and Barrows, C. H.,** The influence of aging on intestinal absorption of vitamins A and D by the rat, *Exp. Gerontol.,* 17, 115, 1982.

15. **Smith, F. R., Raz, A., and Goodman, D. S.,** Radioimmunoassay of human plasma retinol-binding protein, *J. Clin. Invest.,* 49, 1754, 1970.

16. **Kershaw, R. C.,** Factors Controlling Plasma Retinol-Binding Protein Concentration, Ph.D. thesis, University of Liverpool, 1977.

17. **Glover, J., Heaf, D. J., and Large, S.,** Seasonal changes in plasma retinol-binding holoprotein concentration in Japanese quail, *Br. J. Nutr.,* 43, 357, 1980.

18. **Fisher, S., Hendricks, D. G., and Mahoney, A. W.,** Nutritional assessment of senior rural Utahns by biochemical and physical measurements, *Am. J. Clin. Nutr.,* 31, 667, 1978.

19. **Harrill, I. and Cervone, N.,** Vitamin status of older women, *Am. J. Clin. Nutr.,* 30, 431, 1977.

20. Ten-State Nutrition Survey, 1968—1970. IV. Biochemical, Department of Health, Education and Welfare Publ. No. (HSM) 72-8132, Center for Disease Control, U.S. Department of Health, Education and Welfare, Atlanta, 1972.

21. First Health and Nutrition Examination Survey, United States, 1971—1972, Publ. (HRA) 74-1219-1, U.S. Department of Health, Education and Welfare, Washington, D.C., 1974.

22. Nutrition Survey of the West Indies, Interdepartmental Committee on Nutrition for National Defense Report, Department of Defense, Washington, D.C., 1962.

23. **Cheraskin, E., Ringsdorf, W. M., and Medford, F. H.,** The "ideal" daily vitamin A intake, *Int. J. Vitam. Nutr. Res.,* 46, 11, 1976.

24. **Wolf, G.,** Is dietary β-carotene an anti-cancer agent? *Nutr. Rev.,* 40, 257, 1982.

25. Consumption of Households in the U.S., Spring, 1965, Household Food Consumption Survey, 1965 to 1966, Consumer and Food Economics Divisions, Agricultural Research Service, U.S. Department of Agriculture, Washington, D.C., 1968, 212.

26. **Guthrie, H. A., Black, K., and Madden, J. P.,** Nutritional practices of elderly citizens in rural Pennsylvania, *Gerontology,* 12, 330, 1972.

27. **Rawson, I. G., Weinberg, E. I., Herold, J. A., and Holtz, J.,** Nutrition of rural elderly in Southwestern Pennsylvania, *Gerontologist,* 24, 1978.

28. **Pao, E. M. and Hill, M. M.,** Diets of the elderly — nutrition labelling and nutrition education, *J. Nutr. Educ.,* 6, 96, 1974.

29. **Todhunter, N. E. and Darby, W. J.,** Guidelines for maintaining adequate nutrition in old age, *Geriatrics,* 33, 49, 1978.

30. **LeBovit, C.,** The food of older persons living at home, *J. Am. Dietet. Assoc.,* 46, 285, 1965.

31. **Yearick, E. S., Wang, M. L., and Pisias, S. J.,** Nutritional status of the elderly: dietary and biochemical findings, *J. Gerontol.,* 35, 663, 1980.

32. **Dibble, M. W., Brin, M., Thiele, V. F., Peel, A., Chen, N., and McMullen, E.,** Evaluation of the nutritional status of elderly subjects with a comparison between fall and spring, *J. Am. Geriatr. Soc.,* 15, 1031, 1967.

33. **Steinkamp, E. C., Cohen, N. L., and Walsh, H. E.,** Resurvey of an aging population — fourteen year follow-up, *J. Am. Dietet. Assoc.,* 46, 103, 1965.

34. **Lyons, J. S. and Trulson, M. F.,** Food practices of older people living at home, *J. Gerontol.,* 11, 66, 1956.

35. **Fry, P. C., Fox, H. M., and Linkswiler, H.,** Nutrient intakes of healthy older women, *J. Am. Dietet. Assoc.,* 42, 218, 1963.

36. Interdepartmental Committee for National Defense, Manual for Nutrition Surveys, Department of Defense, Washington, D.C., 1957.

37. **Brin, M., Dibble, M. V., and Peele, A.,** Some preliminary findings on the nutrition status of the aged in Onondago County, New York, *Am. J. Clin. Nutr.,* 17, 240, 1965.

38. **Gillum, H. L., Morgan, A. F., and Sailer, F.,** Nutritional status of the aging. V. Vitamin A and carotene, *J. Nutr.,* 55, 655, 1955.

39. **Moore, T.,** *Vitamin A,* Elsevier, New York, 1957.

40. **Fell, H. B. and Mellanby, E.,** The effect of hypervitaminosis A on embryonic limb-bones cultivated *in vitro, J. Physiol. (London),* 116, 320, 1952.

41. **DeLuca, H. F.**, Vitamin D, metabolism and function, *Monographs in Endocrinology*, Springer-Verlag, Berlin, 1979, 1.

42. **Norman, A. W.**, The vitamin D endocrine system: steroid metabolism, hormone receptors, and biological response (calcium-binding proteins), *Endocr. Rev.*, 3, 331, 1982.

43. **Holick, M. F., Richtand, N. M., McNeil, S. C., Holick, S. A., Frommer, J. E., Henley, J. W., and Potts, J. T., Jr.**, Isolation and identification of previtamin D, from the skin of rats exposed to ultraviolet irradiation, *Biochemistry*, 18, 1003, 1979.

44. **Bouillon, R., Van Baelen, H., Rombauts, W., and DeMoor, P.**, The purification and characterisation of the human-serum binding protein for the 25-hydroxycholecalciferol (transcalciferol). Identity with group-specific component, *Eur. J. Biochem.*, 66, 285, 1976.

45. **Bhattacharyya, M. H. and DeLuca, H. F.**, The regulation of rat liver calciferol-25-hydroxylase, *J. Biol. Chem.*, 248, 2969, 1973.

46. **Jones, G.**, Assay of vitamins D_2 and D_3 and 25-hydroxyvitamins D_2 and D_3 in human plasma by high-performance liquid chromatography, *Clin. Chem.*, 24, 287, 1978.

47. **Fraser, D. R. and Kodicek, E.**, Unique biosynthesis by kidney of a biologically active vitamin D metabolite, *Nature (London)*, 228, 764, 1970.

48. **Holick, M. F., Schnoes, H. K., DeLuca, H. F., Gray, R. W., Boyle, I. T., and Suda, T.**, Isolation and identification of 24,25-dihydroxycholecalciferol, a metabolite of vitamin D, made in the kidney, *Biochemistry*, 11, 4251, 1972.

49. **Haussler, M. R. and McCain, T. A.**, Basic and clinical concepts related to vitamin D metabolism and action, *N. Engl. J. Med.*, 297, 974, 1977.

50. **Price, P. A. and Baukal, S. A.**, $1,25$-$(OH)_2D_3$ increases synthesis of the vitamin K-dependent bone protein by osteosarcoma cells, *J. Biol. Chem.*, 255, 11660, 1980.

51. **Abe, E., Miyaura, C., Sakagami, H., Takeda, M., Konno, K., Yamazaki, T., Yoshiki, S., and Suda, T.**, Differentiation of mouse myeloid leukemia cells induced by $1\alpha,25$-dihydroxyvitamin D_3, *Proc. Natl. Acad. Sci. U.S.A.*, 78, 4990, 1981.

52. **Abe, E., Miyaura, C., Tanaka, H., Shiina, Y., Kuribayashi, T., Suda, S., Nishii, Y., DeLuca, H. F., and Suda, T.**, $1\alpha,25$-Dihydroxyvitamin D_3 promotes fusion of mouse alveolar macrophages both by a direct mechanism and by a spleen cell-mediated indirect mechanism, *Proc. Natl. Acad. Sci. U.S.A.*, 80, 5583, 1983.

53. **Weisman, Y., Schen, R. J., Eisenberg, Z., Edelstein, Y., and Harell, A.**, Inadequate status and impaired metabolism of vitamin D in the elderly, *Isr. J. Med. Sci.*, 17, 19, 1981.

54. **Somerville, P. J., Lien, J. W. K., and Kaye, M.**, The calcium and vitamin D status in an elderly female population and their response to administered supplements of vitamin D_3, *J. Gerontol.*, 32, 659, 1977.

55. **MacLaughlin, J. A. and Holick, M. F.**, *American Clinical Proceedings*, Washington, abstract, 1981.

56. **Skinner, R. K.**, 25-Hydroxylation of vitamin D in the elderly, in *Vitamin D, Basic Research and Its Clinical Application*, Norman, A. W., Ed., de Gruyter, New York, 1979, 1011.

57. **Rushton, C.**, Vitamin D hydroxylation in youth and old age, *Age Ageing*, 7, 91, 1978.

58. **Corless, D., Beer, M., Boucher, B. J., Gupta, S. P., and Cohen, R. D.**, Vitamin D status in long stay geriatric patients, *Lancet*, 1, 1404, 1975.

59. **Omdahl, J. L., Garry, P. J., Hunsaker, L. A., Hunt, W. C., and Goodwin, J. S.**, Nutritional status in a healthy elderly population: vitamin D, *Am. J. Clin. Nutr.*, 36, 1225, 1982.

60. **Guggenheim, K., Kravitz, M., Tal, R., and Kaufman, N. A.**, Biochemical parameters of vitamin D nutriture in old people in Jerusalem, *Nutr. Metab.*, 23, 172, 1979.

61. **Manolagas, S. C., Culler, F. L., Howard, J. E., Brickman, A. S., and Deftos, L. J.**, The cytoreceptor assay for $1,25$-$(OH)_2D_3$ and its application to clinical studies, *J. Clin. Endocrinol. Metab.*, 56, 751, 1983.

62. **Horst, R. L., DeLuca, H. F., and Jorgensen, N. A.**, The effect of age on calcium absorption and accumulation of $1,25$-$(OH)_2D_3$ in the intestinal mucosa of rats, *Metab. Bone Dis. Rel. Res.*, 1, 29, 1978.

63. **Gray, R. W. and Gambert, S. R.**, Effect of age on plasma $1,25$-$(OH)_2$ vitamin D in the rat, *Age*, 5, 54, 1982.

64. **Bullamore, J. R., Wilkinson, R., Gallagher, J. C., Nordin, B. E. C., and Marshall, D. H.**, Effects of age on calcium absorption, *Lancet*, 2, 535, 1970.

65. **Markestad, T.**, Plasma concentrations of $1,25$-$(OH)_2D$, $24,25$-$(OH)_2D$ and $25,26$-$(OH)_2D$ in the first year of life, *J. Clin. Endocrinol. Metab.*, 57, 755, 1983.

66. **Gallagher, J. C., Riggs, B. L., Eisman, J. A., Hamstra, A., Arnand, S. B., and DeLuca, H. F.**, Intestinal calcium absorption and serum vitamin D metabolites in normal subjects and osteoporotic patients, *J. Clin. Invest.*, 64, 729, 1979.

67. **Kumar, R., Cohen, W. R., Silva, P., and Epstein, F. H.**, Elevated 1,25-dihydroxyvitamin D levels in normal pregnancy and lactation, *J. Clin. Invest.*, 63, 342, 1979.

68. **Reddy, G. S., Norman, A. W., Willis, D. M., Golzman, D., Guyda, H., Solomon, S., Philips, D. R., Bishop, J. E., and Mayer, E.**, Regulation of vitamin D metabolism in normal human pregnancy, *J. Clin. Endocrinol. Metab.*, 56, 363, 1983.

69. **Parfitt, A. M., Chir, B., Gallaher, J. C., Heaney, R. P., Johnston, C. C., Neer, R., and Whedon, G. D.,** Vitamin D and bone health in the elderly, *Am. J. Clin. Nutr.,* 36, 1014, 1982.

70. **Lukert, B. P., Carey, M. A., McCarty, B., Thieman, S., and Goodnight, L.,** Nutrition calcium homeostasis and bone density in the elderly, *Abstr. 12th Int. Congr. Nutr.,* 77, 1981.

71. **Vir, S. C. and Love, A. H. G.,** Vitamin D status of elderly at home and institutionalized in hospital, *Int. J. Vit. Nutr. Res.,* 48, 123, 1978.

72. **Lester, E., Skinner, R. K., and Wills, M. R.,** Seasonal variation in serum 25-hydroxyvitamin D in the elderly in Britain, *Lancet,* 1, 979, 1977.

73. **Lawson, D. E. M., Paul, A. A., Black, A. E., and Cole, T. J.,** Relative contributions of diet and sunlight to vitamin D state in the elderly, *Br. Med. J.,* 2, 303, 1979.

74. **Hodkinson, H. M., Bryson, E., Klenerman, I., Clark, M. B., and Wooton, R.,** Sex, sunlight, season, diet and the vitamin D status of elderly patients, *J. Clin. Exp. Gerontol.,* 1, 13, 1979.

75. **Lund, B. and Sorenson, O. H.,** Measurement of 25-OH vitamin D in serum and its relation to sunshine, age and vitamin D intake in the Danish population, *Scand. J. Clin. Lab. Invest.,* 39, 23, 1979.

76. **Garcia-Pascual, B., Peytremann, A., Courvoisier, B., and Lawson, D. E. M.,** A simplified competitive protein-binding assay for 25-hydroxycalciferol, *Clin. Chim. Acta,* 68, 99, 1976.

77. **Schmidt-Gayk, H., Goossen, J., Lendie, F., and Seidel, D.,** Serum 25-hydroxycalciferol in myocardial infarction, *Atherosclerosis,* 26, 55, 1977.

78. **Devgun, M. S., Paterson, C. R., Johnson, B. E., and Cohen, C.,** Vitamin D nutrition in relation to season and occupation, *Am. J. Clin. Nutr.,* 34, 1501, 1981.

79. **Fraser, D., Kooh, S. W., and Scriver, C. R.,** Rickets, osteomalacia and vitamin D, *Med. N. Am.,* 12, 1247, 1981.

80. **Yendt, E. R.,** Pharmacological activities of vitamin D, in *International Encyclopedia of Pharmacology and Therapeutics,* Sect. 51, Pergamon Press, Oxford, 1970, 139.

81. **DeLuca, H. F.,** The kidney as an endocrine organ involved in the function of vitamin D, *Am. J. Med.,* 58, 39, 1975.

Chapter 9

FAT-SOLUBLE VITAMINS

Part II

VITAMIN E

Ching K. Chow

TABLE OF CONTENTS

I. INTRODUCTION

Aging is generally regarded as a state of life accompanied by progressive structural and compositional changes, and loss of functional capacity and adaptability of an organism or organ system, which leads to a decreased survival capability and eventual death. Life expectancy at birth in most countries has increased considerably since the turn of this century.[1] As nutrients are essential for all fundamental processes, it is conceivable that the nutritional status of individuals may modify the cellular reactions that regulate or are involved in the biological process of aging. Diet/nutrition is also known to play a role in the pathogenesis of a variety of diseases. Therefore, improvement in diet or nutrition may have contributed significantly to the increase in life expectancy.

A number of dietary components have been linked to or suggested to be associated with the acceleration or retardation of the aging process. Vitamin E, the major biological antioxidant, has received considerable attention in research on aging. This is partially due to the proposed theory that free radical-initiated lipid peroxidation causes tissue damage which may partly be responsible for initiating or accelerating the aging process in vivo.[2-6] The focus of this chapter will be the recent research on vitamin E in protecting cellular constituents from free radical-initiated lipid peroxidation tissue damage that may have an impact on the aging process.

II. FREE RADICAL-INITIATED TISSUE DAMAGE

Free radicals have long been recognized as intermediates in many biological redox reactions essential for the maintenance of life. Biological materials, particularly cell membranes, contain relatively high concentrations of polyunsaturated lipids. In the presence of a free radical initiator and oxygen, lipids may be oxidized. This process, known as lipid peroxidation, has been implicated as a general biological degenerative reaction, and may be an important in vivo process.[7-9]

The cell membrane separates the highly ordered functional activities of metabolism within the cell from the relative randomness of the external environment. The membrane contains a relatively high content of polyunsaturated fatty acid residues and often has heme and flavins as a part of the basic structure. The media on both sides of the membrane contain both oxygen and trace metals. Under these circumstances, the membrane is potentially more susceptible to free radical-initiated lipid peroxidation damage than cytoplasmic components.

The process of lipid peroxidation in biological systems may be associated with the oxidation of essential polyunsaturated fatty acids, and the formation of toxic hydroperoxides, epoxides, aldehydes, and other secondary products. The loss of essential fatty acids may result in a disturbance in the fine structure of biological membranes and thus affect the permeability and function of the membrane.

Any disruption of the integrity of membrane structure will subsequently affect important biological processes in the cells. In addition to lipid peroxidation, proteins interspaced in the bilayer may possibly be altered by free radicals originating in the lipid phase of the membrane. Structural proteins are arranged in a filamentous network on the cytoplasmic face of the membrane. This network is responsible for supporting and maintaining the intact bilayer. Because of the proximity of these structural proteins to the membrane, these proteins may also be modified by free radicals, hydroperoxides, and secondary products present in the phospholipid bilayer, or form cross-linking compounds with lipid peroxidation products, e.g., malonaldehyde.

Extensive oxidation may lead to rupture of cell membranes and concomitant release of destructive lysosomal enzymes. The hydroperoxides and secondary products formed, if not removed, may react with and inactivate essential proteins, enzymes, and nucleic acids.[10,11]

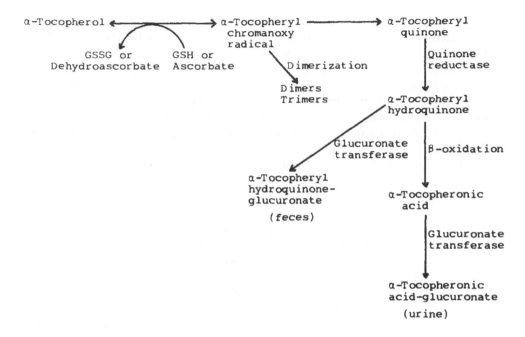

FIGURE 1. Possible mechanism of aging process initiated by free radical lipid peroxidation and its inhibition by vitamin E.

Cumulative damage may eventually lead to the death and/or turnover of the cell (Figure 1). The process of free radical-initiated lipid peroxidation has been suggested to play a role in initiating or accelerating the aging process.[2-6]

III. FREE RADICAL REACTIONS AND AGING

The free radical theory of aging originated by Harman[2,3] is based on the premise that random free radical reactions are a major factor in the breakdown of essential biological systems. It is assumed that free radical reactions either initiated by enzymic or nonenzymic means go on continuously throughout the cells and tissues. If the reactive free radical products generated are not properly controlled, they may cause deleterious effects on essential cellular constituents. The free radical reactions may eventually decrease the ability of an organ to respond adaptively to environmental changes, accelerate the aging process, and thereby shorten the life span of an organism. According to this theory, degenerative changes in normal cellular functions caused by free radical-initiated tissue damage may result either from excessive production of free radicals or from weakened cellular antioxidant defense systems. Therefore, the aging process may be described as the sum of the ever-present deleterious reactions throughout the cells and tissues caused by free radical reactions, and longevity is a reflection of the ability to cope with such reactions.[2,3]

The free radical theory of aging has received considerable support from experimental evidence that products of free radical reactions may form cross-linking substances, which are similar to those observed in aged tissues. Free radicals formed may undergo additional reactions to yield covalent bonds and novel radicals. Hydroperoxides and free radicals may also undergo scission reactions to generate more radicals, or be cleaved to yield a variety of products, including alkanes and aldehydes. One of the aldehydes, malonaldehyde, has been shown to form cross-linking products with amines of proteins, phospholipids, and nucleic acids.[12,13]

A cross-linking reaction between the amino group of macromolecules and aldehydes formed from lipid peroxidation in vitro has also been demonstrated. This finding suggests that lipid peroxidation products may be partly responsible for cross-linking in vivo. These Schiff base products have been shown to be fluorescent, and the accumulation of fluorescent pigments, sometimes termed "aging pigments", "lipofuscin", or "ceroid", has been shown to be directly related to the age of animals.[13,15] The Schiff base products of condensation of malonaldehyde with the primary amines of proteins, nucleic acids, or phospholipids have been shown to exhibit fluorescent properties similar to those observed for native lipofuscin.[8,13] Peroxidation of subcellular fractions of organelles (lysosome, microsome, or mitochondria) has also been shown to form Schiff base products with spectra properties similar to those of native lipofuscin.[13,16] The Schiff base products, however, are not related directly to the polyunsaturated fatty acid composition of tissues. Unusual cross-linking reactions increase with age and may contribute to the loss of cellular integrity known to be associated with advancing age. Rickert and Forbes[17] have shown that collagen isolated from the lungs of elderly persons has an increased ratio of insoluble to acid-soluble collagen compared to that of younger persons. The finding supports the view that abnormal cross-links increase with age. Cell membranes are also known to increase their rigidity with advancing age,[18] an effect probably mediated by free radical lipid peroxidation reactions.

Sagai and Ichinose[19] measured the lipid peroxidation products, ethane, ethylene, butane, and pentane, in respired gases of rats and observed that the amounts of the hydrocarbons exhaled were greater in the older rats (up to 32 months) than those in younger rats. The concentrations of these hydrocarbons in the exhaled gases of experimental animals have been shown to be increased by conditions that enhance lipid peroxidation, and decreased by dietary vitamin E.[8,9,20] The work of Sagai and Ichinose[19] provide another line of evidence that in vivo lipid peroxidation may be associated with acceleration in aging.

IV. CELLULAR ANTIOXIDANT DEFENSE SYSTEMS

In view of the potential damage that may be caused by free radicals and other reactive oxygen species, it is important that the cell contains antioxidant defense systems. Among various enzymic and nonenzymic antioxidant defense systems, the scavenging of free radicals by vitamin E and superoxide dismutase, and the reduction or detoxification of hydroperoxides formed by glutathione (GSH) peroxidase, appear to be the key ones.[9] The action of GSH peroxidase in reducing membrane-associated lipid hydroperoxides may require the prior action of phospholipase to release fatty acetyl hydroperoxides.[21,22] The restoration of GSH involves the reaction of oxidized glutathione (GSSG) through the action of GSSG reductase in which oxidized nicotinamide adenosine dinucleotide phosphate is generated. The latter in turn leads to the activation of the hexose monophosphate shunt through glucose-6-phosphate dehydrogenase to provide a continued supply of reduced nicotinamide adenine dinucleotide phosphate. Catalase is effective only at high concentrations of hydrogen peroxide, whereas GSH peroxidase is effective at physiologic concentrations of hydrogen peroxide. It has been proposed that superoxide dismutase protects the cell against deleterious effects of superoxide anion radicals. Furthermore, a number of enzymic and nonenzymic systems have been shown to be involved in the overall cellular antioxidant defense.[9] Among the important antioxidant defense systems known, however, vitamin E appears to be the most important and dominant. Following a quantitative comparison of the vitamin E to total antioxidant concentrations, Burton et al.[23] concluded that vitamin E is the major, and probably the only, lipid-soluble free radical chain-breaking antioxidant in adult human blood plasma and red blood cells.

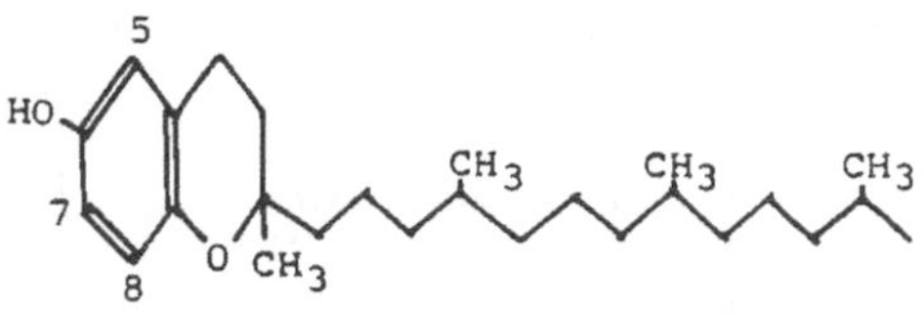

Toco structure

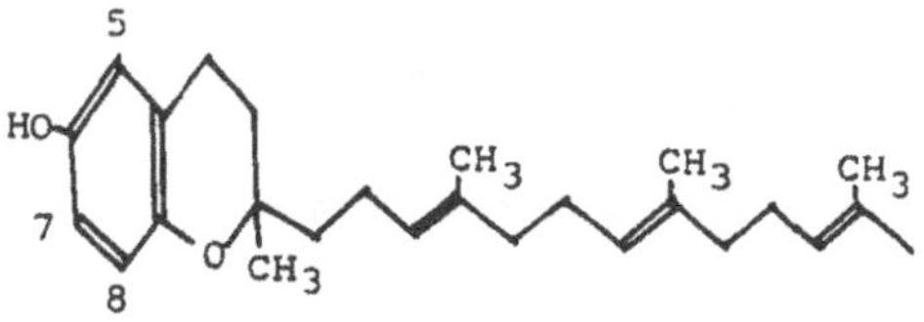

Tocotrienol structure

Position of methyls	Toco structure	Tocotrienol structure
5,7,8	α-tocopherol (α-T)	α-tocotrienol (α-T-3)
5,8	β-tocopherol (β-T)	β-tocotrienol (β-T-3)
7,8	γ-tocopherol (γ-T)	γ-tocotrienol (γ-T-3)
8	δ-tocopherol (δ-T)	δ-tocotrienol (δ-T-3)

FIGURE 2.   Structural formula of tocopherols.

V. BIOLOGICAL FUNCTIONS OF VITAMIN E

Vitamin E is the term suggested for all toco- and tocotrienol derivatives exhibiting qualitatively the biological activity of α-tocopherol. All eight naturally occurring compounds known in the tocopherol series are derivatives of 6-chromanol (Figure 2). This series is made up of four compounds with a toco structure which bears a saturated isoprenoid C_{16} side chain, and four compounds with a tocotrienol structure bearing three double bonds in the isoprenoid side chain.

The main criteria used in the biological evaluation of the tocopherols have been the resorption-gestation assay, the muscular dystrophy score, the erythrocyte hemolysis test, and determination of tocopherol levels in plasma and tissues (mainly liver). While the term "vitamin E" refers to at least eight tocopherol structures possessing vitamin E activity, α-tocopherol predominates in many species and is significantly more potent than any other naturally occurring tocopherols known. Therefore, a determination of α-tocopherol content is usually a good approximation of the total vitamin E activity of the food source.

The primary role of vitamin E in preventing free radical-initiated lipid peroxidation damage in tissues is accepted by most investigators in the field. This action, however, does not adequately explain some biochemical abnormalities observed in vitamin E deficiency. The presence of other cellular antioxidant defense systems, especially the GSH peroxidase system in the parenchymal tissues, may account at least partly for the discrepancy.[9]

Increasing evidence indicates that vitamin E may exert its biological function within cellular membranes. The localization of vitamin E within the membrane as a complex with the polyunsaturated fatty acids of phospholipids[23] may thus inhibit the peroxidation of membrane-bound polyunsaturated fatty acids during electron transport and various other functions.[25] Vitamin E has been shown to be a part of the inner mitochondrial membrane of duck liver[26] and is localized exclusively in the membrane portion of human red blood

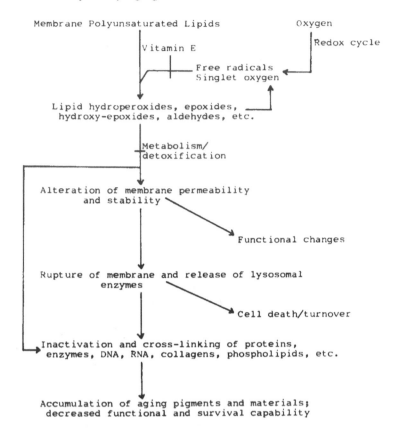

FIGURE 3.   Regulation and metabolic fate of tocopherol.

cells.[27] In addition to erythrocyte membrane, vitamin E deficiency has been shown to markedly affect the ultrastructure and integrity of mitochondrial, endoplasmic reticular, and nuclear membranes of duck liver.[26]

The protective effect of vitamin E against lipid peroxidation tissue damage has been attributed at least partly to its ability to scavenge free radicals. In the process of quenching free radicals, vitamin E is first oxidized to the tocopheryl chromanyl radical (Figure 3). At this stage the tocopherol radical can be reduced to α-tocopherol by a not yet completely defined system or group of systems. Systems, including such reducing agents as ascorbic acid and GSH, and specific tocopherol-regenerating enzymes/proteins have been suggested to be involved in tocopherol regeneration.[28,29] Further oxidation of the tocopherol radical to a quinone, while quenching another molecule of radical, is an irreversible step (Figure 3).

Studies on the metabolism of α-tocopherol quinone and its hydroquinone showed no conversion to α-tocopherol in rat liver.[30] The compound was metabolized partially by reduction to the hydroquinone, conjugation with glucuronic acid, secretion in the bile, and elimination in the feces. Another portion of the quinone may have been degraded through a β-oxidation pathway in the kidney to α-tocopheronic acid, followed by conjugation and elimination in the urine (Figure 3). In addition to tocopheronic acid, tocopheryl quinone and tocopheryl hydroquinone, small amounts of dimer and trimer of α-tocopherol have been isolated in animal tissue.[31,32]

Vitamin E deficiency has been shown to alter the activities of a large number of enzymes in the tissues of various species of animals.[33,34] This leads to the suggestion that vitamin E may have a specific role in regulating the synthesis of enzymes/proteins. The alterations of enzymic activities may have a significant biological consequence in changing normal biological processes. One of the consequences may be related to the alteration of redox status or the balance between factors that exert antioxidant potential and those that favor peroxidation reactions. Increased activity of xanthione oxidase observed in vitamin E-deficient rats and rabbit liver,[34] for example, has been shown to result from increased levels of enzyme proteins rather than activation of the preexisting enzyme. The increased levels of xanthione oxidase due to vitamin E deficiency may lead to an increased production of superoxide radicals, and promote free radical-initiated tissue damage. On the other hand, the activities of GSH peroxidase and metabolically related enzymes have been shown to increase significantly in several tissues of vitamin E-deficient rats[35] and in the lungs of ozone-exposed rats.[36] The apparent compensation reaction may thus afford certain protection against oxidative damage resulting from vitamin E deficiency or other types of oxidative stresses.

VI. AGE-RELATED REQUIREMENT OF VITAMIN E

Vitamin E requirement for maintaining normal function and cellular integrity may change with age. Emerson and Evans[37] reported the necessity of increasing the dose of wheat germ oil from 0.5 to 4.0 g to restore fertility in female rats as they aged from 3 to 12 months, and much higher doses were needed after 15 months of age. Based on a resorption-gestation assay, Fuhr et al.[38] estimated that the vitamin E requirement of female rats increased roughly tenfold from the first half of life to the second. Also, using the resorption-gestation assay, Ames[39] found that the vitamin E requirement of rats aged 43 to 45 weeks was 11 times greater than that at age 9 to 11 weeks, and at 59 weeks of age the requirement rose to 67 times that of young rats. He also observed that the amount of vitamin E necessary to prevent erythrocyte hemolysis was 3 times higher at 71 to 72 weeks of age than at 9 to 11 weeks. On the other hand, based on the ability of vitamin E to prevent necrotizing myopathy, Gabriel et al.[40] concluded that the requirement of vitamin E did not change significantly at 12, 24, 48, and 68 weeks of age. However, it is not known whether the requirement for maintenance of muscle is altered when rats are older than 68 weeks.

According to the free radical theory of aging, levels of antioxidants or prooxidants may be related to the retardation or acceleration of the aging process. Thus, information on the levels of various antioxidant defense systems as a function of age has been a subject of many investigations. The activity of superoxide dismutase, for example, has been shown to decrease with increasing age in rats.[41] It has also been shown that the capacity of GSH utilization via GSH S-transferase in mosquitoes is diminished with aging.[42] This finding suggests that the GSH-linked detoxification pathway may be impaired with senescence. Lower levels of GSH were found in the tissues of old mice as compared to the mature animals.[43] Lower levels of selenium, which is an integral part of GSH peroxidase, and mercaptans have been found in the blood of elderly persons.[44,45] Significant decreases in levels of vitamin D in plasma, platelets, and leukocytes have also been observed in humans as a function of age.[46] Vitamin E nutritional status, however, has not been shown to decline with advancing age in humans.[47,48] In rats, levels of vitamin E in tissues have been shown to increase with age.[49]

VII. VITAMIN E AND AGING PARAMETERS

As the most important biological antioxidant known, the anti-aging role of vitamin E has been a subject of many research investigations. The experimental findings, however, have not been conclusive. Dietary vitamin E has been shown to be inversely related to the concentrations of lipid-soluble fluorescent products measured in animal tissue, as well as ethane and pentane evolved.[8,35,36,50,51] Hirahara et al.[52] have shown that vitamin E supplementation (180 ppm) acts to reduce the decrease of total plasma protein, plasma albumin, and the albumin to globulin ratio with the process of aging in male Wistar rats. On the other hand, some reports fail to support an anti-aging role for vitamin E at the cellular level. Supplementation of high levels (up to 2500 ppm) of vitamin E in the diet, for example, has not been shown to have any effect on the age change discernible in connective tissue parameters of C_3H/He mice.[53]

In a systematic study to examine the effect of the types of dietary fat (coconut oil and safflower oil) with 20 or 200 mg vitamin E per kilogram diet in Wistar male rats during development and aging, Porta and co-workers[54-57] have shown no differences in maximum life span between dietary groups. However, they observed that the 50% survival time of rats fed the safflower oil at high levels of vitamin E was significantly longer than all other groups. Based on serum biochemical parameters and pathological changes observed, the authors attributed this beneficial effect to postponement of the onset and reduction of incidence of malignant neoplasms, and not to any particular influence on the incidence or severity of chronic nephropathy, which developed in nearly all experimental animals. Porta et al.[54] also showed that the body weight of rats consuming unsaturated fat diets increased faster until the 12th month of the experiment than those fed saturated fat diets. However, neither the age-dependent changes in organ weight nor the majority of biochemical parameters studied in serum (i.e., lipid and protein fractions), in brain, liver, and heart (i.e., total protein, DNA, RNA, and total collagen) were in general significantly affected by the types of dietary fat or by the levels of vitamin E. Unfortunately, the information regarding the influence of dietary vitamin E and fat on the same parameters of experimental animals studied beyond 24 months of age has not yet become available. Such information should provide much more insight regarding the possible role of vitamin E and lipid peroxidation in modifying the aging process.

Deprivation of dietary vitamin E is known to result in deficiency symptoms in experimental animals, and a prolonged deficiency state may lead to premature death. Supplementation with vitamin E has been shown to have a modest beneficial effect on life span in experimental animals under certain conditions. Harman,[45] for example, has shown that addition of 0.25% (250 ppm) α-tocopherol acetate, 0.25% santoquin, or 1% NaH_2PO_4 to the diet of New Zealand black male mice starting shortly after weaning, increases the average life span by 7.1, 32.1, and 1.2%, respectively, in comparison with the control life span of 16.8 months. This strain of mice loses T cell suppressor function early in life and develops autoimmune manifestations which mimic those seen in old mice of normal strains. Thus, the findings suggest that free radical reactions may play a role in the deleterious reactions of the immune systems with age.

Supplementing an adequate diet with vitamin E, however, has not been shown to improve or lengthen the life span of normal animals. Tappel et al.[20] tested the effects on mice of mixtures of vitamin E (up to 2740 ppm) and other nutrients and antioxidants that might provide maximum protection against oxidative deterioration. After adult mice were fed test diets for about 1 year, no significant effects on their walking capacity and coordination, kidney function, muscle membrane function, or mortality were observed, although a difference in age pigments was seen.

VIII. VITAMIN E AND ERYTHROCYTE AGING

Erythrocytes have been utilized as a model for aging research by many investigators, in part due to their easy accessibility and well-defined life span. While it is well known that vitamin E-deficient erythrocytes are more susceptible to lysis by hemolytic agents, no apparent structural differences between the membrane proteins of vitamin E-deficient and normal cells have been revealed. Shapiro and Mott[58] have provided experimental evidence that the sites of membrane damage in vitamin E-deficient erythrocytes are on the cytoplasmic face of the membrane, closer to the source of the reactive radical species.

Human adults maintained on a low vitamin E diet for a long period of time have not shown clinical manifestations of vitamin E deficiency, although erythrocytes of both adults and infants with low serum tocopherol levels are susceptible to oxidizing agents in vitro, and erythrocyte survival time is shorter in these adults than in normal persons.[59] The life span of erythrocytes in full-term infants is approximately two thirds that of the erythrocytes of normal adults. The erythrocytes of premature infants have an even shorter life span than do those of *full term infants*.[60] Deficiency of vitamin E *in premature infants at birth may be one of the factors involved in this accelerated senescence*.[61] This has been shown by partial inhibition of accelerated red blood cell destruction when vitamin E was administered to premature infants. Infants, especially those of low birth weight, have been repeatedly shown to have low serum tocopherol levels.[62,63]

A congenital defect of one or more enzymes involved in cellular antioxidant defense may adversely affect the life span of the red cells. Erythrocyte GSH synthetase deficiency, which is a less common cause of chronic hemolytic anemia, may occur with or without oxoprolinuria, depending upon the specific nature of the enzyme deficiency.[64] Individuals with this disorder, for example, exhibit a significant reduction in red cell survival and moderately severe anemia.[65] As in the case of patients deficient in glucose-6-phosphate dehydrogenase, the red cells of patients deficient in GSH synthetase are more susceptible to oxidative stress. Treatment of these patients with vitamin E has been shown to improve both red cell life span and polymorphonuclear cell function.[66] Vitamin E treatment, however, has not been shown to improve the red cell survival or membrane protein pattern of patients deficient in glucose-6-phosphate dehydrogenase.[66,67]

In older individuals, erythrocytes have been shown to be more fragile with a broader and more asymmetric *distribution of fragilities, while the rate of hemolysis is significantly slower than that of the young individuals*.[68] The increase in fragility with age is attributed to the increase in mean cell volume, while the decrease in rate of hemolysis with age appears to be related to a change in the properties of the *erythrocyte membranes. It is not known, however, whether the status of vitamin E is an important determining factor in this respect.* Walker and Nickel[69] have shown that neither the type of dietary fat nor levels of vitamin E affect age-related changes or enzymes in rat erythrocytes. A lack of effect of vitamin E on erythrocyte survival has also been reported by Jacob and Lux.[70]

IX. VITAMIN E AND AGING OF HUMAN DIPLOID CELLS IN CULTURE

Aging changes observed in in vitro systems have been shown to be relevant to those changes observed in vivo.[71,72] Human diploid cells in culture present a unique model system for examining the effects of free radicals and other agents involved in biological redox reactions, including vitamin E. Cultured cells have a definite life span in terms of the population doublings they can achieve. After the cultures are subcultivated a certain number of times, they grow slower and accumulate debris, and eventually the population is lost. This degeneration has been interpreted as a manifestation of aging at the cellular level.

Since Hayflick and Moorhead[72] proposed in 1961 that the finite life span of human diploid fibroblasts in culture may serve as model for cellular aging, several investigators have applied *this experimental model to test various theories of aging, including the free radical theory.* Using fluorescence microscopy, for example, Deamer and Gonzales[73] have demonstrated a large increase in fluorescent materials in cultures near the end of their in vitro life span.

In an attempt to provide evidence *that free radical-mediated reactions are important in* aging processes of human cells, Packer and Smith[74] studied the effect of vitamin E on the life spans of cultured normal human diploid cells, and reported that the addition of 100 μg of d,l-α-tocopherol per milliliter of medium enabled them to culture the WI-38 cells for more than twice the normal life span of approximately 50 population doublings. This exciting finding was interpreted as convincing evidence in support of the free radical theory of aging. Subsequent studies by other investigators, however, did not produce the same results. Balin et al.,[75] for example, did not observe any effect of vitamin E at the concentration range of 10 to 100 μg/mℓ of medium, which was thought to extend the life span of the WI-38 cells at reduced, ambient, or elevated oxygen tension. They concluded that neither oxygen tension nor free radical reaction plays a significant role in limiting the life span of WI-38 cell growth in vitro under ambient oxygen tensions (PO₂ 137 mmHg).

Sakagami and Yamaka[76] also reported the failure of vitamin E to extend the life span of a human fetal lung fibroblast cell line, UT20Lu, in culture. Similarly, Corwin and Humphrey[78] have found either no effect or a mild inhibitory effect of d,l-α-tocopherol on growth in vitro of V79 lung fibroblast from Chinese hamster fibroblast. Later, Packer and Smith[77] reported that they could not reproduce their earlier findings on the effect of vitamin E on the life span of WI-38 cells.

Despite these setbacks, interest in the possibility that vitamin E may be needed for normal aging or slowing of the aging process remains strong. While no spectacular findings have been found, encouraging results are available. Using human lipoprotein fraction enriched with vitamin E (5 to 15 μg/mℓ) to supplement the standard feeding medium of fetal rat brain cultures, Halks-Miller et al.,[79] for example, have shown that the longevity of neurons was increased from 27 to 49 days. It is suggested that the membrane-rich cells, especially for oligodendroglia, need to have an antioxidant such as vitamin E available in order to maintain the integrity of their membranes.

X. CONCLUSIONS

More than 50 years have passed since vitamin E was originally described as a fat-soluble dietary substance necessary for reproduction in rats. The Food and Nutrition Board, National Research Council, included vitamin E as an essential nutrient for man in its Recommended Daily Allowance published in 1965.[80] Although vitamin E deficiency can be induced rather easily in many laboratory animals,[81] it has remained unique among the vitamins in lacking an unequivocal deficiency state in man or an established specific role in cellular metabolism. Due to the lack of definite correlation between clinical deficiency syndromes and vitamin E deficiency in man, vitamin E has been termed "a vitamin in search of a disease".

On the other hand, there is continued interest in the potential role of vitamin E as an important agent in human cellular metabolism. In addition to its ability to alleviate deficiency symptoms in experimental animals, dietary vitamin E has been shown to modify the toxicity or cellular response/susceptibility to a variety of compounds. Vitamin E has also been recommended for the treatment of aging and for such conditions as impotence and muscular dystrophy without the backing of sound experimental or clinical studies. Due in part to considerable tissue storage, it is difficult to produce a vitamin E deficiency under experimental conditions in adult man, yet deficiencies or subclinical deficiencies do occur in association with intestinal malabsorption syndromes of various etiologies.[81] Findings that indicate a need to provide vitamin E supplementation in some infants and those adults with fat absorption problems clearly indicate a role of vitamin E in human nutrition.

If free radical reactions are indeed involved in the initiation and/or acceleration of aging, the aging process, whether it is normal or diseased, or life span might be altered by the status of cellular antioxidant defenses. While several lines of evidence indicate that free radical-initiated degenerative reactions may play a role in the accelerating of the aging process and that vitamin E supplementation may slow down such processes, others do not lead to the same conclusion. The conflicting results can be attributed in part to differences in animal species, strains, sex, levels of vitamin E and other components in the diet, and the time period employed in the experiments.

Despite the fact that definitive evidence supporting the free radical theory of aging and the anti-aging function of vitamin E has yet to be provided, interest in this area remains strong. This is partly due to the possibility that the aging process may be slowed down and the life span maximized by dietary means. A better understanding of the role of vitamin E in human nutrition, including whether aged persons require more vitamin E to maintain normal functions than younger people, is urgently needed.

REFERENCES

1. Vital Statistics of the United States, 1973, Vol. 2, U.S. Department of Health, Education and Welfare, Public Health Service, Health Resources Administration, National Center for Health Statistics, Rockville, Md., 1975, Section 5.
2. **Harman, D.,** Aging: a theory based on free radical and radiation chemistry, *J. Gerontol.,* 11, 298, 1956.
3. **Harman, D.,** Free radical theory of aging: nutritional implications, *Age,* 1, 145, 1978.
4. **Harman, D.,** Free radical theory of aging: beneficial effect of antioxidants on the life span of male NZB mice; role of free radical reactions in the deterioration of the immune system with age and in the pathogenesis of systemic lupus erythematosus, *Age,* 3, 64, 1980.
5. **Leibovitz, B. E. and Siegel, B. V.,** Aspects of free radical reaction in biological systems: aging, *J. Gerontol.,* 35, 45, 1980.
6. **Tappel, A. L.,** Will antioxidant nutrients slow aging processes? *Geriatrics,* 23, 97, 1968.
7. **Pryor, W. A.,** Free radical reactions and their importance in biological systems, *Fed. Proc. Fed. Am. Soc. Exp. Biol.,* 32, 1862, 1973.
8. **Tappel, A. L.,** Vitamin E and free radical peroxidation of lipids, *Ann N.Y. Acad. Sci.,* 203, 12, 1972.
9. **Chow, C. K.,** Nutritional influence on cellular antioxidant defense systems, *Am. J. Clin. Nutr.,* 32, 1066, 1979.
10. **Roubal, W. T. and Tappel, A. L.,** Polymerization of protein induced by free radical lipid peroxidation, *Arch. Biochem. Biophys.,* 113, 150, 1966.
11. **Roubal, W. T. and Tappel, A. L.,** Damage to protein, enzymes and amino acid by peroxidizing lipids, *Arch. Biochem. Biophys.,* 113, 5, 1966.
12. **Chio, K. S. and Tappel, A. L.,** Synthesis and characterization of the fluorescent products derived from malonaldehyde and amino acids, *Biochemistry,* 8, 2821, 1969.
13. **Tappel, A. L.,** Lipid peroxidation damage to cell components, *Fed. Proc. Fed. Am. Soc. Exp. Biol.,* 32, 1870, 1973.
14. **Payes, B.,** Enzymatic repair of x-ray-damaged DNA. 1. Labelling of the precursor of a malonaldehyde-like material in x-irradiated DNA and the enzymatic excision of labelled lesions, *Biochim. Biophys. Acta,* 366, 251, 1974.
15. **Shimaski, H., Nozawa, T., Privett, O. S., and Anderson, W. R.,** Detection of age-related fluorescent substances in rat tissues, *Arch. Biochem. Biophys.,* 183, 443, 1977.
16. **Chio, K. S., Reiss, U., Fletcher, B. L., and Tappel, A. L.,** Peroxidation of subcellular organelle: formation of lipofuscin-like fluorescent pigments, *Science,* 166, 1535, 1969.
17. **Rickert, W. S. and Forbes, W. F.,** Changes in collagen with age. VI. Age and smoking related changes in human lung connective tissue, *Exp. Gerontol.,* 11, 89, 1976.
18. **Nagy, I. Z.,** A membrane hypothesis of aging, *J. Theoret. Biol.,* 75, 189, 1973.
19. **Sagai, M. and Ichinose, T.,** Age-related changes in lipid peroxidation as measured by ethane, ethylene, butane and pentane in respired gases of rats, *Life Sci.,* 27, 731, 1980.
20. **Tappel, A., Fletcher, B., and Deamer, D.,** Effect of antioxidants and nutrients on lipid peroxidation fluorescent products and aging parameters in mouse, *J. Gerontol.,* 28, 415, 1973.
21. **Sevanian, A., Muakkassah-Kelly, S. F., and Montestruque, S.,** The influence of phospholipase A_2 and glutathione peroxidase on the elimination of membrane lipid peroxide, *Arch. Biochem. Biophys.,* 223, 441, 1983.
22. **Grossmann, A. and Wendel, A.,** Non-reactivity of the selenoenzyme glutathione peroxidase with enzymatically hydroperoxidized phospholipids, *Eur. J. Biochem.,* 135, 549, 1983.
23. **Burton, G. W., Joyce, A., and Ingold, K. U.,** Is vitamin E the only lipid-soluble chain breaking antioxidant in human blood plasma and erythrocyte membranes? *Arch. Biochem. Biophys.,* 221, 281, 1983.
24. **Lucy, J. A.,** Functional and structural aspects of biological membranes: a suggested structural role of vitamin E in the control of membrane permeability and stability, *Ann. N.Y. Acad. Sci.,* 203, 4, 1972.

25. **McCay, P. B., Pfeifer, P. M., and Stipe, W. H.,** Vitamin E and free radical peroxidation of lipids, *Ann. N.Y. Acad. Sci.,* 203, 62, 1972.

26. **Molenaar, I., Vos, J., Jager, F. C., and Hommer, F. A.,** The influences of vitamin E deficiency on biological membrane. An ultrastructural study on the intestinal epithelial cells of ducklings, *Nutr. Metab.,* 12, 358, 1970.

27. **Chow, C. K.,** Distribution of tocopherols in human plasma and red blood cells, *Am. J. Clin. Nutr.,* 28, 756, 1975.

28. **Maiorino, M., Ursini, F., Leonelli, M., Finato, N., and Gregolin, C.,** A pig heart peroxidation inhibiting protein with glutathione peroxidase activity on phospholipid hydroperoxides, *Biochem. Int.,* 5, 575, 1982.

29. **Niki, E., Tsuchiyi, J., Tanimura, R., and Kamiya, Y.,** Regeneration of vitamin E from α-chromanoxyl radical by glutathione and vitamin C, *Chem. Lett.,* 1982, 789, 1982.

30. **Chow, C. K., Draper, H. H., Csallany, A. S., and Chiu, M.,** The metabolism of C-14-α-tocopheryl quinone and C-14-α-tocopheryl hydroquinone, *Lipids,* 2, 390, 1967.

31. **Draper, H. H., Csallany, A. S., and Chiu, M.,** Isolation of a trimer of α-tocopherol from mammalian liver, *Lipids,* 2, 47, 1967.

32. **Draper, H. H., Csallany, A. S., and Shah, S. N.,** Isolation and synthesis of a new metabolite of α-tocopherol, *Biochim. Biophys. Acta,* 59, 527, 1962.

33. **Olson, R. E.,** Creatine kinase and myofibrillar protein in hereditary muscular dystrophy and vitamin E deficiency, *Am. J. Clin. Nutr.,* 27, 1117, 1974.

34. **Catignani, G. L., Chytil, R., and Darby, W. J.,** Vitamin E deficiency: immunochemical evidence for increased accumulation of liver xanthione oxidase, *Proc. Nat. Acad. Sci. U.S.A.,* 71, 1966, 1974.

35. **Chow, C. K., Reddy, K., and Tappel, A. L.,** Effect of dietary vitamin E on the activities of glutathione peroxidase system in rat tissues, *J. Nutr.,* 103, 618, 1973.

36. **Chow, C. K. and Tappel, A. L.,** An enzymatic protective mechanism against lipid peroxidation damage to lungs of ozone-exposed rats, *Lipids,* 7, 518, 1972.

37. **Emerson, G. A. and Evans, H. M.,** Restoration of fertility in successively older E-low female rats, *J. Nutr.,* 18, 501, 1939.

38. **Fuhr, A., Johnson, R. E., Kaunitz, H., and Slanetz, C. A.,** Increased tocopherol requirements during the rat's menopause, *Ann. N.Y. Acad. Sci.,* 52, 83, 1949.

39. **Ames, S. R.,** Age, parity and vitamin A supplementation and the vitamin E requirement of female rats, *Am. J. Clin. Nutr.,* 27, 1017, 1974.

40. **Gabriel, E., Machlin, L. J., Filipski, R., and Nelson, J.,** Influence of age on the vitamin E requirement for resolution of necrotizing myopathy, *J. Nutr.,* 110, 1372, 1980.

41. **Reiss, U. and Gershon, D.,** Rat liver superoxide dismutase purification and age-related modifications, *Eur. J. Biochem.,* 63, 617, 1976.

42. **Hazelton, G. A. and Lang, C. A.,** Glutathione S-transferase activities in the yellow-fever mosquito, *Aedes aegypiti* (Louisville), during growth and aging, *Biochem. J.,* 210, 281, 1983.

43. **Hazelton, G. A. and Lang, C. A.,** Glutathione contents of tissues in the aging mouse, *Biochem. J.,* 188, 25, 1980.

44. **Thompson, C. P., Rea, H., Robinson, M. F., and Chapman, O. W.,** Low blood selenium concentrations and glutathione peroxidase activities in elderly people, *Proc. Univ. Otago Med. School,* 55, 18, 1977.

45. **Harman, D.,** The free radical theory of aging: the effect of age on serum mercaptan levels, *J. Gerontol.,* 15, 38, 1960.

46. **Attwood, E. C., Robey, E., Kramer, J. J., Ovenden, N., Snape, S., Ross, J., and Bradley, F.,** A survey of the haematological, nutritional, and biochemical state of the rural elderly with particular reference to vitamin C, *Age Ageing,* 7, 46, 1978.

47. **Chen, L. H., Hsu, S. J., Huang, P. C., and Chen, J. S.,** Vitamin E status of Chinese population in Taiwan, *J. Am. Clin. Nutr.,* 30, 728, 1977.

48. **Barnes, K. J. and Chen, L. H.,** Vitamin E status of the elderly in central Kentucky, *J. Nutr. Elderly,* 1, 41, 1981.

49. **Weglicki, W. B., Luna, Z., and Nair, P. P.,** Sex and tissue specific differences in concentrations of α-tocopherol in mature and senescent rats, *Nature (London),* 221, 185, 1969.

50. **Hafeman, D. G. and Hoekstra, W. G.,** Lipid peroxidation *in vivo* during vitamin E and selenium deficiency in the rat as monitored by ethane evolution, *J. Nutr.,* 107, 666, 1977.

51. **Dillard, C. J., Litov, R. E., and Tappel, A. L.,** Effects of dietary vitamin E, selenium and polyunsaturated fats on *in vivo* lipid peroxidation in the rats as measured by pentane production, *Lipids,* 13, 396, 1978.

52. **Hirahara, F., Takai, Y., and Iwao, H.,** The effects of vitamin E on protein metabolism in the aging process of rats, *Nutr. Rep. Int.,* 20, 261, 1979.

53. **Blackett, A. D. and Hall, D. A.,** The action of vitamin E on the aging of connective tissues in the mouse, *Mech. Age. Dev.,* 14, 305, 1980.

54. **Porta, E. A., Joun, N. S., and Nitta, R. T.,** Effects of the type of dietary fat at two levels of vitamin E in Wistar male rats during development and aging. I. Life span, serum biochemical parameters and pathological changes, *Mech. Age Dev.,* 13, 1, 1980.

55. **Porta, E. A., Nitta, R. T., Kia, L., Joun, N. S., and Nguyen, L.,** Effects of the type of dietary fat at two levels of vitamin E in Wistar male rats during development and aging. II. Biochemical and morphological parameters of the brain, *Mech. Age Dev.,* 13, 319, 1980.

56. **Porta, E. A., Keopuhiwa, L., Joun, N. S., and Nitta, R. T.,** Effects of the type of dietary fat at two levels of vitamin E in Wistar male rats during development and aging. III. Biochemical and morphological parameters of the liver, *Mech. Age Dev.,* 15, 297, 1981.

57. **Porta, E. A., Sablan, H. M., Joun, N. S., and Chee, G.,** Effects of the type of dietary fat at two levels of vitamin E in Wistar male rats during development and aging. IV. Biochemical and morphological parameters of the heart, *Mech. Age Dev.,* 18, 159, 1982.

58. **Shapiro, S. S. and Mott, D. J.,** Alterations of enzymes in the red blood cell membrane in vitamin E deficiency, *Ann. N.Y. Acad. Sci.,* 393, 263, 1982.

59. **Horwitt, M. K., Century, B., and Zeman, A. A.,** Erthyrocyte survival time and reticulocyte levels after tocopherol depletion in man, *Am. J. Clin. Nutr.,* 12, 99, 1963.

60. **O'Brien, R. T. and Pearson, H. A.,** Physiologic anemia of the newborn infant, *J. Pediatr.,* 79, 132, 1971.

61. **Gross, S. J., Landaw, S. A., and Oski, F. A.,** Vitamin E and neonatal hemolysis, *Pediatrics,* 59, 995, 1977.

62. **Nitowsky, H. M., Cornblath, M., and Gordon, H. H.,** Studies of tocopherol deficiency in infants and children. II. Plasma tocopherol and erythrocyte hemolysis in hydrogen peroxide, *J. Dis. Child.,* 92, 164, 1956.

63. **Wright, S. W., Filer, L. J., Jr., and Mason, K. E.,** Vitamin E blood levels in premature and full term infants, *Pediatrics,* 7, 386, 1951.

64. **Spielberg, S. P., Garrick, M. D., Corash, L. M., Butler, J. D., Tietze, H., Rogers, L. V., and Schulman, J. D.,** Biochemical heterogeneity in glutathione synthetase deficiency, *J. Clin. Invest.,* 61, 1417, 1978.

65. **Mohler, D. N., Majerus, P. W., Minnich, V., Hess, C. E., and Garrick, M. D.,** Glutathione synthetase deficiency as a cause of hereditary hemolytic disease, *N. Engl. J. Med.,* 283, 1253, 1970.

66. **Corash, L. M., Sheetz, M., Bieri, J. G., Bartsocas, C., Moses, S., Bashan, N., and Schulman, J. D.,** Chronic hemolytic anemia due to glucose-6-phosphate dehydrogenase deficiency or glutathione synthetase: the role of vitamin E in its treatment, *Ann. N.Y. Acad. Sci.,* 393, 348, 1982.

67. **Johnson, G. J., Vatassery, G. T., Finkel, B., and Allen, D. W.,** High dose vitamin E does not decrease the rate of chronic hemolysis in glucose-6-phosphate dehydrogenase deficiency, *N. Engl. J. Med.,* 308, 1014, 1983.

68. **Araki, K. and Rifkind, J. M.,** Age dependent changes in osmotic hemolysis of human erythrocytes, *J. Gerontol.,* 35, 499, 1980.

69. **Walker, B. L. and Nickel, M. I.,** Dietary vitamin E and age-related changes in rat erythrocyte enzyme activities, *Nutr. Rep. Int.,* 19, 151, 1979.

70. **Jacob, H. S. and Lux, S. E.,** Degradation of membrane phospholipids and thiols in peroxide hemolysis: studies in vitamin E deficiency, *Blood,* 32, 549, 1968.

71. **Daniel, C. W. and Young, L. J. T.,** Influence of cell division on an aging process. Lifespan of mouse mammary epithelium during serial propagation *in vivo, Exp. Cell. Res.,* 65, 27, 1971.

72. **Hayflick, L. and Moorhead, P. S.,** The serial cultivation of human diploid cell strains, *Exp. Cell Res.,* 25, 585, 1961.

73. **Deamer, D. W. and Gonzales, J.,** Autofluorescent structures in cultured WI-38 cells, *Arch. Biochem. Biophys.,* 165, 421, 1974.

74. **Packer, L. and Smith, J. R.,** Extension of the lifespan of cultured normal human diploid cells by vitamin E, *Proc. Natl. Acad. Sci. U.S.A.,* 71, 4763, 1974.

75. **Balin, A. K., Goodman, D. B. P., Rasmuseen, H., and Cristofalo, V. J.,** The effect of oxygen and vitamin E on the lifespan of human diploid cells in vitro, *J. Cell Biol.,* 74, 58, 1977.

76. **Sakagami, H. and Yamada, M.,** Failure of vitamin E to extend the lifespan of a human diploid cell in culture, *Cell Struct. Function,* 2, 219, 1977.

77. **Packer, L. and Smith, J. R.,** Extension of the lifespan of cultured normal human diploid cells by vitamin E: a re-evaluation, *Proc. Natl. Acad. Sci. U.S.A.,* 74, 1640, 1977.

78. **Cowin, L. M. and Humphrey, L. P.,** Vitamin E: substrate dependent growth effect on cells in culture, *Proc. Soc. Exp. Biol. Med.,* 141, 609, 1972.

79. **Halks-Miller, M., Kane, J. P., Beckstead, J. H., and Smuckler, E. A.,** Vitamin E enriched lipoproteins increase longevity of neurons *in vitro, Dev. Brain Res.,* 2, 439, 1982.

80. *Recommended Dietary Allowance,* 7th ed., National Academy of Sciences, Washington, D.C., 1965.

81. **Bieri, J. G. and Farrell, P. M.,** Vitamin E, *Vit. Horm.,* 34, 31, 1976.

Chapter 10

WATER-SOLUBLE VITAMINS

Judy A. Driskell

TABLE OF CONTENTS

I. INTRODUCTION

Vitamins which are soluble in water are usually associated with body fluids. Vitamin C and the members of the vitamin B complex are classified as water-soluble vitamins. Generally, water-soluble vitamins are excreted in the urine when their serum concentrations exceed tissue saturation. Water-soluble vitamins should be supplied daily in the diet. It was once believed that water-soluble vitamins were nontoxic; however, evidence has been reported during the last decade that megadoses of some of these vitamins may have adverse effects upon the body.

Water-soluble vitamins are frequently found together in the same foods and several of them function in the same pathways or even in the same enzyme systems. Hence, deficiency conditions caused by the lack of a single vitamin are rare; they usually are mixed deficiencies. Symptoms which are usually observed include anemia, dermatitis, neurological disorders, and digestive abnormalities. Deficiencies of water-soluble vitamins often develop quite rapidly.

The Recommended Dietary Allowances (RDA) for the elderly are quite similar to those of younger adults. The allowances given for the elderly are mainly extrapolated from allowances for younger adults.[1] Little is known as to the effects of aging on requirements for the water-soluble vitamins. Data collected in the 1977—1978 Nationwide Food Consumption Survey[2] indicated that the average intakes of elderly males and females in the 48 states met the RDA for thiamin, riboflavin, niacin, vitamin B$_{12}$, and vitamin C, but that their average intakes of vitamin B$_6$ were below the RDA, actually from 58 to 79% of the RDA.

II. VITAMIN C

Ascorbic acid is the reduced form of vitamin C and dehydroascorbic acid is the oxidized form. Most of the vitamin exists as ascorbic acid; however, dehydroascorbic acid is produced spontaneously from ascorbic acid upon oxidation. Both forms seem to be equally physiologically active in the body. Vitamin C may be classified as a monosaccharide. Ascorbic acid is not a necessary dietary component for all species. Humans have inadequate quantities of the enzyme L-gulonolactone oxidase,[3] which is needed for the synthesis of ascorbic acid from gulonic acid, which may come from glucose or galactose.

A. Functions
Vitamin C appears to be essential for the normal functioning of all cellular and subcellular units in higher plants and animals. Vitamin C plays a role in hydrogen ion transfer as well as in several oxidation-reduction reactions. It functions as an antioxidant. It is needed for

the formation of bone, teeth, and cartilage. It facilitates the absorption of iron[4] and may play a role in the conversion of folic acid to its active forms.[5] The vitamin is involved in the distribution of iron in the iron storage compounds ferritin and hemosiderin.[6] Ascorbic acid is needed for the hydroxylation of proline in collagen synthesis,[7] the hydroxylation of tryptophan to 5-hydroxytryptophan (which may be converted to serotonin),[8] the synthesis of epinephrine and norepinephrine,[9] tyrosine and phenylalanine metabolism,[10] and the synthesis of anti-inflammatory steroids by the adrenal glands.[11] Vitamin C deficiency has been associated with impaired wound healing.[12] Ascorbic acid may affect cholesterol metabolism;[13] however, researchers report both positive and negative correlations between plasma ascorbic acid and cholesterol levels.[14,15] Some evidence exists that the vitamin also functions in detoxification of some poisonous substances via its participation in hydroxylation reactions.[16] The precise biochemical functions of vitamin C are not well defined; however, the vitamin does seem to be involved in many reactions in the body.

B. Metabolism

Vitamin C is readily and rapidly absorbed in the upper small intestine; generally, very little fecal loss occurs. The vitamin is absorbed at an efficiency of 80 to 90% when the intake is 100 mg/day or less;[17] the vitamin is less efficiently absorbed at higher intakes.[18] In about 4 hr the absorbed ascorbic acid is equilibrated with that in all the tissues in the body. The glandular tissues (salivary glands, pancreas, adrenals, spleen, and testes) tend to have high concentrations of the vitamin; however, large quantities are also found in the brain, pituitary, liver, and eyes. Healthy adult males have a body pool slightly above 1500 mg.[19,20] The vitamin is catabolized at about 3% daily during depletion.[19-21] Vitamin C is filtered by the glomerulus of the kidney; the tubules actively reabsorb ascorbic acid as long as the plasma concentration is within normal ranges. After the renal threshold is reached, the vitamin is excreted in the urine in the form of ascorbic acid, oxalic acid, diketogulonic acid, and ascorbate-2-sulfate.

Vitamin C nutriture may be evaluated by measuring the ascorbic acid levels in whole blood, plasma, serum, leukocytes, or the buffy coat (leukocytes and platelets) with the ascorbic acid concentration of the leukocytes being the highest; however, relatively large quantities of blood are needed for white blood cell determinations. Most often, vitamin C nutriture is determined by measuring the ascorbic acid concentration of serum or plasma. Methods for the assessment of vitamin C nutriture were discussed in Chapter 6.

Several factors may influence the metabolism of vitamin C. Emotional or environmental stress appears to increase the quantity of vitamin C intake needed to maintain normal plasma levels.[22,23] The use of cigarettes[24,25] and of oral contraceptives[26] tends to lower plasma ascorbic acid levels. Men have been observed to have lower plasma and leukocyte concentrations of the vitamin than women, perhaps because of ovarian hormone activity.[27]

C. Age-Related Changes in Metabolism

Plasma and leukocyte ascorbic acid levels tend to be lower in elderly men and women than in younger adults; these levels progressively decline with advancing age according to Burr et al.[28] These researchers found no correlations between vitamin C concentrations and the presence of anemias, body weights, skinfold thickness, or chronic illnesses. Some illnesses which occur in the elderly have been reported to lower leukocyte ascorbic acid levels.[29] Treatment of elderly men with tetracycline may also depress leukocyte vitamin C levels.[30] Blood vitamin C levels of the aged can be elevated with ascorbic acid supplements.[23,31]

The hypovitaminosis C which frequently is observed in aged individuals may increase the risk of hypercholesterolemia.[15] The poor vitamin C status frequently observed in this age group may also be associated with a defect in collagen proline hydroxylation, thus affecting the repair of connective tissue.[32]

D. Requirements in Aging

More research is needed regarding the vitamin C requirements of the elderly. The Recommended Dietary Allowance[1] for men and women 51 years of age and above living in the U.S. is 60 mg/day, the same as for younger adults. The recommended intake of ascorbic acid for adult men and women given by FAO/WHO[33] is 30 mg/day.

Several investigations of vitamin C nutriture of the elderly living in the U.S. have been conducted over the last couple of decades. The mean vitamin C intakes of males and females 65 years of age and above in the Nationwide Food Consumption Survey[2] of Spring 1977 were 150 to 167% of the 1980 RDA; most of the reported vitamin C intake was from citrus fruits and tomatoes, noncitrus fruits, and vegetables. Approximately one fourth of the individuals 65 years and over in the 1977 survey consumed less than 70% of the 1980 RDA.[34] About one fourth of the elderly subjects included in the Health and Nutrition Examination Survey of 1971—1972[35] reported consuming less than 30 mg ascorbic acid daily. Black, white, and Spanish-American men from low income ratio states included in the Ten-State Nutrition Survey of 1968 to 1970[36] frequently were found to have nutritional problems with relation to vitamin C; around one third of the high- and low-income elderly subjects of all sex-race groups reported consuming less than 30 mg/day. The mean intakes of men and women 65 years of age and above who were included in the Food Consumption Survey of Spring 1965[37] were between 57 and 66 mg/day.

Garry et al.[38] reported that about 5% of their 270 free-living, middle-income, healthy men and women over 60 years old living in the Albuquerque, N.M. area, were observed to receive less than 75% of the 60-mg RDA for vitamin C. The vitamin C intakes of 196 males and females at least 65 years of age in Belfast, Northern Ireland[39] were estimated and over one fourth of these subjects reported receiving less than 20 mg ascorbic acid daily.

Some researchers have determined the vitamin C nutriture of the elderly using plasma, serum, or leukocyte levels of the vitamin as the status criterion. The serum ascorbic acid levels of 68 elderly women were found to be within acceptable limits.[40] The plasma ascorbic acid concentrations of 29% of 196 elderly subjects in Belfast, Northern Ireland[39] were found to be low. The circulating ascorbic acid levels of noninstitutionalized elderly subjects were observed to be lower than those of elderly in institutions or younger control subjects.[41] A group of 18 elderly hospitalized psychogeriatric patients had lower plasma ascorbic acid values than a group of 10 healthy elderly subjects who lived in their own residences; the psychogeriatric patients may have been at risk of ascorbic acid deficiency.[42] The majority of the 93 acute geriatric subjects being admitted to a hospital were found to have leukocyte vitamin C levels that were deemed to be abnormal.[43] Burr et al.[44] found that men who ate alone, but not women, tended to have a poorer vitamin C status than those who ate with others. The vitamin C nutriture of 547 participants of the Title VII Nutrition Program for the Elderly was positively associated with the frequency of participation in the program.[45] Although many investigators have found the elderly to have low leukocyte or plasma vitamin C levels, clinical evidence of deficiency generally has not been observed in these subjects.[43,46]

E. Problems of Deficiency and Excess

Deficiency of vitamin C results in scurvy, a disease which is characterized by widespread capillary hemorrhaging. Adults need from 8 to 10 mg of vitamin C daily to prevent scurvy. However, this quantity does not allow for acceptable tissue reserves. The first signs of scurvy appear when the plasma vitamin C levels are from 0.13 to 0.24 mg/dℓ.[47] Well-defined scurvy is relatively rare in the U.S.

Clinical symptoms of scurvy have been well documented;[47-51] this disease affected sailors many years ago. Early symptoms of scurvy include weakness, fatigue, listlessness, anorexia, as well as aching of the bones, joints, and muscles. Dermatitis is observed; mental changes frequently occur. Perifollicular hyperkeratotic papules appear on the thighs and buttocks and later on the arms and trunk; hairs become buried in the follicles followed by erythema and purpura in these follicles. Petechiae in the skin coalesce, developing ecchymoses. Hemorrhages occur deep in the muscles. Gums become hemorrhagic and necrotic; periodontal infection may occur. Subperiosteal bleeding occurs; there is also bleeding into the joints, the peritoneal cavity, and on into the pericardial sac. Massive hemorrhaging is also observed in the adrenals. Alterations in folate metabolism and impaired iron absorption occur; anemia is frequently observed. Wound healing is defective. Infectious diseases are also much more common among individuals with scurvy than with the normal population.[52]

Overt manifestations of scurvy in the elderly are rare. Leung and Guze[53] diagnosed two elderly alcoholic men as having scurvy. An elderly man living alone was found to have severe scurvy.[54] Taylor[55] has suggested that the vascular sublingual abnormalities that frequently are observed in the elderly may be due to vitamin C inadequacies.

Vitamin C intakes of 60 to 75 mg/day allow for serum ascorbic acid levels of about 0.75 mg/dℓ.[47,56] Higher serum levels are attainable when vitamin intakes are higher. Maximal serum ascorbic acid concentration seems to be about 1.4 mg/dℓ.[57] Larger than normal body reserves of vitamin C may be obtained via the consumption of about 200 mg of the vitamin daily;[58] however, as is stated in the 1980 RDA publication,[1] "efforts to attain such pool sizes are unnecessary in view of the decreased efficiency of absorption and increased rate of excretion of unmetabolized ascorbic acid at these higher intakes."

Many people, including the elderly, take large intakes (1 g or more daily) of ascorbic acid; the publications of Pauling[59,60] have advocated these dosages. Other researchers in the early 1970s also reported large intakes of the vitamin to be beneficial in reducing the severity and frequency of some of the symptoms of the common cold;[61-66] upon completing double-blind studies, some of these same scientists[58,67] later reported these large intakes to have considerably smaller effects or no effects.

Disagreement exists among investigators as to whether large doses of ascorbic acid may[69] or may not[47,70] lower serum cholesterol in hypercholesterolemic subjects. Ascorbic acid supplementation may increase serum levels of IgA and IgM in humans.[71] Vitamin C may be of some benefit in the treatment of cancer.[72]

Until recent years, large doses of ascorbic acid have been considered to be nontoxic. Individuals usually have nausea followed by osmotic diarrhea when they first take large doses of the vitamin. Samborskaya and Ferdman[73] reported that large amounts of ascorbic acid given to pregnant women result in abortion, probably due to increased estrogen production.

There are several reports in the literature of possible adverse effects on humans taking large quantities of the vitamin; these publications have been mainly based on case histories. Humans have been reported to excrete increased quantities of oxalic acid[74] and uric acid[75] in the urine following the ingestion of large amounts of the vitamin. High doses of vitamin C have been reported to destroy vitamin B_{12}.[76] Individuals taking large doses may absorb excessive quantities of food iron.[77] Males taking large quantities of the vitamin have impaired leukocyte bactericidal activity.[78] Individuals may also become conditioned to consuming large doses and develop scurvy when they resume intakes near 60 mg/day.[79]

III. THIAMIN

In 1912 Funk[80] reported the isolation of an amine substance from rice polishings which cured beriberi; he named this substance "vitamine", an amine essential for life; later the "e" was dropped from the name as used in the U.S. In 1936 Williams and Cline[81] synthesized the substance and gave it the name thiamin or vitamin B_1 (also thiamine and vitamin B-1). Thiamin is composed of both a pyrimidine and a thiazole ring.

A. Functions

Thiamin functions in growth, energy production, promotion of a normal appetite and digestion, and maintenance of a functionally normal nervous and muscular system. The vitamin, in the form of thiamin pyrophosphate (TPP), serves as a coenzyme in several reactions in the body. Thiamin pyrophosphate functions in the oxidative decarboxylation[82] of α-keto acids to aldehydes; as such, the coenzyme is involved in the conversion of pyruvate to acetyl coenzyme A and α-ketoglutarate to succinyl coenzyme A; both of these reactions also require lipoic acid, nicotinamide adenine dinucleotide (NAD), and coenzyme A, and are part of the tricarboxylic acid cycle (TCA; also known as the Krebs cycle and the citric acid cycle). This cycle is involved in energy metabolism. The coenzyme also functions in the pentose phosphate shunt (also known as the hexose monophosphate shunt and the phosphogluconate pathway) in the transfer of an α-keto group from xylulose-5-phosphate and ribose-5-phosphate to form sedoheptulose-7-phosphate and glyceraldehyde-3-phosphate.[83,84] The pentose phosphate shunt provides for the interconversions of monosaccharides containing from 3 to 7 carbons (including ribose for synthesis of DNA, RNA, and other nucleotides), the synthesis of reduced nicotinamide adenine dinucleotide phosphate (NADPH), and the oxidation of glucose to pyruvate without the involvement of adenosine triphosphate (ATP).

This vitamin may in some manner influence the action of the neurotransmitters acetylcholine and/or serotonin and thus affect the neural membranes.[85] Stimulation of the nerve fibers results in the release of free thiamin and thiamin monophosphate. Neural disorders are observed in beriberi. Hence, thiamin must in some fashion affect the functioning of the central nervous system; however, its specific function has not yet been elucidated.

B. Metabolism

Thiamin is absorbed rapidly, primarily in the proximal small intestine, by active transport and passive diffusion. The maximum absorption of the vitamin following a single dose has been reported to be about 8.3 mg for healthy men.[86] Thiamin absorption is reduced in alcoholics.[87] Thiamin is found in the tissues primarily as a pyrophosphate; however, the vitamin may also exist in the body as free thiamin, a triphosphate, and a monophosphate. Thiamin concentration is highest in cardiac muscle and is also found in kidney, liver, brain, and skeletal muscle. The adult body contains not more than 30 mg of thiamin.[88] The vitamin is excreted in the urine as thiamin or as one of its numerous metabolites.[89]

Substances which are thiamin antagonists occur normally in some foods. Fresh and saltwater fish contain thiaminase which inactivates thiamin; if large quantities of fish are eaten raw, a deficiency of thiamin may develop.[90] However, cooking inactivates the enzyme. Compounds with antithiamin activity may also be present in tea[91] and coffee.[92]

The parameters most commonly utilized in the assessment of thiamin nutriture are urinary excretion of the vitamin and pyrophosphate stimulation of erythrocyte transketolase activities. Methods for assessment of thiamin nutriture are discussed in Chapter 6.

C. Age-Related Changes in Metabolism

Aged individuals may utilize thiamin less efficiently than younger adults.[93,94] Thomson[95] presented evidence that thiamin malabsorption seems not to be the cause of deficits of the vitamin in the elderly. Increasing age may[96] or may not[97,98] affect thiamin requirements. Low thiamin intakes of the elderly may be associated with the prevalence of cardiovascular and neural disorders.[99]

D. Requirements in Aging

More research is needed regarding the thiamin requirements of the elderly. The Recommended Dietary Allowances[1] for men and women 51 years of age and above living in the U.S. are 1.2 and 1.0 mg thiamin daily, respectively. A thiamin allowance of 0.5 mg/

1000 kcal has been recommended for adults; however, the elderly should consume 1 mg/day even if their energy intake is below 2000 kcal. The recommended intakes of thiamin for adult men and women given by FAO/WHO[33] are 1.2 and 0.9 mg/day, respectively.

Several investigations of thiamin nutriture of the elderly living in the U.S. have been conducted over the last couple of decades. The mean thiamin intakes of males and females 65 years of age and above in the Nationwide Food Consumption Survey[2] of Spring 1977 were 101 to 117% of the 1980 RDA; most of the reported thiamin intakes were from grain products, meat, poultry, and fish; 17% of the subjects included in the 1977 survey (all ages) reported consuming less than 70% of the 1980 RDA for thiamin.[34] About 40% of the elderly men and women included in the Ten-State Nutrition Survey of 1968 to 1970[36] reported consuming less than two thirds of the 1980 RDA for thiamin. The mean thiamin intakes of men and women 65 years of age and over which were included in the Spring 1965 Food Consumption Survey[37] were 0.84 to 1.17 mg/day.

Of the men and women who were included in a survey of 270 free-living, middle-income, healthy individuals over 60 years of age living in the Albuquerque, N.M. area, 3 and 13%, respectively, reported consuming less than 75% of the 1980 RDA for thiamin.[38] About 7% of the 196 institutionalized and noninstitutionalized elderly in a Belfast, Northern Ireland study consumed less than 0.7 mg (women) or 0.8 mg (men) of thiamin daily.[39]

Brin et al.[100] and Dibble et al.[101] have reported subclinical deficiencies of thiamin in the elderly as assessed using transketolase measurements. Vir and Love[39] reported that 17% of their 51 male and 13% of their 145 female institutionalized and noninstitutionalized subjects were deficient in thiamin as judged by transketolase measurements. Harrill and Cervone[40] reported that about 15% of their female subjects between 62 and 99 years of age had urinary thiamin levels indicative of deficient or low status. Elderly men who ate alone usually had a poorer thiamin status than those who ate with others; this difference was not observed in women.[44]

Little evidence of thiamin malnutrition was observed in 93 acute geriatric admissions to a hospital in the U.K.[43] About 23% of 153 geriatric patients seemed to be deficient in thiamin as assessed using the transketolase parameter;[102] 30% of the 127 geriatric patients in another study[103] were found to be deficient in the vitamin using this parameter. Long-stay hospitalized psychogeriatric patients were found frequently to have transketolase measurements indicative of deficiency.[42] The thiamin status of geriatric patients found to be deficient in the vitamin as assessed by transketolase measurements improved after supplementation with the vitamin.[43,102]

E. Problems of Deficiency and Excess

Deficiency of thiamin results in beriberi. In that thiamin is required for energy metabolism, particularly carbohydrate oxidation, the requirement for the vitamin usually is related to the food energy intake and the carbohydrate intake; the presence of large amounts of fat in the diet reduces the requirement for the vitamin. Adults need intakes in excess of 0.12 mg/1000 kcal/day to prevent the development of clinical signs of beriberi.[104,105] Thiamin deficiency is relatively rare in the U.S., probably due to the enrichment of white flour and rice with the vitamin; however, beriberi is sometimes seen in alcoholics.[87] Renal patients who have had long-term dialysis treatment are sometimes deficient in thiamin.[106] Beriberi is most frequently observed in individuals whose staple food is polished nonenriched rice.

Clinical symptoms of beriberi have been well documented[107-109] and mainly involve the nervous and circulatory systems. There are two forms of beriberi — dry beriberi or chronic polyneuropathy and wet beriberi or acute/subacute high-output cardiac failure. Early symptoms of beriberi include anorexia, weakness, shortened attention span, calf tenderness, burning and numbness in the extremities, altered tendon reflexes, increased irritability, anemia, and frequently, edema. This is followed by peripheral paralysis, opthalmoplegia, tachycardia, and enlarged heart. As polyneuropathy progresses there is marked muscle weakness, foot and wrist drop occur, and the patient is cachectic and more prone to infections. Individuals with wet beriberi become quite edematous, their capillaries dilate, and they usually are short of breath; cardiac failure follows.

Thiamin intakes of 0.33 to 0.35 mg/1000 kcal/day seem to satisfy the minimum requirement for thiamin as judged by urinary excretion measurements.[110,111] However, more than 0.5 mg/1000 kcal may be needed to attain tissue saturation.[105,112] Normal erythrocyte transketolase activities were observed when daily intakes of the vitamin were 0.4 mg/1000 kcal[112] or 0.5 mg/1000 kcal;[113,114] adults had maximal activity at an intake of 1.1 mg/1000 kcal.[112]

The ingestion of large amounts (100 to 500 mg) of thiamin seems to produce no toxic effects.[115] Occasionally, cases have been reported in which men were hypersensitive to large doses of thiamin taken parenterally; they had reactions resembling anaphylactic shock. A feeling of slight warmth and burning, sweating, weakness, restlessness, nausea, dyspnea, hypotension, and tachycardia were initially observed; in more severe cases angioneurotic edema, cyanosis, hemorrhage in the gastrointestinal tract, pulmonary edema, and collapse occurred.[116] Generally, the patients recovered fairly quickly. Sudden death almost immediately following intravenous or intramuscular injection of thiamin has been reported in about five individuals; the symptoms were those of anaphylactic shock.[117]

IV. RIBOFLAVIN

Riboflavin is an orangish-yellow compound which has a green fluorescence. In 1932, Warburg and Christian[118] isolated an "active yellow respiratory enzyme" from yeast which was yellow in aqueous solutions and fluoresced; this compound is often referred to as the "old yellow enzyme". Lyxoflavin, currently known as riboflavin, was synthesized in 1935.[119] Structurally, riboflavin consists of an isalloxazine ring with a ribityl side chain. Riboflavin is also called vitamin B_2.

A. Functions

Riboflavin is essential for growth, tissue repair, general good health and vigor, and integrity of the skin, eyes, and nerves. Riboflavin primarily functions in oxidation-reduction reactions as the reactive portion of the prosthetic groups flavin mononucleotode (FMN) and flavin adenine dinucleotide (FAD), which are collectively known as flavoproteins. Flavin adenine dinucleotide functions as the prosthetic group in D-amino acid oxidases[120] and flavin mononucleotide or dinucleotide, in L-amino acid oxidases;[121] these two enzymes oxidatively deaminate D- and L-amino acids, respectively. There are several metalloflavoproteins, including the molybdenum- and iron-containing xanthine oxidase,[122] which are needed for the conversion of purine to uric acid, the molybdenum-containing aldehyde oxidase,[123] the copper-containing butryl-CoA dehydrogenase,[124] and the iron-containing nicotinamide adenine dinucleotide cytochrome reductase.[125] Other flavoproteins include pyruvate oxidase, oxalate oxidase, succinate dehydrogenase, pyridoxaminephosphate oxidase, and acyl-CoA dehydrogenase. Flavoproteins also function in electron transport.

B. Metabolism

Riboflavin is readily absorbed in the small intestine and is phosphorylated by flavokinase (also called riboflavin kinase), which is located in the intestinal mucosa to form flavin mononucleotide. Flavin mononucleotide may be phosphorylated by FMN adenylyltransferase to form flavin adenine dinucleotide, the primary form found in tissues. Free riboflavin is found in the retina, milk, plasma, and semen. Flavin mononucleotide and flavin adenine dinucleotide are necessary for the synthesis of flavoproteins. Comparatively large quantities of the vitamin are found in liver and kidney with lesser amounts in muscles. The majority of the flavin adenine dinucleotide is bound to flavoproteins and released only when proteolysis occurs.[126] Riboflavin is excreted in the urine primarily as the free vitamin; it also appears in the feces where it may have been derived from bacterial synthesis. The urinary excretion of riboflavin varies with the intake and the degree to which tissue stores are saturated.

Riboflavin nutriture generally is assessed by measuring the urinary excretion of the vitamin or the activity coefficient (degree of stimulation with flavin adenine dinucleotide added in vitro) of erythrocyte glutathione reductase activity (sometimes called EGR activity coefficient). Methods for assessment of riboflavin nutriture were discussed in Chapter 6.

C. Age-Related Changes in Metabolism

The urinary riboflavin levels of subjects over the age of 70 years were similar to those of younger adults under conditions of depletion and repletion;[127] only five to six subjects were included in each group in this study. Garry et al.[128] measured the erythrocyte glutathione reductase activity coefficients of 667 subjects between 20 and 87 years of age and found that inadequate riboflavin nutriture appeared to be more of a problem for younger than older (those over 60 years) adults.

D. Requirements in Aging

More research is needed regarding the riboflavin requirements of the elderly. The Recommended Dietary Allowances[1] for men and women 51 years of age and above living in the U.S. are 1.4 and 1.2 mg/day, respectively. A riboflavin allowance of 0.6 mg/1000 kcal has been recommended for people of all ages; however, a minimum intake of 1.2 mg is recommended independent of energy intake. The recommended intakes of riboflavin for adult men and women given by FAO/WHO are 1.8 and 1.3 mg/day, respectively; here again, an intake of 0.6 mg/1000 kcal is recommended.[33]

Several investigations of riboflavin nutriture of the elderly living in the U.S. have been conducted over the last couple of decades. The mean riboflavin intakes of males and females 65 years of age and above in the Nationwide Food Consumption Survey[2] of Spring 1977 were 117 to 132% of the 1980 RDA; most of the reported riboflavin intakes were from grain products, milk and milk products, meat, poultry, and fish. Of the subjects included in the 1977 survey (all ages) 12% reported consuming less than 70% of the 1980 RDA for the vitamin.[34] Between 10 and 25% of the elderly included in the Health and Nutrition Examination Survey of 1971 to 1974[35] reported consuming less than two thirds of the 1980 RDA for riboflavin. About one sixth to one third of the men and women 60 years of age and over included in the Ten-State Nutrition Survey of 1968 to 1970[36] reported consuming less than two thirds of the 1980 RDA for riboflavin; 2 to 6% of the elderly males and females included in a survey in Albuquerque, N.M. reported consuming less than 75% of the 1980 RDA for the vitamin.[38]

Erythrocyte glutathione reductase activity coefficients indicative of riboflavin deficiency were observed in 7% of the 196 aged (65 years and above) subjects residing at home or in institutions in Belfast, Northern Ireland;[129] however, Chen and Fan-Chiang,[130] using the same parameter, found 27 to 34% of their institutionalized and noninstitutionalized subjects 60 years of age and above to be deficient in the vitamin. Less than acceptable urinary levels of riboflavin were observed in 15 of 46 elderly men and women in three nursing homes in Colorado.[131] Only 1 of 359 subjects in a study involving participants in the Title VII Nutrition Program for the Elderly[45] had an activity coefficient indicative of deficiency.

There was little evidence of riboflavin inadequacy in 93 acute geriatric admissions to a hospital.[43] Of 153 geriatric patients, 18 had erythrocyte glutathione reductase activity coefficients indicative of riboflavin deficiency; the deficiencies were corrected by supplementation.[98] Of a group of geriatric patients in another study,[100] 16 also had activities indicative of riboflavin deficiency.

E. Problems of Deficiency and Excess

Lesions of riboflavin deficiency have been produced in adults that received less than about 0.25 mg/1000 kcal.[132,133] Riboflavin intakes of 0.5 mg/1000 kcal or less do not maintain tissue reserves as indicated by urinary excretion measurements[134] or erythrocyte glutathione reductase activity coefficients.[135]

There are few characteristic symptoms of riboflavin deficiency. Riboflavin deficiency is frequently observed in association with pellagra, the niacin deficiency condition. Early symptoms of riboflavin deficiency include soreness and burning of the lips, tongue, and mouth, with increased discomfort of the aforementioned on eating and swallowing. Symptoms of more severe riboflavin deficiency include angular stomatitis, glossitis, cheilosis, seborrheic dermatitis around the nose and scrotum or vulva, corneal vascularization, anemia, photophobia, conjunctivitis, and lacrimation.[132,136,137]

Riboflavin intakes above 0.6 mg/1000 kcal maintain tissue reserves of the vitamin as indicated by the urinary excretion measurements[134] and activity coefficients.[135] All of the male subjects who participated in a study[138] had activity coefficients within normal ranges when they consumed 1.5 mg of riboflavin daily.

The author is not aware of any documented reports of riboflavin toxicity in humans. The low solubility of the vitamin may be responsible for this apparent lack of toxicity. However, there seems to be no reason for healthy humans of any age to consume large amounts of this vitamin.

V. NIACIN

Niacin is a generic name used for nicotinic acid and nicotinamide. Niacin is known as the "pellagra-preventing factor". Pellagra, the niacin deficiency disease, was endemic in the southern part of the U.S. at the turn of the century; it was prevalent in population groups where corn was a staple food. Pellagra was believed to be caused by a toxic substance present in corn or an infectious agent. Goldberger and Wheeler[139] fed prisoners a diet similar to that consumed by pellagrins and produced pellagra in the prisoners, thus showing that the disease was diet related. Pellagra was also proven not to be infectious.[140] Nicotinic acid and nicotinamide are both white crystalline compounds which have a pyridine ring.

A. Functions

Niacin plays a role in growth, glycolysis, tissue respiration, fat synthesis, amino acid and protein metabolism, and functioning of the nervous system. Niacin functions in several oxidation-reduction reactions in the body as the active portion of the coenzymes nicotinamide adenine dinucleotide (NAD, formerly called Coenzyme I or coenzymase and later, diphosphopyridine nucleotide, DPN) and nicotinamide adenine dinucleotide phosphate (NADP, formerly called Coenzyme II and later, triphosphopyridine nucleotide, TPN). The oxidized forms of the coenzymes are known as NAD and NADP and the reduced, as NADH and NADPH. These coenzymes function in the removal of hydrogen from certain substrates as part of dehydrogenases and the transfer of hydrogen or electrons to another coenzyme or substrate which is then reduced. Nicotinamide adenine dinucleotide and its phosphate function in many enzymatic reactions in the cells such as in the synthesis of adenosine triphosphate in the electron transport chain, glycolysis, pyruvate metabolism, pentose metabolism, lipid metabolism, and amino acid and protein metabolism. Reactions involving niacin coenzymes have been reviewed.[141,142]

B. Metabolism

Niacin as well as its precursor, tryptophan, are readily absorbed into the small intestine. Nicotinic acid is phosphoribosylated to nicotinate mononucleotide and then reacts with adenosine triphosphate and glutamine to become nicotinamide adenine dinucleotide; this compound may be further phosphorylated to form nicotinamide adenine dinucleotide phosphate. Niacin is present in every cell of the body. The body's reserve of niacin is not large; the liver, heart, and muscle serve as reserves. Niacin is excreted primarily as *N*-methylnicotinamide and *N*-methyl-2-pyridone-5-carboxamide.[143,144]

The amino acid tryptophan can be converted to niacin;[145] this reaction takes place in the liver and requires the participation of pyridoxal phosphate, a vitamin B_6 coenzyme. Humans can convert approximately 60 mg of tryptophan to 1 mg nicotinic acid.[146,147] Hence, niacin requirements and allowances are given in terms of niacin equivalents (NE).[1]

Niacin nutriture may be evaluated by measuring the *N*-methylnicotinamide or its 2-pyridone which is excreted in the urine. Methods for the assessment of niacin nutriture were discussed in Chapter 6.

C. Age-Related Changes in Metabolism

Intestinal hypermotility which is often observed in the elderly may cause niacin deficiency. Whanger[148] has reported that tissue levels of niacin are somewhat decreased in the aged. Circulating nicotinate levels have been found to be severely depressed in the 473 elderly living in nursing homes or at home as compared to those of 204 younger adults surveyed;[41] vitamin intake data for these subjects were not reported. There is some indication that niacin metabolism may be affected by age but the data are inconclusive.

D. Requirements in Aging

More research is needed regarding the niacin requirements of the elderly. The Recommended Dietary Allowances[1] for men and women 51 years of age and above living in the U.S. are 16 and 13 niacin equivalents (mg) daily, respectively. The recommended intakes of niacin for adult men and women given by FAO/WHO[33] are 19.8 and 14.5 mg/day.

Several investigations of niacin intakes of the elderly living in the U.S. have been conducted over the last couple of decades. The mean niacin intakes of males and females 65 years of age and above in the Nationwide Food Consumption Survey[2] of Spring 1977 were 106 to 130% of the 1980 RDA; most of the preformed niacin came from grain products, meat, poultry, and fish; 9% of the individuals (all ages) included in the 1977 survey consumed less than 70% of the 1980 RDA.[34] Between 10 and 25% of the males and females above 55 years of age included in the Health and Nutrition Examination Survey of 1971 to 1974[35] reported consuming less than two thirds of the 1980 RDA. Around 20 to 35% of the males and females 60 years of age and above included in the Ten-State Nutrition Survey[36] had niacin intakes less than two thirds of the 1980 RDA. None of the 270 free-living, middle-income, healthy men and women over 60 years of age included in a study in Albuquerque, N.M. consumed less than the 1980 RDA for the vitamin.[38]

Few niacin assessment studies have been conducted utilizing biochemical parameters. Harrill and Cervone[40] found that only 1 of their 68 elderly women had a urinary *N*-methylnicotinamide level indicative of niacin inadequacy. Stiedemann et al.[132] found the urinary *N*-methylnicotinamide:creatinine ratio of only one of their 46 elderly men and women in nursing homes to be lower than normal. Over 50% of the 93 acute geriatric admissions to a hospital in the U.K.[43] had lower than normal urinary *N*-methylnicotinamide levels; however, the authors did say that the urinary collections of some patients may have been incomplete. Baker et al.[41] found circulating nicotinate levels in the elderly to be frequently indicative of deficiency. Niacin deficiency may be somewhat prevalent in the elderly.

E. Problems of Deficiency and Excess

Pellagra, the deficiency disease due to inadequacies of niacin and tryptophan, is seldom seen in the U.S. today. Pellagra remains to be a problem in India and Africa. The symptoms of pellagra have been described.[149-151] Early symptoms of pellagra include weakness, apathy, lassitude, anorexia, indigestion, achlorhydria, anemia, and general aches and pains. These symptoms are followed by the 3 Ds — dermatitis, diarrhea, and dementia; the deficiency chiefly affects the skin, gastrointestinal tract, and nervous tissues. Sometimes the 4 Ds are used, with the last D being death. The dermatitis is bilaterally symmetrical and appears on the parts of the body exposed to sunlight, heat, or friction; typically the hands, forearms, skin around the neck, and upper thorax are affected as well as the areas around the body openings (stomatitis, proctitis, vulvovaginitis, and scrotum). Mental symptoms include irritability, headaches, depression, loss of memory, psychosis, delirium, and catatonia.

Pellagra has been produced in adults consuming less than 7.5 mg niacin daily.[139,150] Intakes of tryptophan also need to be considered. The deficiency has been observed in adults who consumed 4.9 niacin equivalents per 1000 kcal and as much as 8.8 niacin equivalents daily.[152,153] A daily intake of 11.3 to 13.3 niacin equivalents seems to prevent depletion of body stores of the vitamin.[152,154] The allowance recommended for adults is 6.6 niacin equivalents and not less than 13 niacin equivalents.[1]

Blood cholesterol, β-lipoprotein, and triglyceride levels are reduced when 3 to 6 g of nicotinic acid are taken;[155,156] increased muscle glycogen utilization and decreased mobilization of fatty acids from adipose tissues during exercise have also been observed when 3 g or more were ingested.[157] Nicotinic acid may be beneficial in the treatment of myocardial infarctions.[158] All of these are effects of nicotinic acid and not of nicotinamide. Unfortunately, these large doses (3 g and above) also produce vascular dilation (flushing reaction) and itching as well as a greater incidence of arrhythmias and gastrointestinal problems.[158] High doses of nicotinic acid should not be used except when recommended by a physician.

VI. VITAMIN B_6

Vitamin B_6 is a collective term for pyridoxine (also called pyridoxol), pyridoxal, and pyridoxamine. Gyorgy[159] in 1934 proposed that a factor which prevented dermatitis acrodynia in rats be called vitamin B_6 (later referred to as B-6). Vitamin B_6 is a 3-hydroxy-2-methylpyridine.

A. Functions

Vitamin B_6 is essential for growth, amino acid and protein metabolism, synthesis of hormones and body regulators, central nervous system functioning, and carbohydrate and lipid metabolism. Vitamin B_6 functions in the body primarily in the form of pyridoxal phosphate.[160] Pyridoxal phosphate is a coenzyme that functions in practically all the reactions involved in amino acid metabolism including transamination,[161,162] decarboxylation,[163] nonoxidative deamination, and desulfhydration. Aminotransferases, which contain the coenzyme, function in the synthesis of nonessential amino acids and in the catabolism of the amino acids. The vitamin also plays a role in protein synthesis.[164] Decarboxylases containing the coenzyme are necessary for the synthesis of γ-aminobutyric acid, 3,4-dihydroxyphenylalanine (DOPA), serotonin, histamine, epinephrine, and norepinephrine.[165] Vitamin B_6 is involved in the conversion of tryptophan to niacin.[166] Pyridoxal phosphate is part of glycogen phosphorylase which catabolizes glycogen.[167] The vitamin may[168,169] or may not[170,171] affect essential fatty acid metabolism, specifically the conversion of linoleic acid to arachidonic acid. The coenzyme is essential for the synthesis of porphyrins via the enzyme δ-aminolevulate synthetase.[172] Vitamin B_6 may function in amino acid absorption or transport[173] and in absorption and utilization of vitamin B_{12}.[174] Vitamin B_6 functions in a large variety of metabolic reactions most of which are involved in the metabolism of protein.

B. Metabolism

Vitamin B_6 is absorbed primarily in the jejunum, but also in the ileum by passive diffusion.[175] The various forms of the vitamin are interconvertible.[176] Pyridoxal kinase phosphorylates pyridoxal, pyridoxine, and pyridoxamine; adenosine triphosphate is required for this reaction. Acid phosphatase is responsible for the dephosphorylation of pyridoxal, pyridoxine, and pyridoxamine phosphates. Pyridoxine dehydrogenase, a nicotinamide adenine dinucleotide phosphate dependent enzyme, is needed for the conversion of pyridoxine to pyridoxal. The flavoprotein pyridoxaminephosphate oxidase functions in the conversion of pyridoxine phosphate and pyridoxamine phosphate to pyridoxal phosphate and of pyridoxamine to pyridoxal. The conversion of pyridoxal phosphate to pyridoxamine phosphate and of pyridoxal to pyridoxamine involves aspartate aminotransferase. Pyridoxine, pyridoxal, and pyridoxamine administered parenterally have equal vitamin B_6 activities in rat tissues; information is lacking as to their activities in man.

Small quantities of the vitamin are stored in the body in the liver, heart, brain, and skeletal muscles. Pyridoxal phosphate is considered to be the most active form of the vitamin in metabolism; however, pyridoxamine phosphate may also serve as a coenzyme. Although small quantities of pyridoxine, pyridoxal, and pyridoxamine as well as their phosphorylated derivatives are excreted into the urine, the major metabolite is pyridoxic acid.[177]

Several vitamin B_6 antagonists exist. Isonicotinic acid hydrazine[178] and cycloserine,[179] two drugs used in the treatment of tuberculosis, are vitamin B_6 antagonists.

Vitamin B_6 nutriture has been assessed by a variety of parameters including urinary excretion of xanthurenic acid after a tryptophan load, urinary pyridoxic acid levels, erythrocyte (or blood or serum) alanine or aspartate aminotransferase activities with or without coenzyme stimulation, and radioisotopic measurement of plasma pyridoxal phosphate concentrations. Methods for the assessment of vitamin B_6 nutriture were discussed in Chapter 6.

C. Age-Related Changes in Metabolism

Elderly men and women have been found to have lower serum aspartate aminotransferase activities and higher xanthurenic acid excretion than younger subjects.[180] A negative correlation of erythrocyte alanine aminotransferase activity (no stimulation; basal) with age has been reported.[181] Lower plasma pyridoxal phosphate levels have been observed in older subjects as compared to younger men.[182,183] A negative correlation of plasma pyridoxal phosphate levels with age has also been reported.[183] Parameters used to assess vitamin B_6 status may be affected by the aging process, but are the parameters affected or is vitamin B_6 metabolism affected by aging? Two different research groups have reported that oral administration of 20 mg/day of the vitamin did not normalize erythrocyte aspartate[99] or alanine[184] aminotransferase stimulations in several of their elderly subjects, thus indicating that the cause of the deficiency was probably not merely low intake of the vitamin. Perhaps the elderly may frequently have defects in absorption of the vitamin or in B_6 vitamin phosphorylation. Some of the aged may also have much higher vitamin B_6 requirements than others in their age group. More research is needed regarding the vitamin B_6 metabolism of the elderly.

Vitamin B_6 deficiency has been observed to have increased incidence in alcoholics. Baker et al.[185] have suggested that this vitamin B_6 deficit may ensue from the inability of the chronic alcoholic to absorb the vitamin from foods.

D. Requirements in Aging

More research is needed regarding the vitamin B_6 requirements of the elderly. The Recommended Dietary Allowances[1] for men and women 51 years of age and above living in the U.S. are 2.2 and 2.0 mg/day, respectively. The quantity of protein that individuals consume has been shown to affect the vitamin B_6 requirement.[186-188] The suggestion has been made that humans should ingest 0.02 mg vitamin B_6 per gram of protein consumed.[189] The FAO/WHO Handbook on Human Nutritional Requirements[33] does not give a recommended intake for vitamin B_6.

Little information is available regarding vitamin B_6 intakes of the elderly. The mean vitamin B_6 intakes of males and females 65 years of age and above in the Nationwide Food Consumption Survey[2] of Spring 1977 were 58 to 72% of the 1980 RDA; most of the vitamin B_6 came from meat, poultry, fish, and grain products. Of the individuals (all ages) included in the 1977 survey, 51% consumed less than 70% of the 1980 RDA.[34] Hampton et al.[190] found that 90% of their 20 women and 47% of their 17 men above the age of 60 living in Virginia reported consuming less than 1.4 mg of vitamin B_6 daily. Garry et al.[38] found that men and women over 60 years of age residing in the Albuquerque, N.M. vicinity, who did not take vitamin supplements had a median vitamin B_6 intake of less than 1.0 mg/day while those who took supplements had a median intake of almost 6 mg/day. Vitamin B_6 was observed to be one of the nutrients most lacking in the diets of 196 aged institutionalized and noninstitutionalized subjects in Belfast, Northern Ireland.[39] The mean vitamin B_6 intakes of hospitalized aged males and females in a study[191] conducted in Northern Ireland was between 0.8 and 1.1 mg/day.

Several investigations of vitamin B_6 nutriture of the aged have been conducted using various assessment parameters. All 20 of the subjects over the age of 70 years included in a study by Crepaldi et al.[192] had abnormal urinary excretion of tryptophan metabolites after a tryptophan load; the abnormal excretory patterns were normalized following vitamin administration. Of 153 geriatric patients, 29 (19%) had erythrocyte aspartate aminotransferase activity coefficients (stimulations) indicative of vitamin B_6 deficiency; these coefficients normalized in 26 of the 29 patients after vitamin B_6 was administered orally;[99] why did the coefficients of the other 3 subjects not normalize? Approximately one fourth of the 37 aged subjects in a Virginia study[190] were found to have coenzyme stimulation of erythrocyte alanine aminotransferase (or activity coefficient) values indicative of subclinical vitamin B_6 deficiency. Vitamin B_6 deficiency, as assessed by coenzyme stimulation of erythrocyte alanine aminotransferase measurements, has been reported in half of the institutionalized and noninstitutionalized aged in a study in Kentucky[130] and in 40% in a similar study in Belfast, Northern Ireland.[184] About one fourth of 102 hospitalized aged were found to be deficient in vitamin B_6 as indicated by erythrocyte alanine aminotransferase stimulations.[193] Low plasma pyridoxal phosphate levels have also been observed to be prevalent in elderly subjects.[180,183] Vitamin B_6 inadequacy appears to be a nutritional problem in the elderly.[194]

Vir and Love[191] gave daily supplements of 2.5 mg vitamin B_6 for 15 days to their hospitalized elderly subjects found to be deficient in the vitamin; some of these subjects continued on a 50-mg vitamin B_6 supplement for another 15 days; the vitamin B_6 status of the subjects improved after the 2.5 mg treatment, but the 50-mg treatment following the 2.5-mg treatment was not beneficial. A recommended allowance for vitamin B_6 intakes of 2.5 to 25 mg/day has been suggested as a "precaution to prevent possible serious pathological conditions".[195]

E. Problems of Deficiency and Excess

Estimates of the vitamin B_6 status and requirements of humans have been primarily based upon the appearance or cure of clinical signs, xanthurenic acid excretion after a tryptophan load, pyridoxic acid excretion, and serum and erythrocyte aspartate and alanine aminotransferase activities with or without coenzyme stimulation. In recent years, researchers have also utilized radiomonitored plasma pyridoxal phosphate levels as an indicator of vitamin B_6 status.

The first evidence that humans required vitamin B_6 was reported in 1939;[196] patients who were receiving poor diets experienced weakness, irritability, insomnia, and difficulty in walking, and these symptoms were relieved upon administration of the vitamin. Vitamin B_6 was also found to be effective in healing cheilosis that did not respond to riboflavin treatment.[197] Subjects fed a diet deficient in vitamin B_6 developed personality changes manifested by irritability, depression, and loss of a sense of responsibility; these subjects also developed stomatitis, seborrheic dermatitis, glossitis, and electroencephalographic abnormalities.[198] Hypochromic microcytic anemia[199] and increased incidence of kidney stones[200] and dental caries are also observed in vitamin B_6 deficiency. Epileptiform convulsions have been observed in vitamin B_6 deficient infants[201-203] and convulsions in adults[204,205] who were deficient in the vitamin.

As stated earlier, the vitamin B_6 requirement is affected by the protein intake. The optimal vitamin B_6 intake of males that consumed 100 g protein daily was 1.75 to 2.0 mg/day, whereas it was 1.25 to 1.5 mg when 30 g protein was consumed, as judged by urinary xanthurenic acid measurements following a tryptophan load.[186] Linkswiler[206] has suggested that the vitamin B_6 requirement for men consuming diets containing 100 to 150 g protein is between 1.5 and 2.0 mg/day and 1.0 to 1.5 mg when the protein intake is below 100 g. A vitamin B_6 intake of 1.5 mg/day was borderline for young women receiving 78 g protein daily, yet 2.2 mg was in excess of requirements as assessed by varying status parameters.[207,208] Based upon average protein intakes as determined in food consumption surveys and utilizing the ratio of 0.02 mg vitamin B_6 per gram of protein consumed,[189] the Recommended Dietary Allowance[1] for vitamin B_6 for adult males was calculated to be 2.2 mg (with 110 g protein intake) and for adult females, 2.0 mg (with 100 g protein intake).

Larger than RDA doses of vitamin B_6 have several suggested therapeutic applications. The most common therapeutic dose is 50 mg.

Some investigators have studied the effects of vitamin B_6 in the control of hyperemesis gravidarum (nausea and vomiting during pregnancy). Willis et al.[209] observed beneficial effects when 50 mg of the vitamin was given to pregnant women; similar findings were reported by other researchers. However, Hasseltine[210] concluded in a controlled study that the use of the vitamin for treatment of hyperemesis gravidarum was not warranted. Emotional factors do enter into such studies and treatments.

The withdrawal seizures in chronic alcoholism have been beneficially treated with 100 mg of the vitamin.[211] Beneficial effects of 25- to 200-mg doses have been observed in the treatment of radiation sickness.[212]

Pyridoxine hydrochloride administration has been shown to correct apparent vitamin B_6 suboptimacy in pregnant women[213] and in females using oral contraceptives.[214] Individuals diagnosed to have Parkinson's disease generally are treated daily with L-DOPA; pyridoxine hydrochloride supplements can counteract the effect of L-DOPA at certain dosage levels.[215]

Vitamin B_6 supplements are commonly employed when isoniazid is used in tuberculosis treatment and when penicillamine is used in the treatment of Wilson's disease.[216] Patients with carpal tunnel syndrome have been shown to be vitamin B_6 deficient; supplementation with 100 mg of the vitamin alleviated symptoms of the syndrome.[217] The vitamin has also been used with some success in the treatment of hyperoxaluria and recurring oxalate kidney stones, chorea, and levodopa-induced dystonia. Pyridoxine hydrochloride supplements are useful in the treatment of primary oxalosis. Various dose levels have been utilized; however, excellent responses were observed when two patients received 1 g/day.[218]

Schumacher et al.[219] found patients with rheumatoid arthritis to have low plasma pyridoxal phosphate levels; these patients then received 50 to 150 mg of pyridoxine hydrochloride daily for 3 months. The plasma pyridoxal phosphate levels of the patients increased but no improvement in the clinical symptoms of rheumatoid arthritis were observed.

A group of disorders known as vitamin B_6-dependency states have been observed in which the metabolism and availability of the vitamin in the body appear to be normal yet a higher than normal requirement for the vitamin exists. If vitamin B_6-dependent individuals are not treated promptly with the vitamin or one of its metabolites, irreversible alterations in brain function leading to mental retardation may occur. Typical vitamin B_6-dependency states include infantile convulsions,[220] vitamin B_6-responsive anemias,[221] congenital cystathioninuria,[222] xanthurenic aciduria, and homocystinuria.[223] The enzyme lesion is known in three of these diseases and in the other two, the mutant enzyme seems in some cases to interact abnormally with pyridoxal phosphate in a manner which can be overcome by high coenzyme concentration.[223] These conditions appear to be inherited metabolic abnormalities.

The toxicity of all forms of vitamin B_6 is quite low in laboratory animals. Toxic effects were not observed in men that received 50 to 200 mg of the vitamin daily for several months.[224]

Neurohistological examination of tissues from rats and dogs that had received 2 to 6 g of vitamin B_6 per kilogram body weight revealed degeneration of the spinal cord and peripheral nerves.[225] Rats that consumed 1.56 g pyridoxine hydrochloride per kilogram diet (14% protein) exhibited different behavioral patterns than animals fed 78 to 780 mg of the vitamin per kilogram of diet.[226] Seven adults were found to have ataxia and severe sensory nervous system abnormalities after taking 2 to 6 g pyridoxine hydrochloride daily for 2 to 40 months.[227]

Doses over about 200 mg vitamin B_6 daily should not be taken unless the individual is under the care of a physician. The progress of patients taking such megadoses should be carefully observed.

VII. PANTOTHENIC ACID

Pantothenic acid is a member of the vitamin B complex. The vitamin was isolated by Williams et al. in 1938.[228] Pantothenic acid is an essential nutrient for humans.

A. Functions

Pantothenic acid functions in the body as a component, but not the functional unit, of coenzyme A[229] (sometimes referred to as CoA); this coenzyme also contains β-mercaptoethylamine, adenine, ribose, and phosphoric acid. The coenzyme functions as an acyl group acceptor and donor. Coenzyme A is involved in the tricarboxylic acid cycle, gluconeogenesis, fatty acid metabolism, and in the synthesis of acetylcholine, sphingosine, porphyrins, and sterols (including steroid hormones).[230]

B. Metabolism

Pantothenic acid is absorbed readily into the intestines and is phosphorylated by adenosine triphosphate to phosphopantothenic acid and is subsequently converted to coenzyme A via several reactions involving adenosine triphosphate and cysteine. Coenzyme A can be synthesized by all cells. High concentrations of the coenzyme are found in the liver, brain, kidneys, adrenals, and heart. Pantothenic acid administered orally is excreted rapidly in the urine.[231]

C. Age-Related Changes in Metabolism

To the knowledge of the author there are no reports of alterations in pantothenic acid metabolism due to age. The quantity of the vitamin required by rats for optimal growth has been reported not to increase during aging up to 2 years of age.[232]

D. Requirements in Aging

Adults fed 5 to 7 mg of the vitamin daily have been reported to have urinary excretions of 2 to 7 mg/day and fecal excretions of 1 to 2 mg/day.[233] Subjects fed 10 mg of the vitamin daily had urinary excretions of 5 to 7 mg/day.[234] The 1980 Recommended Dietary Allowances[1] publication has recommended an estimated safe and adequate daily dietary intake of 4 to 7 mg/day for adults; there is no separate age category for the elderly. The *FAO/WHO Handbook on Human Nutritional Requirements*[33] does not list a recommended intake for pantothenic acid.

Meals representing usual dietary intakes excluding beverages of residents of a Utah nursing home were analyzed for pantothenic acid content; the vitamin content of these meals was observed to be 2.22 mg/1000 kcal.[235] The average pantothenic acid intake of institutionalized and noninstitutionalized men and women 65 years of age and above as calculated via food consumption records was 5.9 mg/day or 2.9 mg/1000 kcal.[236]

E. Problems of Deficiency and Excess

Dietary pantothenic acid deficiencies have not been clinically recognized in man probably because of the vitamin being present in a wide variety of foods; good sources include meats, eggs, whole grains, and legumes. Pantothenic acid may also be synthesized by the intestinal microflora.[1] Pantothenic acid deficiency may be present in multiple nutrient deficiencies seen in chronic malnutrition[237] and in alcoholism.[238]

Deficiency symptoms have been produced in humans via the feeding of a semisynthetic diet which was virtually free of pantothenic acid for a 10-week period.[234] Symptoms include headache, fatigue, sleep disturbances, nausea, personality changes, mental depression, abdominal distress, numbness and tingling of hands and feet, muscle cramps, impaired coordination, increased sensitivity to insulin, and decreased antibody formation.[234,239,240] These symptoms were reversed when pantothenic acid was added to the diet.

Pantothenic acid seems to be relatively nontoxic. No toxicity symptoms were observed when men were given 10 g of the vitamin daily for 6 weeks;[241] however, occasional diarrhea and water retention problems were observed when humans received daily doses of 10 to 20 g.[242]

Pantothenic acid deficiency does produce graying of hair in black rats; however, researchers have not observed the vitamin to be of benefit in preventing the graying of hair in humans. The vitamin may also improve the capability of well-nourished subjects to withstand stress.[243]

VIII. BIOTIN

Biotin is a member of the vitamin B complex. This vitamin was earlier known as the curative factor for "egg white injury".[244] Biotin has an ureido ring containing a sulfur atom and a valeric acid side chain.

A. Functions

Biotin is involved in carbohydrate, fat, and to some degree, protein metabolism. Biotin functions in carboxylation reactions; the vitamin is tightly attached via the carboxyl group of its valeric acid side chain to the ϵ-N-lysine moiety of the apoenzyme.[245] Carboxylases transport carboxyl groups as well as fix carbon dioxide in tissues. Carboxylases containing biotin include acetyl-CoA carboxylase (converts acetyl-CoA to malonyl-CoA; important in fatty acid synthesis), propionyl-CoA carboxylase (converts propionyl-CoA to methylmalonyl-CoA; an intermediate in the synthesis of succinyl-CoA), methylcrontonyl-CoA carboxylase (converts methylcrontonyl-CoA to methylglutaconyl-CoA; also a reaction in the conversion of leucine to acetoacetate and acetyl-CoA), geranoyl-CoA carboxylase (converts geranoyl-CoA to orthophosphate and isohexenylglutaconyl-CoA), pyruvate carboxylase (converts pyruvate to oxaloacetate; important in gluconeogenesis and replenishment of oxaloacetate in tricarboxylic acid cycle), and methylmalonyl-CoA carboxyltransferase (converts methylmalonyl-CoA and pyruvate to propionyl-CoA and oxaloacetate; important in tricarboxylic acid cycle).[246] Propionyl-CoA carboxylase requires biotin as well as vitamin B[12] for its functioning.[247]

Biotin functions in the synthesis of carbamyl phosphate which is important in the synthesis of pyrimidines and some amino acids.[248] Biotin may play a role in the deamination of threonine, aspartic acid, and serine; however, its involvement is believed to be indirect.

B. Metabolism

Biotin is absorbed in the upper small intestine. The vitamin is stored in minute quantities in the liver, kidneys, brain, and adrenals. The vitamin is excreted in the urine primarily in the form of free biotin. Generally, the biotin content of the urine and feces is considerably greater than the dietary intake. It is currently believed that the biotin which is synthesized by the intestinal microflora can be absorbed and does make a significant contribution to the biotin needs of humans; hence the dietary requirement of humans for biotin is uncertain.[1] Long-term use of antibiotics and sulfa drugs may decrease biotin synthesis by the intestinal microflora.

Biotin deficiency can be produced in humans by feeding large amounts of raw egg whites; these egg whites contain the glycoprotein avidin and avidin binds biotin and prevents its absorption.[249] Cooking denatures the avidin and the denatured avidin does not bind with biotin.

C. Age-Related Changes in Metabolism

To the knowledge of the author there are no reported alterations in biotin metabolism due to aging.

D. Requirements in Aging

In that the biotin synthesized by the intestinal microflora can be absorbed, requirements for the vitamin cannot be established. Syndenstricker[249] found that injected doses of 150 to 300 μg of biotin were needed to relieve the biotin deficiency symptoms under conditions of avidin ingestion. The urinary biotin excretion of two adults after a 200-μg dose of the vitamin was increased by 50 to 100 μg over a 24-hr period, thus indicating that the subjects retained 100 to 150 μg.[250] The 1980 Recommended Dietary Allowances[1] publication lists an estimated safe and adequate daily dietary intake of 100 to 200 μg of biotin for adults of all ages. If half of the dietary biotin is absorbed and half excreted, then the lower level of intake is estimated to provide for tissue biotin replenishment. The *WHO/FAO Handbook on Human Nutritional Requirements*[33] does not list a recommended intake for biotin.

E. Problems of Deficiency and Excess

Biotin deficiency has rarely been reported in human adults except when these adults were eating large quantities of raw egg whites. Deficiency symptoms include anorexia, nausea, glossitis, mental depression, pallor, substernal pain, scaly dermatitis, and desquamation of the lips.[249,250] Infants under the age of 6 months have been reported to have seborrheic dermatitis due to nutritional biotin deficiency;[251] perhaps the intestinal microflora are somewhat incapable of biotin synthesis.

Biotin appears to be rather nontoxic. Infants with seborrheic dermatitis have responded to daily biotin injections of 5 mg with seemingly no adverse effects.[251] This quantity is probably in excess of physiological needs. The American Academy of Pediatrics[252] has recommended that infant formulas provide at least 15 μg of biotin per 1000 kcal. Beneficial effects have been observed when mothers of breast-fed infants with seborrheic dermatitis received daily biotin injections of 5 to 10 mg.[253] Individuals with rarely observed inborn errors of metabolism affecting propionyl-CoA carboxylase and methylcrotonyl-CoA carboxylase activities may respond to pharmacologic doses of the vitamin.[254]

IX. FOLIC ACID AND VITAMIN B$_{12}$

There is evidence that folic acid and vitamin B$_{12}$ inadequacies are frequently observed in the elderly. Folic acid and vitamin B$_{12}$ are discussed in Volume II, Chapter 10.

X. CONCLUSIONS

Frequently nutrient requirements are stated in terms of 1000 kcal consumed. If the nutrient requirements of the aged are considered in terms of energy intakes, are these requirements really any different than those of the other age groups that may be consuming less than 2000 or 2500 kcal/day? Researchers do not agree on the answer to that question. There is a progressive reduction in the voluntary intake of kilocalories and thus of other nutrients by the elderly due to reduced physical activity and loss of muscular tissues.

The Heart Disease Epidemiological Study group in Framingham, Mass. has collected medical information from their male and female subjects for almost 30 years; the subjects still living today are over 55 years of age. Jette and Branch[255] reported that these Framingham elders exhibited substantial physical ability. Advancing age is accompanied by an increased loss in physical ability but not of the magnitude suggested by "conventional wisdom".

Reduced physical activity and socioeconomic conditions among the elderly may lead to a reduction in the intake of foods that may supply vitamins. Various diseases may also indirectly induce vitamin inadequacies. The elderly may feel that they no longer perform a "useful role" in society and they may become anorexic. The aged often experience loss of their teeth, ill-fitting dentures, and declining and altered smell and taste sensations. Medications and alcohol intake may also affect the appetite of indivduals as well as the absorption and metabolism of nutrients. Chronic alcoholism is prevalent in the elderly. Physical and mental factors do affect the vitamin intakes and the vitamin nutriture of the elderly.

The results of many investigations have indicated that the elderly frequently are deficient or marginally deficient in several of the water-soluble vitamins. The water-soluble vitamin inadequacies observed most frequently are those of vitamin B$_6$, folic acid, ascorbic acid, and thiamin. Vitamin B$_{12}$ inadequacy is fairly prevalent in the aged primarily because of the lack of adequate intrinsic factor for its absorption.

More research investigating the nutritional requirements of the elderly is needed. These requirements are known to be affected by degenerative diseases as well as by nutrient and drug interactions.

REFERENCES

1. National Research Council, National Academy of Sciences, *Recommended Dietary Allowances*, 9th ed., National Academy of Sciences, Washington, D.C., 1980.
2. U.S. Department of Agriculture, Food and Nutrient Intakes of Individuals in 1 Day in the United States, Spring 1977, Nationwide Food Consumption Survey 1977—78, Preliminary Report No. 2, U.S. Department of Agriculture, Washington, D.C., 1982.
3. **Burns, J. J.**, Biosynthesis of *L*-ascorbic acid; basic defect in scurvy, *Am. J. Med.*, 26, 740, 1959.
4. **Monsen, E. R., Hallberg, M., Layrisse, D. M., Hegsted, D. M., Cook, J. D., Mertz, W., and Finch, C. A.**, Estimation of available dietary iron, *Am. J. Clin. Nutr.*, 31, 134, 1978.
5. **Stokes, P. L., Melikian, V., Leeming, R. L., Portman-Graham, H., Blair, J. A., and Cooke, W. T.**, Folate metabolism in scurvy, *Am. J. Clin. Nutr.*, 28, 126, 1975.
6. **Lipschitz, P. A., Bothwell, T. H., Seftel, H. C., Wapnick, A. A., and Charlton, R. W.**, The role of ascorbic acid in the metabolism of storage iron, *Br. J. Haematol.*, 20, 155, 1971.
7. **Peterkofsky, B. and Udenfriend, S.**, Enzymatic hydroxylation of proline in microsomal polypeptides leading to formation of collagen, *Proc. Natl. Acad. Sci. U.S.A.*, 53, 335, 1965.

8. **Cooper, J. R.,** The role of ascorbic acid in the oxidation of tryptophan to 5-hydroxy-tryptophan, *Ann. N.Y. Acad. Sci.,* 92, 208, 1961.

9. **Levin, E. Y., Levenberg, B., and Kaufman, S.,** The enzymatic conversion of 3,4-dihydroxyphenylalanine to norepinephrine, *J. Biol. Chem.,* 235, 2080, 1960.

10. **La Du, B. N. and Zannoni, V. G.,** The role of ascorbic acid in tyrosine metabolism, *Ann. N.Y. Acad. Sci.,* 92, 175, 1961.

11. **Stone, K. J. and Townsley, B. H.,** The effect of L-ascorbate on catecholamine biosynthesis, *Biochem. J.,* 131, 611, 1973.

12. **Schwartz, P. L.,** Ascorbic acid in wound healing — a review, *J. Am. Dietet. Assoc.,* 56, 497, 1970.

13. **Ginter, E.,** Ascorbic acid in cholesterol and bile acid metabolism, *Ann. N.Y. Acad. Sci.,* 258, 410, 1975.

14. **Bates, C. J., Burr, M. K., and St. Leger, A. S.,** Vitamin C, high density lipoproteins and heart disease in elderly subjects, *Age Ageing,* 8, 177, 1979.

15. **Greco, A. M. and La Rocca, L.,** Correlation between chronic hypovitaminosis in old age and plasma levels of cholesterol and triglycerides, *Int. J. Vitam. Nutr. Res.,* 23, 129, 1982.

16. **Zannoni, V. G. and Sato, P. H.,** Effects of ascorbic acid on microsomal drug metabolism, *Ann. N.Y. Acad. Sci.,* 258, 119, 1975.

17. **Kallner, A., Hartman, D., and Hornig, D.,** On the absorption of ascorbic acid in man, *Int. J. Vitam. Nutr. Res.,* 47, 383, 1977.

18. **Mayersohn, M.,** Ascorbic acid absorption in man — pharmacokinetic implications, *Eur. J. Pharmacol.,* 19, 140, 1972.

19. **Baker, E. M., Hodges, R. E., Hood, J., Sauberlich, H. E., March, S. C., and Canham, J. E.,** Metabolism of ^{14}C- and ^{3}H-labeled L-ascorbic acid in human scurvy, *Am. J. Clin. Nutr.,* 24, 444, 1971.

20. **Hodges, R. E., Hood, J., Canham, J. E., Sauberlich, H. E., and Baker, E. M.,** Clinical manifestations of ascorbic acid deficiency in man, *Am. J. Clin. Nutr.,* 24, 432, 1971.

21. **Baker, E. M., Hodges, R. E., Hood, J., Sauberlich, H. E., and March, S. C.,** Metabolism of ascorbic-1-^{14}C in experimental human scurvy, *Am. J. Clin. Nutr.,* 22, 549, 1969.

22. **Sauberlich, H. E. and Baker, E. M.,** Studies in human nutrition, in Annual Research Progress Report, U.S. Army Medical Research and Nutrition Laboratory, San Francisco, June 30, 1967, 180.

23. **Irwin, M. I. and Hutchins, B. K.,** A conspectus of research on vitamin C requirements of man, *J. Nutr.,* 106, 823, 1976.

24. **Brook, M. and Grimshaw, J. J.,** Vitamin C concentration of plasma and leukoytes as related to smoking habit, age, and sex of humans, *Am. J. Clin. Nutr.,* 21, 1254, 1968.

25. **Keith, R. E. and Driskell, J. A.,** Effects of chronic cigarette smoking on vitamin C status, lung function, and resting exercise cardiovascular metabolism in humans, *Nutr. Rep. Int.,* 21, 907, 1980.

26. **Rivers, J. M. and Devine, M. M.,** Plasma ascorbic acid concentrations and oral contraceptives, *Am. J. Clin. Nutr.,* 25, 684, 1972.

27. **Loh, H. S. and Wilson, C. W. M.,** Relationship of human ascorbic-acid metabolism to ovulation, *Lancet,* 1, 110, 1971.

28. **Burr, M. L., Elwood, P. C., Holes, D. J., Hurley, R. J., and Hughes, R. E.,** Plasma and leukocyte ascorbic acid levels in the elderly, *Am. J. Clin. Nutr.,* 27, 144, 1974.

29. **Wilson, T. S., Datta, S. B., Murrell, J. S., and Andrews, C. T.,** Relation of vitamin C levels to mortality in a geriatric hospital: a study of the effect of vitamin C administration, *Age Ageing,* 2, 163, 1973.

30. **Windsor, A. C. M., Hobbs, C. B., Treby, D. A., and Cowper, R. A.,** Effect of tetracycline on leukocyte ascorbic acid levels, *Br. Med. J.,* 1, 214, 1972.

31. **Burr, M. L., Hurley, R. J., and Sweetnam, P. M.,** Vitamin C supplementation of old people with low blood levels, *Gerontol. Clin.,* 17, 236, 1975.

32. **Bates, C. J.,** Proline and hydroxyproline excretion and vitamin C status in elderly human subjects, *Clin. Sci. Mol. Med.,* 52, 535, 1977.

33. Food and Agriculture Organization of the United Nations, *Handbook on Human Nutritional Requirements,* FAO/WHO, Rome, 1974.

34. **Pao, E. M. and Mickle, S. J.,** Problem nutrients in the United States, *Food Technol.,* 35(9), 58, 1981.

35. United States Department of Health, Education and Welfare, Dietary Intake Source Data, United States, 1971—74, DHEW Publ. No. (PHS) 79-1221, 1979.

36. United States Department of Health, Education, and Welfare, Ten-State Nutrition Survey 1968—1970, DHEW Publ. No. (HSM) 72-8130 to 8134, 1972.

37. United States Department of Agriculture, Food Intake and Nutritive Values of Diets of Men, Women, and Children in the United States, Spring 1965: A preliminary Report, USDA, Washington, D.C., 1969.

38. **Garry, P. J., Goodwin, J. S., Hunt, W. C., Hooper, E. M., and Leonard, A. G.,** Nutritional status in a healthy elderly population: dietary and supplemental intakes, *Am. J. Clin. Nutr.,* 36, 319, 1982.

39. **Vir, S. C. and Love, A. H. G.,** Nutritional status of institutionalized and noninstitutionalized aged in Belfast, Northern Ireland, *Am. J. Clin. Nutr.,* 32, 1934, 1974.

40. **Harrill, I. and Cervone, N.,** Vitamin status of older women, *Am. J. Clin. Nutr.,* 30, 431, 1977.
41. **Baker, H., Frank, O., Thind, I. S., Jaslow, S. P., and Louria, D. B.,** Vitamin profiles in elderly persons living at home or in nursing homes, versus profile in healthy young subjects, *J. Am. Geriatr. Soc.,* 27, 444, 1979.
42. **Basu, T. K., Jordan, S. J., Jenner, M., and Williams, D. C.,** Blood values of some vitamins in long-stay psycho-geriatric patients, *Int. J. Vitam. Nutr. Res.,* 46, 61, 1976.
43. **Morgan, A. G., Kelleher, J., Walker, B. E., Losowsky, M. S., Droller, H., and Middleton, R. S.,** A nutritional survey in the elderly: blood and urine vitamin levels, *Int. J. Vitam. Nutr. Res.,* 45, 448, 1975.
44. **Burr, M. L., Milbank, J. E., and Gibbs, D.,** The nutritional status of the elderly, *Age Ageing,* 11, 89, 1982.
45. **Kohrs, M. B., Nordstrom, J., Plowman, E. L., O'Hanlon, P., Moore, C., Davis, C., Abrahams, O., and Eklund, D.,** Association of participation in a nutritional program for the elderly with nutritional status, *Am. J. Clin. Nutr.,* 33, 2643, 1980.
46. **Vir, S. C. and Love, A. H.,** Vitamin C status of institutionalised and non-institutionalised aged, *Int. J. Vitam. Nutr. Res.,* 48, 274, 1978.
47. **Hodges, R. E., Hood, J., Canham, J. E., Sauberlich, H. E., and Baker, E. M.,** Clinical manifestations of ascorbic acid deficiency in man, *Am. J. Clin. Nutr.,* 24, 432, 1971.
48. **Lind, J.,** A treatise of the scurvy, A. Millar, London, 1753. Republished by Stewart, C. P., and Guthrie, D., Eds., *Lind's Treatise on Scurvy,* University of Edinburgh, Press, Edinburgh, 1953.
49. **Walker, A.,** Chronic scurvy, *Br. J. Dermatol.,* 80, 625, 1968.
50. **Kinsman, R. A. and Hood, J.,** Some behavioral effects of ascorbic acid deficiency, *Am. J. Clin. Nutr.,* 24, 455, 1971.
51. **Woodruff, C. W.,** Ascorbic acid — scurvy, *Progr. Food Nutr. Sci.,* 1, 493, 1975.
52. **Scrimshaw, N. S., Taylor, C. E., and Gorden, J. E.,** *Interactions of Nutrition and Infection,* World Health Organization, Geneva, 1968.
53. **Leung, F. W. and Guze, P. A.,** Adult scurvy, *Ann. Emerg. Med.,* 10, 652, 1981.
54. **Connelly, T. J., Becker, A., and McDonald, J. W.,** Bachelor scurvy, *Int. J. Dermatol.,* 21, 209, 1982.
55. **Taylor, G. F.,** Nutrition and old age, *Commun. Health,* 3, 244, 1972.
56. **Dodds, M. J. and MacLeod, F. L.,** Blood plasma ascorbic acid levels on controlled intakes of ascorbic acid, *Science,* 106, 67, 1947.
57. **Friedman, G. J., Sherry, S., and Ralli, E. P.,** The mechanism of the excretion of vitamin C by the human kidney at low and normal plasma levels of ascorbic acid, *J. Clin. Invest.,* 19, 685, 1940.
58. **Baker, E. M., Saari, J. C., and Tolbert, B. M.,** Ascorbic acid metabolism in man, *Am. J. Clin. Nutr.,* 19, 371, 1966.
59. **Pauling, L.,** *Vitamin C and the Common Cold,* W. H. Freeman, San Francisco, 1970.
60. **Pauling, L.,** The significance of the evidence about ascorbic acid and the common cold, *Proc. Natl. Acad. Sci. U.S.A.,* 68, 2678, 1971.
61. **Anderson, T. W., Reid, D. B. W., and Beaton, G. H.,** Vitamin C and the common-cold: a double-blind trial, *Can. Med. Assoc. J.,* 107, 503, 1972.
62. **Anderson, T. W., Suranyi, G., and Beaton, G. H.,** The effect on winter illness of large doses of vitamin C, *Can. Med. Assoc. J.,* 111, 31, 1974.
63. **Anderson, T. W., Beaton, G. H., Corey, P. N., and Spero, L.,** Winter illness and vitamin C: the effect of relatively low doses, *Can. Med. Assoc. J.,* 112, 823, 1975.
64. **Wilson, C. W. M., Loh, H. S., and Foster, F. G.,** The beneficial effect of vitamin C on the common cold, *Eur. J. Clin. Pharmacol.,* 6, 26, 1973.
65. **Coulehan, J. L., Reisinger, K. S., Rogers, K. D., and Bradley, D. W.,** Vitamin C prophylaxis in a boarding school, *N. Engl. J. Med.,* 290, 6, 1974.
66. **Karlowski, T. R., Chalmers, T. C., Frenkel, L. D., Kapikian, A. Z., Lewis, T. L., and Lynch, J. M.,** Ascorbic acid for the common cold: a prophylactic and therapeutic trial, *JAMA,* 231, 1038, 1975.
67. **Anderson, T. W.,** Large-scale trials of vitamin C, *Ann. N.Y. Acad. Sci.,* 258, 498, 1975.
68. **Coulehan, J. L., Eberhard, S., Kapner, L., Taylor, F., Rogers, K., and Garry, P.,** Vitamin C and acute illness in Navajo school children, *N. Engl. J. Med.,* 295, 973, 1976.
69. **Ginter, E., Cerna, O., Budlovsky, J., Balaz, V., Hruba, F., Roch, V., and Sasko, E.,** Effect of ascorbic acid on plasma cholesterol in humans in a long-term experiment, *Int. J. Vitam. Nutr. Res.,* 47, 123, 1977.
70. **Peterson, V. E., Crapo, P. A., Weininger, J., Ginsberg, H., and Olefsky, J.,** Quantification of plasma cholesterol and triglyceride levels in hypercholesterolemic subjects receiving ascorbic acid supplements, *Am. J. Clin. Nutr.,* 28, 584, 1975.
71. **Prinz, W., Bortz, R., Bregin, B., and Hersch, M.,** The effect of ascorbic acid supplementation on some parameters of the human immunological defence system, *Int. J. Vitam. Nutr. Res.,* 47, 248, 1977.
72. **Cameron, E. and Campbell, A.,** The orthomolecular treatment of cancer. II. Clinical trial of high-dose ascorbic acid supplements in advanced human cancer, *Chem. Biol. Interact.,* 9, 285, 1974.

73. **Samborskaya, E. P. and Ferdman, T. D.,** The mechanism of termination of pregnancy by ascorbic acid, *Byull. Eksp. Biol.,* 62(8), 96, 1966.

74. **Lamden, M. P. and Chrystowski, G. A.,** Urinary oxalate excretion by man following ascorbic acid ingestion, *Proc. Soc. Exp. Biol. Med.,* 85, 190, 1954.

75. **Stein, H. G., Hasan, A., and Fox, I. H.,** Ascorbic acid-induced uricosuria. A consequence of megavitamin therapy, *Ann. Intern. Med.,* 84, 385, 1976.

76. **Herbert, V. and Jacob, E.,** Destruction of vitamin B₁₂ by ascorbic acid, *JAMA,* 230, 241, 1974.

77. **Cook, J. D. and Monsen, E. R.,** Vitamin C, the common cold, and iron absorption, *Am. J. Clin. Nutr.,* 30, 235, 1977.

78. **Shilotri, P. G. and Bhat, K. S.,** Effect of mega doses of vitamin C on bactericidal activity of leukocytes, *Am. J. Clin. Nutr.,* 30, 1077, 1977.

79. **Schrauser, G. N. and Rhead, W. J.,** Ascorbic acid abuse: effects of long-term ingestion of excessive amounts on blood levels and urinary excretion, *Int. J. Vitam. Nutr. Res.,* 43, 201, 1973.

80. **Funk, J.,** Die vitamine, *J. State Med.,* 20, 341, 1912.

81. **Williams, R. R. and Cline, J. K.,** Synthesis of vitamin B₁, *J. Am. Chem. Soc.,* 58, 1504, 1936.

82. **Lohmann, K. and Schuster, P.,** Untersuchungen über die cocarboxylase, *Biochem. Z.,* 294, 188, 1937.

83. **Horecker, B. L. and Smyrniotis, P. Z.,** The coenzyme function of thiamine pyrophosphate in pentose phosphate metabolism, *J. Am. Chem. Soc.,* 75, 1009, 1953.

84. **Racker, E. G., de La Haba, G., and Leder, I. G.,** Thiamine pyrophosphate, a coenzyme of transketolase, *J. Am. Chem. Soc.,* 75, 1010, 1953.

85. **Dreyfus, P. M.,** Thiamin and the nervous system: an overview, *J. Nutr. Sci. Vitaminol.,* 22(Suppl.), 13, 1976.

86. **Thomson, A. D. and Leevy, C. M.,** Observations on the mechanism of thiamine hydrochloride absorption in man, *Clin. Sci.,* 43, 153, 1972.

87. **Leevy, C. M. and Baker, H.,** Vitamins and alcoholism, *Am. J. Clin. Nutr.,* 21, 1325, 1968.

88. **Jansen, B. C. P.,** Thiamine, in *Vitamins: Chemistry, Physiology, Pathology,* Sebrell, W. H. and Harris, R. L., Eds., Academic Press, New York, 1954, 425.

89. **Pollack, H., Ellenberg, M., and Dolger, H.,** Study of the excretion of thiamine and its degradation products in humans, *J. Nutr.,* 10(Suppl.), 21, 1941.

90. **Murata, K.,** Thiaminase, in *Review of Japanese Literature on Beriberi and Thiamine, Vitamin B Research Committee of Japan,* Shimazono, N. and Katsura, E., Eds., Igaku Shoin, Tokyo, 1965, 220.

91. **Vimokesant, S. L., Nakornchi, S., Dhanamitta, S., and Hilker, D. M.,** Effect of tea consumption on the thiamin status in man, *Nutr. Rep. Int.,* 9, 371, 1974.

92. **Somogyi, J. C. and Nageli, V.,** Antithiamine effect of coffee, *Int. J. Vitam. Nutr. Res.,* 46, 148, 1976.

93. **Horwitt, M. K., Liebert, E., Kreisler, O., and Wittman, P.,** *Investigations of Human Requirements for B-Complex Vitamins,* NRC Bull. no. 116, National Academy of Science, Washington, D.C., 1948.

94. **Oldham, H. G.,** Thiamine requirements of women, *Ann. N.Y. Acad. Sci.,* 98, 542, 1962.

95. **Thomson, A. D.,** Thiamine absorption in old age, *Gerontol. Clin.,* 8, 354, 1966.

96. **Markkanen, T., Herkinheimo, R., and Dahl, M.,** Transketolase activity of red blood cells from infancy to old age, *Acta Haematol.,* 42, 148, 1969.

97. **Darke, S. J.,** Requirements for vitamins in old age. Symp. Swedish Nutr. Foundation: X, in *Nutrition in Old Age,* Carlson, L. A., Ed., Almqvist & Wiksell, Uppsala, Sweden, 1972, 107.

98. **Horwitt, M. K.,** Dietary requirements of the aged, *J. Am. Dietet. Assoc.,* 29, 443, 1958.

99. **Schlenker, E. D., Feurig, J. S., Stone, L. H., Ohlson, M. A., and Michelson, O.,** Nutrition and health of older people, *Am. J. Clin. Nutr.,* 26, 1111, 1973.

100. **Brin, M., Dibble, M. V., Peel, A., McMullen, E., Bourquin, A., and Chen, N.,** Some preliminary findings on the nutritional status of the aged in Onondaga County, New York, *Am. J. Clin. Nutr.,* 17, 240, 1965.

101. **Dibble, M. W., Brin, M., Thiele, V. F., Peel, A., Chen, N., and McMullen, E.,** Evaluation of the nutritional status of elderly subjects, with a comparison between fall and spring, *J. Am. Geriatr. Soc.,* 15, 1031, 1967.

102. **Hoorn, R. K. J., Flikweert, J. P., and Westerink, D.,** Vitamin B-1, B-2 and B-6 deficiencies in geriatric patients, measured by coenzyme stimulation of enzyme activities, *Clin. Chim. Acta,* 61, 151, 1975.

103. **Pollitt, N. T. and Salkeld, R. M.,** Vitamin B status of geriatric patients, *Nutr. Metab.,* 21(Suppl. 1), 24, 1977.

104. **Elsom, K. O., Reinhold, J. G., Nicholson, J. T. L., and Chornock, C.,** Studies of the B vitamins in the human subject: the normal requirement for thiamine; some factors influencing its utilization and excretion, *Am. J. Med. Sci.,* 203, 569, 1942.

105. **Williams, R. D., Mason, H. L., Smith, B. F., and Wilder, R. M.,** Induced thiamine (vitamin B₁) deficiency and the thiamine requirement of man; further observations, *Arch. Intern. Med.,* 69, 721, 1942.

106. **Raskin, N. H. and Fishman, R. A.,** Neurologic disorders in renal failure, *N. Engl. J. Med.,* 294, 204, 1976.

107. **Williams, R. R.,** *Toward the Conquest of Beriberi,* Harvard University Press, Cambridge, Mass., 1961.
108. **Inouye, K. and Katsura, E.,** Etiology and pathology of beriberi, in *Review of Japanese Literature on Beriberi and Thiamine, Vitamin B Research Committee of Japan,* Shimazono, N. and Katsura, E., Eds., Igaku Shoin, Tokyo, 1965, 1.
109. **Platt, S.,** Thamine deficiency in human beriberi and in Wernicke's encephalopathy, in *Thiamine Deficiency,* Wolstenholme, G. E. W. and O'Connor, M., Eds., Little, Brown, Boston, 1967, 135.
110. **Melnick, D.,** Vitamin B₁ (thiamine) requirement of man, *J. Nutr.,* 24, 139, 1942.
111. **Ziporin, Z. Z., Nunes, W. T., Powell, R. C., Waring, P. P., and Sauberlich, H. E.,** Thiamine requirement in the adult human as measured by urinary excretion of thiamine metabolites, *J. Nutr.,* 85, 297, 1965.
112. **Reuter, H., Gasmann, B., and Bohm, M.,** Thiamine requirement in humans, *Int. J. Vitam. Nutr. Res.,* 37, 315, 1967.
113. **Bamji, M. S.,** Transketolase activity and urinary excretion of thiamin in the assessment of thiamin-nutrition status of Indians, *Am. J. Clin. Nutr.,* 23, 52, 1970.
114. **Haro, E. N., Brin, M., and Faloon, W. W.,** Fasting in obesity: thiamine depletion as measured by erythrocyte transketolase changes, *Arch. Intern. Med.,* 117, 175, 1966.
115. **Jolliffe, N.,** Treatment of neuropsychiatric disorders with vitamins, *JAMA,* 117, 1496, 1941.
116. **Jaros, S. H., Wnuck, A. L., and de Beer, E. J.,** Thiamine intolerance, *Ann. Allergy,* 10, 291, 1952.
117. **Weigard, C. C.,** Reactions attributed to administration of thiamin chloride, *Geriatrics,* 5, 274, 1950.
118. **Warburg, O. and Christian, W.,** Uber ein neus oxydationsferment und sein absorption-spektrum, *Biochem. Z.,* 254, 438, 1932.
119. **Kuhn, R., Reinemund, K., Weygand, F., and Strobele, R.,** Uber die synthese des lactoflavins (vitamin B₂), *Chem. Ber.,* 68, 1765, 1935.
120. **Warburg, O. and Christian, W.,** Isolation of the prosthetic group of the d-amino acid oxidase, *Biochem. Z.,* 298, 150, 1938.
121. **Blanchard, M., Green, D. E., Nocito, V., and Ratner, S.,** Isolation of *L*-amino acid oxidase, *J. Biol. Chem.,* 161, 583, 1945.
122. **Mackler, B., Mahler, H. R., and Green, D. E.,** Studies on metalloflavoproteins. Xanthine oxidase, a molybdoflavoprotein, *J. Biol. Chem.,* 210, 149, 1954.
123. **Mahler, H. R., Mackler, B., Green, D. E., and Bock, R. M.,** Studies on metalloflavoproteins. III. Aldehyde oxidase: a molybdoflavoprotein, *J. Biol. Chem.,* 210, 465, 1954.
124. **Mahler, H. R.,** Butyryl co-A dehydrogenase, a cupro-flavo protein, *J. Am. Chem. Soc.,* 75, 3288, 1953.
125. **Mahler, H. R. and Elowe, D. G.,** Studies on metalloflavoproteins. II. The role of iron in diphosphopyridine nucleotide cytochrome c reductase, *J. Biol. Chem.,* 210, 165, 1954.
126. **Slater, E. C., Ed.,** *Flavins and Flavoproteins,* Elsevier, Amsterdam, 1966.
127. **Horwitt, M. K.,** Dietary requirements of the aged, *J. Am. Dietet. Assoc.,* 29, 443, 1958.
128. **Garry, P. J., Goodwin, J. S., and Hunt, W. C.,** Nutritional status in a healthy elderly population: riboflavin, *Am. J. Clin. Nutr.,* 36, 902, 1982.
129. **Vir, S. C. and Love, A. H. G.,** Riboflavin status of institutionalised and non-institutionalised aged, *Int. J. Vitam. Nutr. Res.,* 47, 336, 1977.
130. **Chen, L. H. and Fan-Chiang, W. L.,** Biochemical evaluation of riboflavin and vitamin B₆ status of institutionalized and noninstitutionalized elderly in Central Kentucky, *Int. J. Vitam. Nutr. Res.,* 51, 232, 1981.
131. **Stiedemann, M., Jansen, C., and Harrill, I.,** Nutritional status of elderly men and women, *J. Am. Dietet. Assoc.,* 73, 132, 1978.
132. **Sebrell, W. H. and Butler, R. E.,** Riboflavin deficiency in man: preliminary note, *U.S. Public Health Rep.,* 53, 2282, 1938.
133. **Horwitt, M. K., Hills, O. W., Harvey, C. C., Liebert, E., and Steinburg, D. L.,** Effects of dietary depletion of riboflavin, *J. Nutr.,* 39, 357, 1949.
134. **Sebrell, W. H., Jr., Butler, R. E., Wooley, J. G., and Isbell, H.,** Human riboflavin requirement estimated by urinary excretion of subjects on controlled intake, *U.S. Public Health Rep.,* 56, 510, 1941.
135. **Bamji, M. S.,** Glutathione reductase activity in red blood cells and riboflavin status in humans, *Clin. Chim. Acta,* 26, 263, 1969.
136. **Sydenstricker, V. P., Sebrell, W. H., Cleckley, H. M., and Kruse, H. D.,** Ocular manifestations of ariboflavinosis; progress note, *JAMA,* 114, 2437, 1940.
137. **Syndensticker, V. P.,** Clinical manifestations of nicotinic acid and riboflavin deficiency (pellagra), *Ann. Intern. Med.,* 14, 1499, 1941.
138. **Tillotson, J. A. and Baker, E. M.,** An enzymatic measurement of the riboflavin status in man, *Am. J. Clin. Nutr.,* 25, 425, 1972.
139. **Goldberger, J. and Wheeler, G. A.,** Experimental pellagra in the human subject brought about by a restricted diet, *Public Health Rep.,* 30, 3336, 1915.
140. **Goldberger, J.,** The transmissibility of pellagra, *Public Health Rep.,* 31, 3159, 1916.

141. **Chaykin, S.,** Nicotinamide coenzymes, *Ann. Rev. Biochem.,* 36, 149, 1967.

142. **Everse, J., Anderson, B., and You, K.-S.,** *The Pyridine Nucleotide Coenzymes.* Academic Press. New York, 1982.

143. **Holman, V. I. M. and deLange, D. J.,** Metabolism of nicotinic acid and related compounds by humans, *Nature (London),* 165, 604, 1950.

144. **Rosenthal, H. L., Goldsmith, G. A., and Sarett, H. P.,** Excretion of N^1-methylnicotinamide and the 6-pyridone of N^1-methylnicotinamide in urine of human subjects. *Proc. Soc. Exp. Biol. Med.,* 84, 208, 1953.

145. **Krehl, W. A., Tepley, L. J., Sarma, P. S., and Elvejhem, C. A.,** Growth retarding effect of corn in nicotinic acid low rations and its counteraction by tryptophan, *Science,* 101, 489, 1945.

146. **Horwitt, M. K., Harvey, C. C., Rothwell, W. S., Cutler, J. L., and Haffron, D.,** Tryptophan-niacin relationships in man, *J. Nutr.,* 60(Suppl. 1), 1, 1965.

147. **Goldsmith, G. A., Miller, O. N., and Unglaub, W. G.,** Efficiency of tryptophan as a niacin precursor in man, *J. Nutr.,* 73, 172, 1961.

148. **Whanger, A. D.,** Vitamins and vigor at 65 plus, *Postgrad. Med.,* 53. 167, 1973.

149. **Elvehjem, C. A.,** Relationship of nicotinic acid to pellagra, *Physiol. Rev.,* 20, 249. 1940.

150. **Goldsmith, G. A., Sarett, H. P., Register, U. D., and Gibbens, J.,** Studies of niacin requirement in man. I. experimental pellagra in subjects on corn diets low in niacin and tryptophan, *J. Clin. Invest.,* 31, 533, 1952.

151. **Frostig, J. P. and Spies, T. D.,** Initial nervous syndrome of pellagra and associated deficiency diseases, *Am. J. Med. Sci.,* 199, 268, 1940.

152. **Goldsmith, G. A., Gibbens, J., Unglaub, W. G., and Miller, O. N.,** Studies of niacin requirement in man. III. Comparative effects of diets containing lime-treated and untreated corn in the production of experimental pellagra, *Am. J. Clin. Nutr.,* 4, 151, 1956.

153. **Goldsmith, G. A.,** Niacin-tryptophan relationships in man and niacin requirement, *Am. J. Clin. Nutr.,* 6, 479, 1958.

154. **Goldsmith, G. A., Rosenthal, H. L., Gibbens, J., and Unglaub, W. G.,** Studies of niacin requirement in man. II. Requirement on wheat and corn diets low in tryptophan, *J. Nutr.,* 56, 371, 1955.

155. **Miller, O. N., Hamilton, J. G., and Goldsmith, G. A.,** Investigation of mechanism of action of nicotinic acid on serum lipid levels in man, *Am. J. Clin. Nutr.,* 8, 480, 1960.

156. **Shawver, J. R., Scarborough, J. S., and Tarnowski, S. M.,** Control of hypercholesteremia and hyperlipemia in a neuropsychiatric hospital, *Am. J. Psychiatry,* 117, 741, 1961.

157. **Darby, W. J., McNutt, K. W., and Todhunter, E. N.,** Niacin, *Nutr. Rev.,* 33, 289, 1975.

158. Coronary Drug Project Research Group, Clofibrate and niacin in coronary heart disease, *JAMA,* 231, 360, 1975.

159. **Gyorgy, P.,** Vitamin B_2 and the pellagra-like dermatitis in rats, *Nature (London),* 133, 498, 1934.

160. **Umbreit, W. W. and Gunsalus, I. C.,** The function of pyridoxine derivative: arginine and glutamic acid decarboxylases, *J. Biol. Chem.,* 159, 333, 1945.

161. **Schlenk, F. and Snell, E. E.,** Vitamin B_6 and transamination, *J. Biol. Chem.,* 157, 425, 1945.

162. **Lichstein, H. C., Gunsalus, I. C., and Umbreit, W. W.,** Function of vitamin B_6 group; pyridoxal phosphate (codecarboxylase) in transamination, *J. Biol. Chem.,* 161, 311, 1945.

163. **Gunsalus, I. C., Bellamy, W. D., and Umbreit, W. W.,** A phosphorylated derivative of pyridoxal as the coenzyme of tyrosine decarboxylase, *J. Biol. Chem.,* 155, 685, 1944.

164. **Driskell, J. A. and Kirksey, A.,** The cellular approach to the determination of pyridoxine requirements in pregnant and nonpregnant rats, *J. Nutr.,* 101, 661, 1971.

165. **Hsu, J. M.,** Interrelations between vitamin B_6 and hormones, *Vitam. Horm.,* 21, 113, 1963.

166. **Wiss, O. and Weber, F.,** Die reindarstellung der kynureninase, *Z. Physiol. Chem.,* 304, 232, 1956.

167. **Cori, C. F. and Illingworth, B.,** The prosthetic group of phosphorylase, *Proc. Natl. Acad. Sci., U.S.A.,* 43, 547, 1957.

168. **Mueller, J. F.,** Vitamin B_6 in fat metabolism, *Vitam. Horm.,* 22, 787, 1964.

169. **Sato, Y.,** A possible role of pyridoxine in lipid metabolism, *Nagoya J. Med. Sci.,* 33, 105, 1970.

170. **Williams, M. A. and Scheier, G. E.,** Effect of methyl arachidonate supplementation on the fatty acid composition of livers of pyridoxine-deficient rats, *J. Nutr.,* 74, 9, 1961.

171. **Johnson, P. V., Kopaczyk, K. C., and Kummerow, F. A.,** Effect of deficiency on fatty acid composition of carcass and brain lipids in the rat, *J. Nutr.,* 74, 96, 1961.

172. **Shemin, D. and Kikuchi, G.,** Enzymatic synthesis of γ-aminolevulinic acid, *Ann. N.Y., Acad. Sci.,* 75, 122, 1958.

173. **Christensen, H. N. and Riggs, T.,** Structural evidences for chelation and Schiff's base formation in amino acid transfer into cells, *J. Biol. Chem.,* 220, 265, 1956.

174. **Ranke, B., Ranke, E., and Chow, B. F.,** The interrelationship between vitamin B_6 and B_{12}, *J. Nutr.,* 71, 411, 1960.

175. **Booth, C. C. and Brain, M. C.,** The absorption of tritium-labelled pyridoxine hydrochloride in the rat, *J. Physiol. (London),* 164, 282, 1962.

176. **Snell, E. E.,** Chemical structure in relation to biological activities of vitamin B$_6$, *Vitam. Horm.*, 16, 77, 1958.
177. **Robinowitz, J. C. and Snell, E. E.,** Vitamin B$_6$ group; urinary excretion of pyridoxal, pyridoxamine, pyridoxine and 4-pyridoxic acid in human subjects, *Proc. Soc. Exp. Biol. Med.*, 70, 235, 1949.
178. **Price, J. M., Brown, R. R., and Larson, F. C.,** Quantitative studies on human urinary metabolites of tryptophan as affected by isoniazid and deoxypyridoxine, *J. Clin. Invest.*, 36, 1600, 1957.
179. **Cohen, A. C.,** Pyridoxine in the prevention and treatment of convulsions and neurotoxicity due to cycloserine, *Ann. N.Y. Acad. Sci.*, 166, 346, 1969.
180. **Ranke, E., Tauber, S. A., Horonick, A., Ranke, B., Goodhart, R. S., and Chow, B. F.,** Vitamin B$_6$ deficiency in the aged, *J. Gerontol.*, 15, 41, 1960.
181. **Jacobs, A., Cavill, I. A. J., and Hughes, J. N. P.,** Erythrocyte transaminase activity: effect of age, sex, and vitamin B$_6$ supplementation, *Am. J. Clin. Nutr.*, 21, 502, 1968.
182. **Hamfelt, A.,** Age variation of vitamin B$_6$ metabolism in man, *Clin. Chim. Acta*, 10, 48, 1964.
183. **Rose, C. S., Gyorgy, P., Butler, M., Andres, R., Norris, A. H., Shock, N. W., Tobin, J., Brin, M., and Spiegel, H.,** Age differences in vitamin B$_6$ status of 617 men, *Am. J. Clin. Nutr.*, 29, 847, 1976.
184. **Vir, S. C. and Love, A. H. G.,** Vitamin B$_6$ status of institutionalised and non-institutionalised aged, *Int. J. Vitam. Nutr. Res.*, 47, 364, 1977.
185. **Baker, H., Frank, O., Zetterman, R. K., Rajan, K. S., Hove, W., and Leevy, C. M.,** Inability of chronic alcoholics with liver disease to use food as a source of folates, thiamin and vitamin B$_6$, *Am. J. Clin. Nutr.*, 28, 1377, 1975.
186. **Baker, E. M., Canham, J. E., Nunes, W. T., Sauberlich, H. E., and McDowell, M. E.,** Vitamin B$_6$ requirement for adult men, *Am. J. Clin. Nutr.*, 15, 59, 1964.
187. **Miller, L. T. and Linkswiler, H. M.,** Effect of protein intake on the development of abnormal tryptophan metabolism by men during vitamin B$_6$ depletion, *J. Nutr.*, 93, 53, 1967.
188. **Canham, J. E., Baker, E. M., Harding, R. S., Sauberlich, H. E., and Plough, I. C.,** Dietary protein — its relationship to vitamin B$_6$ requirements and function, *Ann. N.Y. Acad. Sci.*, 166, 16, 1969.
189. Bureau of Nutritional Sciences, Dietary Standard for Canada, Department of National Health and Welfare, Ottawa, Canada, 1975.
190. **Hampton, D. J., Chrisley, B. M., and Driskell, J. A.,** Vitamin B$_6$ status of the elderly in Montgomery County, VA, *Nutr. Rep. Int.*, 16, 743, 1977.
191. **Vir, S. C. and Love, A. H. G.,** Vitamin B$_6$ status of the hospitalized aged, *Am. J. Clin. Nutr.*, 31, 1383, 1978.
192. **Crepaldi, G., Allergri, G., De Antoni, A., Costa, C., and Muggeo, M.,** Relationship between tryptophan metabolism and vitamin B$_6$ and nicotinamide in aged subjects, *Acta Vitamin. Enzymol.*, 29, 140, 1975.
193. **Vir, S. C. and Love, A. H. G.,** Nutritional evaluation of B groups of vitamins in institutionalised aged, *Int. J. Vitam. Nutr. Res.*, 47, 211, 1977.
194. **Driskell, J. A.,** Vitamin B$_6$ status of the elderly, in *Human Vitamin B$_6$ Requirements*, National Academy of Sciences, Washington, D.C., 1978, 252.
195. **Gyorgy, P.,** Developments leading to metabolic role of vitamin B$_6$, *Am. J. Clin. Nutr.*, 24, 1250, 1971.
196. **Spies, T. D., Bean, W. B., and Ashe, W. F.,** Note on use of vitamin B$_6$ in human nutrition, *JAMA*, 112, 1414, 1939.
197. **Smith, S. G. and Martin, D. W.,** Cheilosis successfully treated with synthetic vitamin B$_6$, *Proc. Soc. Exp. Biol. Med.*, 43, 660, 1940.
198. **Sauberlich, H. E., Canham, J. E., Baker, E. M., Raica, N., Jr., and Herman, Y. F.,** Human vitamin B$_6$ nutriture, *J. Sci. Ind. Res.*, 29, 528, 1970.
199. **Dinning, J. S. and Day, P. L.,** Vitamin B$_6$ and erythropoiesis in rat, *Proc. Soc. Exp. Biol. Med.*, 92, 115, 1956.
200. **Gershoff, S. N., Faragalla, F. F., Nelson, D. A., and Andrus, S. B.,** Vitamin B$_6$ deficiency and oxalate nephrocalcinosis in the cat, *Am. J. Med.*, 27, 72, 1959.
201. **Snyderman, S. E., Holt, L. E., Jr., Corretero, R., and Jacobs, K. G.,** Pyridoxine deficiency in the human infant, *Am. J. Clin. Nutr.*, 1, 200, 1953.
202. **Coursin, D. B.,** Convulsive seizures in infants with pyridoxine deficient diet, *JAMA*, 154, 406, 1954.
203. **Molony, C. J. and Parmelle, A. H.,** Convulsion in young infants as a result of pyridoxine (vitamin B$_6$) deficiency, *JAMA*, 154, 405, 1954.
204. **Canham, J. E., Nunes, W. T., and Eberlin, E. W.,** Electroencephalographic and central nervous system manifestation of B$_6$ deficiency and induced B$_6$-dependency in normal human adults, in *Proceedings VI International Congress on Nutrition*, E & S Livingstone, Edinburgh, 1964.
205. **Sauberlich, H. E.,** Human requirements for vitamin B$_6$, *Vitam. Horm.*, 22, 807, 1964.
206. **Linkswiler, H. M.,** Vitamin B$_6$ requirements of men, in *Human Vitamin B$_6$ Requirements*, National Academy of Sciences, Washington, D.C., 1978, 279.
207. **Shin, H. K. and Linkswiler, H.,** Tryptophan and methionine metabolism of adult females as affected by vitamin B$_6$ deficiency, *J. Nutr.*, 104, 1348, 1974.

208. **Brown, R. R., Rose, D. P., Leklem, J. E., Linkswiler, H., and Anand, R.,** Urinary 4-pyridoxic acid, plasma pyridoxal phosphate, and erythrocyte aminotransferase levels in oral contraceptive users receiving controlled intakes of vitamin B_6, *Am. J. Clin. Nutr.*, 28, 10, 1975.

209. **Willis, R. S., Winn, W. W., Morris, A. T., Newsome, A. A., and Massey, W. E.,** Clinical observations in treatment of nausea and vomiting in pregnancy with vitamin B_1 and B_6: a preliminary report, *Am. J. Obstet. Gynecol.*, 44, 265, 1942.

210. **Hasseltine, H. C.,** Pyridoxine failure in nausea and vomiting of pregnacy, *Am. J. Obstet. Gynecol.*, 51, 82, 1946.

211. **Lunde, F.,** Pyridoxine deficiency in chronic alcoholism, *J. Nerv. Mental Dis.*, 131, 77, 1960.

212. **Maxfield, J. R., McIlwain, A. J., and Robertson, J. E.,** Treatment of radiation sickness with vitamin B_6 (pyridoxine hydrochloride), *Radiology*, 41, 383, 1943.

213. **Dempsey, W. B.,** Vitamin B_6 and pregnancy, in *Human Vitamin B_6 Requirements*, National Academy of Sciences, Washington, D.C., 1978, 202.

214. **Rose, D. P.,** Oral contraceptives and vitamin B_6, in *Human Vitamin B_6 Requirements*, National Academy of Sciences, Washington, D.C., 1978, 193.

215. **Ebadi, M.,** Vitamin B_6 and biogenic amines in brain metabolism, in *Human Vitamin B_6 Requirements*, National Academy of Sciences, Washington, D.C., 1978, 129.

216. **Jaffe, I. A., Altman, D., and Merryman, P.,** The antipyridoxine effect of penicillamine in man, *J. Clin. Invest.*, 43, 1869, 1964.

217. **Ellis, E., Folkers, K., Watanabe, T., Kaji, M., Saji, S., Caldwell, J. W., Temple, C. A., and Wood, F. S.,** Clinical results of a cross-over treatment with pyridoxine and placebo of the carpal tunnel syndrome, *Am. J. Clin. Nutr.*, 32, 2040, 1979.

218. **Will, E. J. and Bijvoet, O. L. M.,** Primary oxalosis: clinical and biochemical response to high-dose pyridoxine therapy, *Metabolism*, 28, 542, 1979.

219. **Schumacher, H. P., Bernhart, F. W., and Gyorgy, P.,** Vitamin B_6 levels in rheumatoid arthritis: effect of treatment, *Am. J. Clin. Nutr.*, 28, 1200, 1975.

220. **Hunt, A. D., Stokes, J., Jr., McCrory, W. W., and Stroud, H. H.,** Pyridoxine dependency: report of case of intractable convulsions in infant controlled by pyridoxine, *Pediatrics*, 13, 140, 1954.

221. **Horrigan, D. L. and Harris, J. W.,** Pyridoxine-responsive anemia: analysis of 62 cases, *Adv. Intern. Med.*, 12, 103, 1964.

222. **Frimpter, G.,** Cystathioninuria: nature of the defect, *Science*, 149, 1095, 1965.

223. **Mudd, S. H.,** Pyridoxine-responsive genetic disease, *Fed. Proc. Fed. Am. Soc. Exp. Biol.*, 30, 970, 1971.

224. **Jolliffe, N.,** Treatment of neuropsychiatric disorders with vitamins, *JAMA*, 117, 1496, 1941.

225. **Antopol, W. and Tarlov, I. M.,** Experimental study of effects produced by large doses of vitamin B_6, *J. Neuropathol. Exp. Neurol.*, 1, 330, 1942.

226. **Driskell, J. A. and Loker, S. F.,** Behavioral patterns of female rats fed high levels of vitamin B_6, *Nutr. Rep. Int.*, 14, 467, 1976.

227. **Schaumburg, H., Kaplan, J., Windeback, A., Vick, N., Rasmus, S., Pleasure, D., and Brown, M. J.,** Sensory neuropathy from pyridoxine abuse: a new megavitamin syndrome, *N. Engl. J. Med.*, 309, 445, 1983.

228. **Williams, R. J., Truesdail, J. H., Weinstock, H. H., Rohrmann, E., Lyman, C. M., and McBurney, C. H.,** Pantothenic acid. II. Its concentration and purification from liver, *J. Am. Chem. Soc.*, 60, 2719, 1938.

229. **DeVries, W. H., Grovier, V. M., Evans, J. S., Gregory, J. D., Novelli, G. D., Soodak, M., and Lipmann, F.,** Purification of coenzyme A from fermentation sources and its further identification, *J. Am. Chem. Soc.*, 72, 4838, 1950.

230. **Goldman, P. and Vagelos, P. R.,** Acyl-transfer reactions (CoA — structure, function), in *Comprehensive Biochemistry*, Vol. 15, Florkin, M. and Stotz, E. H., Eds., Elsevier, New York, 1964, 71.

231. **Silber, R. H. and Unna, K.,** Studies on the urinary excretion of pantothenic acid, *J. Biol. Chem.*, 142, 623, 1942.

232. **Mills, C. A.,** B vitamin requirements with advancing age, *Am. J. Physiol.*, 153, 31, 1948.

233. **Fox, H. M. and Linkswiler, H.,** Pantothenic acid excretion on three levels of intake, *J. Nutr.*, 75, 451, 1961.

234. **Fry, P. C., Fox, H. M., and Tao, H. G.,** Metabolic response to a pantothenic acid deficient diet in humans, *J. Nutr. Sci. Vitaminol.*, 22, 339, 1976.

235. **Walsch, J. H., Wyse, B. W., and Hansen, R. G.,** Pantothenic acid content of a nursing home diet, *Ann. Nutr. Metab.*, 25, 178, 1981.

236. **Scinivasan, V., Christensen, N., Wyse, B. W., and Hansen, R. G.,** Pantothenic acid nutritional status in the elderly — institutionalized and noninstitutionalized, *Am. J. Clin. Nutr.*, 34, 1736, 1981.

237. **Gershberg, H., Rubin, S. H., and Ralli, E. P.,** Urinary pantothenate, blood glucose and inorganic serum phosphate in patients with metabolic disorders treated with doses of pantothenate, *J. Nutr.*, 39, 107, 1949.

238. **Leevy, C. M., Baker, H., ten Hove, W., Frank, O., and Cherrick, G. R.,** B-Complex vitamins in liver disease of the alcoholic, *Am. J. Clin. Nutr.*, 16, 339, 1965.

239. **Hodges, R. E., Bean, W. B., Ohlson, M. A., and Bleiler, B.,** Human pantothenic acid deficiency produced by omega-methylpantothenic acid, *J. Clin. Invest.*, 38, 1421, 1959.

240. **Glusman, N.,** The syndrome of "burning feet" (nutritional melalgia) as a manifestation of nutritional deficiency, *Am. J. Med.*, 3, 211, 1947.

241. **Ralli, E. P. and Dumm, M. E.,** Relation of pantothenic acid to adrenal cortical function, *Vitam. Horm.*, 11, 133, 1953.

242. **Harris, R. S. and Lepkovsky, S.,** Pantothenic acid, in *The Vitamins: Chemistry, Physiology, Pathology*, Vol. 2, Sebrell, W. H., Jr. and Harris, R. S., Eds., Academic Press, New York, 1954, 591.

243. **Ralli, E. P.,** The effect of certain nutritional factors on the reactions produced by acute stress in human subjects, *Nutr. Symp. Ser.*, 5, 78, 1952.

244. **Gyorgy, P.,** The curative factor (vitamin H) for egg white injury, with particular reference to its presence in different foodstuffs and in yeast, *J. Biol. Chem.*, 131, 733, 1939.

245. **Lane, M. D. and Lynen, F.,** The biochemical function of biotin. VI. Chemical structure of the carboxylated active site of propionyl carboxylase, *Proc. Natl. Acad. Sci. U.S.A.*, 49, 379, 1963.

246. **Lynen, F.,** The role of biotin-dependent carboxylations in biosynthetic reactions, *Biochem. J.*, 102, 381, 1967.

247. **Dupont, J. and Mathias, M. M.,** Bio-oxidation of linoleic acid via methyl-malonyl-CoA, *Lipids*, 4, 478, 1969.

248. **Wellner, V. P., Santos, J. I., and Meister, A.,** Carbamyl phosphate synthetase: a biotin enzyme, *Biochemistry*, 7, 2848, 1968.

249. **Sydenstricker, V. P., Singel, S. A., Briggs, A. P., DeVaughn, N. M., and Isbell, H.,** Observations of the "egg white injury" in man and its cure with a biotin concentrate, *JAMA*, 118, 1199, 1942.

250. **Baugh, C. M., Malone, J. W., and Butterworth, C. E., Jr.,** Human biotin deficiency: a case history of biotin deficiency induced by raw egg consumption in a cirrhotic patient, *Am. J. Clin. Nutr.*, 21, 173, 1968.

251. **Bonjour, J. P.,** Biotin in man's nutrition and therapy — a review, *Int. J. Vitam. Nutr. Res.*, 47, 107, 1977.

252. American Academy of Pediatrics, Committee on Nutrition, Commentary on breast-feeding and infant formulas, including proposed standards for formulas, *Pediatrics*, 57, 278, 1976.

253. **Nisenson, A.,** Seborrheic dermatitis of infants: treatment with biotin injections for the nursing mother, *Pediatrics*, 44, 1014, 1969.

254. **Barnes, N. D., Hull, D., Balgobin, L., and Gompertz, D.,** Biotin-responsive propionic acidaemia, *Lancet*, 2, 244, 1970.

255. **Jette, A. M. and Branch, L. G.,** The Framingham disability study. II. Physical disability among the aging, *Am. J. Public Health*, 71, 1211, 1981.

Chapter 11

TRACE MINERALS

Janet L. Greger

TABLE OF CONTENTS

I. INTRODUCTION

More than 50 elements have been identified in mammalian tissues. Most of these are present in trace amounts. Several of these trace elements are known to be essential for life. The importance of five of these essential trace elements — zinc, copper, manganese, chromium, and selenium — to human nutrition, especially in regard to the nutritional needs of the aged, will be reviewed in this chapter. Iron and fluoride, other trace elements, will be discussed in the chapters on anemia and osteoporosis, respectively.

II. ZINC

A. Zinc Intake of the Elderly

In 1977 officials in the U.S. Department of Agriculture (USDA) surveyed more than 9600 Americans in regard to their food intake.[1] USDA officials did not calculate the zinc intake of individuals participating in this survey. However, useful information pertinent to the topic of zinc intake can be gleaned from this survey. In this large survey, the average man and woman older than 75 years of age consumed 26 and 15%, respectively, less energy daily than the average young (23 to 34 years of age) man and woman. Energy intakes of individuals between the ages of 65 to 74 years of age were also considerably less (20% for men and 11% for women) than those of young adults.

Not only did the elderly consume less food than younger adults, but they also made different food choices. Generally, the intakes of grain products by the elderly were similar to those of young adults. Egg consumption was greater by the elderly than by young adults. However, the average man and woman over 75 years of age consumed 31 and 17%, respectively, less meat, fish, and poultry than young (23 to 34 years of age) adults. The decline in the intake of meat, fish, and poultry by individuals between the ages of 65 to 74 was less dramatic.

These differences in food intake by the elderly may reflect differences in income. USDA officials estimated that 34% of the individuals over 65 years of age who participated in this survey were from households that had less than $6000 annual income; only 10% of the younger (19 to 64 years of age) adults surveyed were from households that had less than $6000 annual income. The differences in food choices among various age groups may also reflect differences in food preferences and in health status, including dental health.

On the basis of the data from this national survey, it can be concluded that the elderly in the U.S. probably consume less zinc than younger adults. This reflects two factors: a decline in total energy intake and a decline in the consumption of meat, fish, and poultry. These flesh foods are the major sources of zinc in the diets of many Americans. Welsh and Marston[2] estimated that 47% of the zinc consumed by Americans in 1980 was supplied by meat, fish, and poultry.

Other surveys on smaller groups of the elderly tend to confirm these observations.[3-11] Investigators have noted that the average intakes of zinc by elderly subjects were between 7 and 10 mg of zinc daily (Table 1). The Recommended Dietary Allowance (RDA) for zinc is 15 mg.[12] Hence, the average elderly individual in these surveys consumed one half to two thirds of the amount of zinc recommended by the Food and Nutrition Board of the National Academy of Sciences in the RDAs.

The fact that most elderly individuals consume less zinc daily than the amount suggested in the RDA does not prove that their zinc intakes are inadequate to meet their needs. It does suggest that a large percentage of the elderly may be at risk of consuming inadequate amounts of zinc. Whether these individuals would benefit from additional zinc intake depends on a variety of factors: physiological or pathological condition of the individual, medications used by the individual, and bioavailability of the dietary zinc.

Table 1
ZINC INTAKE OF ELDERLY INDIVIDUALS

Age (years)	Gender	Number	Type of diet	Zinc intake (mean ± SEM) (mg/day)	Ref.
67	M	17	Free living	8.2	3
67	F	20	Free living	7.2	3
52—89	M	18	Free living	10.2 ± 0.7	4
52—89	F	26	Free living	9.9 ± 0.7	4
67—96	F	36	Free living	7.1 ± 1.0	5
72—87	M	10	Free living	11.8 ± 0.9	6
70—78	F	14	Free living	10.4 ± 0.5	6
53 ± 15(SD)	F	49	Free living vegetarian	9.2 ± 0.4	7
60—97	F & M	173	Free living	7.3 ± 0.3	8
>62	M	9	Free living	7	9
>62	F	49	Free living	7	9
75 ± 10(SD)	M	31	Institutional	8.0 ± 0.4	10
76 ± 6(SD)	F	34	Institutional	8.6 ± 0.5	10
69—96	M	13	Institutional	8.1 ± 1.0	6
40—82	F	53	Institutional	7.4 ± 0.5	6
>60	M	54	Institutional	9.6 ± 0.5	11
>60	F	108	Institutional	7.8 ± 0.3	11

B. Zinc Requirements of the Elderly

The zinc requirements of humans of any age are not well defined. Although urinary losses of zinc by healthy individuals are small and fairly constant, the loss of zinc in the feces can vary greatly.[13,14] Usually, adults will achieve balance (i.e., intake ≥ fecal and urinary losses) in regard to zinc when they consume somewhere between 8 and 16 mg of zinc daily.[15-31] The exact level at which balance is achieved depends on the individual and a variety of dietary factors.

Factors which depress the utilization of zinc by livestock and laboratory animals tend to depress the absorption of zinc by human subjects.[13,14,32,33] However, the levels and sometimes forms of these factors in the diets of humans and livestock differ; thus, their relative importance sometimes differs. It is well established that the incorporation of large amounts of dietary fiber and phytate into diets will depress zinc absorption in livestock and humans.[13,25,27,33] However, Sandstead and co-workers[30] observed that the incorporation of moderate amounts of soft white wheat bran and corn bran into the diets of human subjects did not result in significantly decreased bioavailability of dietary zinc.

The addition of large amounts of phosphate salts, especially in the presence of high dietary levels of calcium, tends to depress zinc utilization by animals.[32] The effect of phosphorus on zinc utilization is sometimes[22,29] but not always observed[18,31] in studies with humans. Differences in the forms of the phosphorus and levels of other dietary factors appear to be important. Some investigators have noted that humans absorb zinc more efficiently when protein is added to the diet;[18,22] others have not.[29]

Levels of other trace elements in the diet can also affect the utilization of zinc by human subjects. Both tin[24] and iron[34] can depress the apparent absorption of zinc by human subjects. Elderly Americans can ingest large amounts of tin and iron from canned foods and nutritional supplements, respectively.

Furthermore, the level of zinc in the diet influences the efficiency of zinc absorption. Individuals generally absorb zinc more efficiently when less zinc is present in the diet.[13,14,16-18,34]

Heaney and associates[35] have stated that the efficiency of calcium absorption decreases after middle age. Whether the efficiency of zinc absorption decreases with age has not been established. Weigand and Kirchgessner[35] observed that young rats absorbed zinc more efficiently than mature rats; they did not compare young adults and old rats. Several groups of investigators have conducted balance studies in regard to zinc with relatively healthy older (50 to 85 years of age) men and women.[15-19,36] The efficiency of utilization of zinc by these older subjects[15-19,36] did not appear to differ markedly from that of younger adults in other studies.[20-31] However, older subjects tended to absorb zinc less efficiently.

Although there is little evidence that the zinc requirements of healthy elderly are greater than the zinc requirements of young adults, there is definitely no evidence that zinc requirements of humans are reduced with age. This is an important point because many elderly consume less zinc than young adults. Furthermore, many elderly individuals use medications that increase the requirements for zinc.

C. Pathological Conditions and Medications that Affect the Requirements for Zinc

Any pathological condition or medication that reduces the absorption of zinc, increases the loss of zinc from endogenous sources (i.e., bile, pancreatic, or intestinal secretions or cells sloughed by the gastrointestinal tract), or increases urinary losses of zinc could effectively increase a patient's requirement for zinc. Of course, these factors would affect the utilization of zinc by individuals of any age. However, the incidence of chronic medical problems and the usage of medications are generally greater among the elderly than among young adults.

Circulating levels of zinc are depressed in patients during a number of pathological conditions.[37] Only those conditions which are known to affect absorption or excretion of zinc are mentioned here.

Any chronic malabsorption syndrome can result in decreased absorption of a variety of nutrients, including zinc.[37,38] These conditions are important because the gut is generally the major organ that controls zinc homeostasis in animals.[13]

Urinary losses of zinc are generally small in normal subjects; they usually account for less than 7% of dietary intake. However, in several pathological conditions urinary losses of zinc are large enough to be of practical significance. Askari et al.[39] monitored the urinary zinc losses of 14 trauma patients (13 had skeletal injuries). These patients excreted upwards of five times as much zinc as normal subjects daily. Hallböök and Hedelin[40] observed that, postoperatively, plasma zinc levels were inversely related to the size of surgical wounds. Similarly, Cohen et al.[41] observed that patients with thermal burns excreted high levels of zinc in their urine and had depressed serum levels of zinc. Patients with progressive systemic sclerosis[42] and conditions that result in the wasting of muscle tissue[43] have also been found to lose larger amounts of zinc in their urine than normal adults.

A number of investigators have noted that patients with alcohol-induced cirrhosis of the liver excreted large amounts of zinc in their urine.[38,44,45] These patients had lower concentrations of zinc in their livers than normal subjects and subjects with nonalcohol-induced liver disease.[46] This excessive loss of zinc in the urine by individuals with alcohol-induced cirrhosis of the liver is at least partially attributable to alcohol consumption per se. McDonald and Margen[47] have observed that alcohol consumption caused normal subjects to excrete more zinc in their urine. Furthermore, Sullivan and Lankford[45] found that urinary zinc losses of most chronic alcoholics returned to normal levels in 1 to 2 weeks following abstention from alcohol. Spencer et al.[18] even noted that alcoholic subjects, during the initial phases of withdrawal of alcohol, absorbed dietary zinc more efficiently than most normal subjects.

Several types of medications, i.e., diuretics, chelating agents, laxatives, and nutritional iron supplements, that are used by many elderly individuals, are known to affect zinc metabolism. The clinical significance of these medications in regard to the elderly has not been studied extensively.

Diuretics, including thiazides, chlorthalidone, and furosemide, increase urinary excretion of zinc.[48] Wester[49] noted that although serum zinc levels of patients taking diureties remained in the normal range, tissue levels of zinc were depressed in patients who received diuretic therapy for 6 months or longer.

Penicillamine is a chelating agent that is used in the treatment of heavy metal poisoning, Wilson's disease, cystinuria, and sometimes, rheumatoid arthritis.[48] Besides increasing urinary excretion of copper, penicillamine has been found to increase urinary losses of zinc. The use of this medication to treat rheumatoid arthritis may have been associated in at least one clinical trial with adverse effects on the nutritional status of patients (only 2 out of 105) in regard to zinc.[48,50]

Many elderly individuals consume bran and iron-vitamin supplements daily to ensure regularity and "good health". Generally, these supplements do little harm and sometimes are beneficial. However, if either of these types of supplements is used excessively, zinc absorption may be impaired. Bran contains both phytate and fiber, factors which are known to depress zinc absorption.[25,27,30] Whenever, an individual consumes much larger quantities of iron than zinc, there is the potential that zinc absorption will also be depressed.[34] The clinical significance of these factors will be greatest among individuals who consume low or marginally low levels of zinc.

D. Clinical Evaluation of Nutritional Status in Regard to Zinc

During the last 20 years many investigators have studied the zinc metabolism of human subjects. However, there is no universally accepted way to monitor the nutritional status of patients in regard to zinc.

Plasma and serum zinc levels are the most commonly used indicators of nutritional status in regard to zinc. Plasma and serum zinc levels generally, but not always, respond to variations in zinc intake.[13] Circulating levels of zinc have been found to be depressed in patients with some gastrointestinal disorders, uremia, alcohol-induced cirrhosis, acute infections, traumatic injuries, and neoplastic disease.[37,38,40-42,44,45,51,52] During the first few days after a myocardial infarction, plasma levels of zinc are depressed.[53] The changes in circulating levels of zinc noted among individuals with these various pathological conditions may reflect losses of zinc from the individuals' total body stores of zinc and/or changes in the distribution of zinc among tissues.

Investigators have not agreed as to what levels of zinc in plasma and serum samples are indicative of zinc deficiency. Some consider only plasma zinc levels below 50 μg/dℓ as indicative of poor nutritional status in regard to zinc; others consider plasma zinc levels of less than 70 or 80 μg/dℓ as indicative of poor nutritional status (Table 2).

Given these conflicting factors, it is not surprising that investigators have reported the incidence of poor nutritional status in regard to zinc, as determined by plasma or serum zinc levels, to range from 2 to 61% of the elderly participants in various surveys.[5-8,54-58] Lindeman et al.[59] have noted that elderly subjects tend to have lower plasma zinc levels than younger subjects; Chooi et al.[60] observed that plasma zinc levels were lower among healthy adults older than 50 years of age. However, other investigators have not been able to confirm these observations.[6,56,57]

Several investigators have also used the concentration of zinc in hair samples from patients as an indication of nutritional status in regard to zinc. Although the levels of zinc in hair samples have been found to be related to the dietary zinc intake of animals, the levels of zinc in hair samples from humans have often not reflected dietary intake.[13,37] Environmental contamination of samples is another problem. Furthermore, hair zinc levels of patients are often not correlated to plasma zinc levels.[13,37]

The levels of zinc found in hair samples from elderly subjects are reported in Table 3. Several investigators have observed lower levels of zinc in the hair of elderly subjects than

Table 2
ZINC LEVELS IN PLASMA AND SERUM SAMPLES
FROM ELDERLY INDIVIDUALS

Age (years)	No.	Mean plasma (P) or serum (S) zinc levels (μg/dℓ)	Subjects with low P or S zinc levels (%)	Ref.
34—82	80	92 (P)	NR[a]	54
41—91	187	110 (S)	10[b]	55
65—95	146	79 (P)	2[c]	56
60—87	91	93 (S)	7[d]	57
67—96	36	78 (S)	61[c]	5
61—99	90	91 (P)	27[c]	6
53 (mean)	49	99 (S)	14[d]	7
60—97	89	92 (S)	6[d]	8
72 (mean)	61	80 (P)	NR[a]	29
65—74 (male)	1014	86 (S)	12[d]	58
65—74 (female)	1203	83 (S)	12[d]	58

[a] Not reported.
[b] <81 μg/dℓ defined as low level.
[c] <50 μg/dℓ defined as low level.
[d] <70 μg/dℓ defined as low level.
[e] <80 μg/dℓ defined as low level.

Table 3
ZINC LEVELS IN HAIR FROM ELDERLY INDIVIDUALS

Age (years)	No.	Mean hair zinc level (μg/g)	Subjects with low hair zinc levels (%) <100 μg/g	<70 μg/g	Ref.
52—89	44	174	11	0	4
75 (mean)	62	173	5	1	10
65—95	146	221	3	0	56
60—87	127	142	29	8	57
55—87	132	157	21	NR[a]	61
53 (mean)	52	187	2	0	7
60—97	103	140	20	3	8
>62	48	142	17	6	9

[a] Not reported.

in the hair of young adults and adolescents.[4,7-10,56,57,61] Between 2 and 29% of the elderly participants in 8 different surveys were found to have moderately low levels (<100 μg zinc per gram of hair) of zinc in their hair. Only a few elderly subjects (0 to 8%) had less than 70 μg zinc per gram of hair. A concentration of less than 70 μg zinc per gram of hair is generally indicative of impaired nutritional status in regard to zinc.

Generally, other methods of evaluating nutritional status of patients in regard to zinc, i.e., leukocyte zinc levels and the activity of metalloenzymes, have not been used to evaluate the nutritional status of elderly subjects in regard to zinc.[29,37,38]

Low intakes of zinc and even reduced levels of zinc in tissues mean little if there are no functional changes. Thus, it is worthwhile to briefly examine the functions of zinc in the body.

E. Functions of Zinc in the Body

Zinc has been identified as a component of more than 80 enzymes and proteins.[62] These include such diverse enzymes as DNA polymerase, RNA polymerase, carboxypeptidase, alcohol dehydrogenase, δ-aminolevulinic dehydratase, superoxide dismutase, and alkaline phosphatase. The diverse functions of these enzymes account for the fact that many and diverse symptoms are noted in zinc deficiency. Some of the zinc deficiency symptoms that have been observed in laboratory animals and man include growth retardation, impaired wound healing, reproductive failure, anorexia, dermatitis, depressed taste acuity, and impaired immune functions.[13,62]

F. Clinical Evidence of Zinc Deficiency Among the Elderly

Some of the symptoms of zinc deficiency, (i.e., slow wound healing, anorexia, dermatitis, depressed taste acuity, impaired immune function) resemble common problems observed among the elderly. Thus, during the last 10 years a number of investigators have attempted to relate the dietary zinc intake and the zinc status of elderly individuals to functional changes in regard to taste acuity,[4,9,10,64-65] wound healing,[54,66-68] and immune function.[5,8,69,70]

Several investigators have noted that elderly subjects had higher taste detection levels for sodium chloride and sucrose than young adults during standardized tests for taste acuity.[4,9,10,63,71,72] (High taste detection levels indicate poor taste acuity.) However, investigators were unable to find correlations between the nutritional status of the elderly in regard to zinc, as indicated by dietary zinc level or hair or plasma levels of zinc, and taste acuity.[4,9,10,63]

Two double-blind zinc supplementation studies have also been conducted to examine the relationship between nutritional status in regard to zinc and taste acuity. Greger and Giessler,[64] in a double-blind study, administered either 15 mg zinc daily or placebo tablets to 49 elderly individuals for 95 days. No significant changes were observed in the taste acuity of subjects, although hair zinc levels increased in the individuals receiving the zinc supplements. Seligson[65] confirmed these observations in another study with 113 elderly individuals.

Zinc is essential for wound healing.[73] However, it is not clear that the nutritional status of most elderly individuals in regard to zinc is a significant factor affecting wound healing.

Husain and Bessant[54] treated 90 patients (34 to 82 years of age) with chronic ulcers on their legs and feet with zinc sulfate for 112 days. The ulcers healed completely in 68% of the patients. No control group was studied.

In two double-blind therapeutic trials, zinc supplements were found to improve the healing of leg ulcers in elderly subjects.[66,67] In one study, the zinc status of the subjects was not evaluated.[66] However, the investigators reported that the rate of wound healing among patients given zinc sulfate was more rapid than among patients given placebo tablets. In the other double-blind study, the investigators found that only those subjects with serum zinc levels of less than 110 μg/100 mℓ responded to large zinc supplements (600 mg zinc sulfate daily).[67]

Weismann et al.[68] found 26 individuals (mean age was 82 years) with "skin manifestations suggestive of chronic zinc deficiency" in an institution for the elderly. Of the subjects, 10 had subnormal (<70 μg/dℓ) plasma zinc levels; they were treated for 4 weeks with zinc sulfate tablets (0.6 g/day). Although the plasma zinc levels of subjects rose, there were no improvements in the skin conditions of the patients.

There have been only a few studies on the importance of zinc nutriture to the immune function of elderly persons. The subject is important because deterioration of T cell immune function is associated with aging.[69]

Stiedemann and Harrill[5] studied the zinc status and immunocompetence of 36 elderly women. They found a low correlation ($r = 0.325$, $p < 0.05$) between serum zinc levels and postimmunization serum HAI antibody titers to influenza A/NJ/8/76 antigen. No relationships were observed between serum zinc levels or dietary zinc levels and serum levels of IgG and IgM.

Two groups of investigators have studied the effects of zinc supplements on the immune function of elderly subjects. Neither study was conducted as a double-blind study. In the first study, Duchateau and associates[70] administered 100 mg zinc daily to 15 elderly subjects (mean age was 81 years) for 1 month. The group given the zinc supplements exhibited significant improvements in regard to several immunological parameters. Wagner[8] found the 22% of the 173 elderly subjects (mean age was 74 years) that she surveyed were nonresponsive to four standard antigens; 5 months after the original survey, 5 of these subjects were still anergic. After 4 weeks of treatment with oral doses of 55 mg zinc daily, these 5 elderly subjects developed positive responses to skin tests.

The effects of zinc supplements on the immunocompetence of elderly individuals need to be evaluated with double-blind trials. Both nutritional and pharmacological doses of zinc should be evaluated with a variety of patients. Several investigators have noted low circulating zinc levels among individuals with malignancies.[38,51,52] The significance of low tissue zinc levels among the elderly needs to be evaluated not only in regard to immunocompetence but also in regard to the development of malignancies.[74]

In general, clinical evidence of impaired nutritional status in regard to zinc among the elderly is limited. More studies are needed before scientists can estimate the true significance of apparently low dietary intakes of zinc by the elderly.

III. COPPER

A. Copper Intake

As was observed in the section on zinc, the total food intake of elderly individuals tends to be less than that of young adults.[1] Holden et al.[75] noted that diets which contained less energy usually contained less zinc and copper. However, it should be noted that the distribution of copper in foods is different than the distribution of zinc. Grain products and fruits and vegetable products provide 30 and 38%, respectively, of the copper consumed by Americans.[76] Milk and milk products and meat, fish, and poultry provide only 4 and 10%, respectively, of the copper consumed by Americans. Thus, many elderly individuals probably consume less copper daily than young adults because of reductions in total food intake. However, the food selections patterns of the elderly would not be expected to affect their copper intake as adversely as their zinc intake.[1]

Only Gibson et al.[77] have reported the copper intake of elderly subjects. They calculated that 36 vegetarian women (mean age was 69 years) and 30 control women (mean age was 60 years) consumed 2.1 ± 0.7 (SD) and 1.6 ± 0.6 mg copper daily, respectively. The copper intake of these elderly subjects appear to be somewhat higher than would be expected on the basis of chemical analyses of diet composites for their copper content.[75,78-84] Table 4 is a summary of these data. During the last 15 years, investigators generally have found that most "Western-style" diets contained less than 1.5 mg copper daily. There may be several reasons for the somewhat high intake of copper by the elderly women in Gibson's survey. One of these is that estimates of copper intake based on food composition tables may be inaccurate because of the limitations of food composition tables. Only a limited number of foods have been analyzed for their copper content.[80,85-87] Furthermore, a variety of factors, including soil conditions, fertilizers, and pesticides, can greatly alter the copper content of individual food items.[85]

The Food and Nutrition Board has estimated that safe and adequate dietary intakes of copper for adults are between 2 and 3 mg/day.[12] Many Americans, including many elderly individuals, would not routinely consume this amount of copper daily.

B. Copper Requirements

Scientists at the USDA laboratory at Grand Forks have found that young men need to consume about 1.3 mg copper in a mixed diet in order to compensate for fecal and urinary

Table 4
ANALYZED COPPER CONTENT OF
DIET COMPOSITES

Type of diet	Copper content	Ref.
Self-selected	0.97	78
Self-selected	0.34	79
Institutional	0.36	80
Self-selected	2.4	81
Dormitory (no liver)	1.5	82
Dormitory (with liver)	7.6	82
Dormitory	3.79	83
Self-selected	1.01	75
Hospital	0.78	84

losses.[88,89] Subjects were found to lose an additional 0.3 mg copper per day in sweat.[90] Thus, the average requirement of a young adult male is about 1.6 mg copper daily. Other laboratories have also observed that healthy young subjects have minimal losses of copper in their urine and appear to absorb copper more efficiently than they absorb many trace elements.[20-22,24,25,31,91]

Two groups of investigators have monitored copper excretion in elderly subjects during metabolic balance studies.[16,19,92] Burke et al.[16] found that 8 of 10 elderly subjects (56 to 83 years of age) were in positive balance in regard to copper when they consumed 2.33 mg copper daily. Turnlund et al.[19] observed that, on the average, 6 elderly men (65 to 74 years of age) were in positive balance in regard to copper when they consumed about 3.3 mg copper daily. The apparent absorption of 8 elderly subjects (65 to 74 years of age) was 19 to 29% when they consumed 3.3 mg copper daily.[92]

It is not possible on the basis of these data to assess whether the efficiency of copper utilization is altered with age. The diets fed in the various studies differed in regard to their content of a variety of dietary factors and nutrients. A number of these substances could alter the apparent absorption of copper.[93]

One factor that can greatly alter copper absorption is zinc. Human subjects generally,[16,93,94] but not always,[91] absorb copper less efficiently when the level of zinc in the diet is increased. Phytate, some sources of dietary fiber, and ascorbic acid can also depress the bioavailability of dietary copper.[88,93,95] Individuals who supplement their diets with large quantities of any of these substances, especially zinc and ascorbic acid, can increase their risk of being in marginal status in regard to copper.[95-97]

Dietary protein also appears to affect the apparent absorption of copper.[19,22,93] Turnlund[19] observed that elderly subjects absorbed copper significantly better when the amount of lactalbumin in the diet was increased. Similarly, Greger and Snedeker[22] observed that young males apparently absorbed copper more efficiently when purified proteins were added to their diets. Sandstead[93] observed that young males required less dietary copper to achieve positive balance in regard to copper when dietary protein levels were increased. However, others have found that levels of dietary protein did not affect copper utilization by humans.[91,98]

C. Disease States and Medications that Alter Copper Metabolism

The metabolism of copper is affected by a variety of disease states and medications. Some of these factors actually alter the absorption and excretion of copper; many of these conditions and medications appear to alter only the distribution of copper in tissues.

Hypocupremia has been observed in patients with protein-energy malnutrition, malabsorption syndromes, active ulcerative colitis, and nephrotic syndromes.[99,100] In protein-energy malnutrition, low circulating levels of copper are believed to reflect hypoproteinemia and the inability of the body to synthesize sufficient ceruloplasmin.[99] Similarly, plasma copper levels are low in patients with Wilson's disease, although the levels of copper in the livers of these patients are very high.[13,99]

Penicillamine, a metal-chelating agent used to treat Wilson's disease, heavy metal poisoning, cystinuria, and rheumatoid arthritis, increases urinary excretion of copper.[48] The administration of corticosteroids and ACTH can also depress plasma copper levels.[100] It might be anticipated that this therapy would induce a copper-deficient state in some patients. However, few, if any, cases have been reported in the literature.[101]

Hypercupremia can also occur in some disease states. Patients with cancer, rheumatoid arthritis, emphysema, congestive heart failure, infections, and psychoses have all been found to have elevated circulating levels of copper.[52,99,100,102-106] Askari et al.[102] have even suggested that serum copper levels may be clinically useful for the evaluation of the efficacy of treatment among cancer patients. The mechanisms causing the elevation of plasma copper levels in patients with these conditions are generally unknown.[99] However, Cohen et al.[103] observed that rats bearing tumors absorbed copper more efficiently than control rats. Garofalo et al.[104] have questioned whether some of the elevations in serum copper levels ascribed to the effects of malignancies should be ascribed to age. Serum levels of copper are elevated in patients for at least 6 days after surgery.[107] Therapy with estrogens and diuretics also tends to elevate circulating levels of copper.[99,100,108]

The major excretory route for copper is through the bile.[13,99] Thus, it is not surprising that patients with diseases in which biliary excretion of copper is blocked often have elevated plasma copper and ceruloplasmin levels and lose more copper in the urine.[99,109]

In normal individuals, more than 90% of the copper in plasma is bound to ceruloplasmin.[13] The percentage of the copper bound to ceruloplasmin can be altered by disease states.[13,99] However, disease states and medications that alter circulating levels of copper often also alter circulating levels of ceruloplasmin.[13,99,100,109]

D. Clinical Evaluation of Nutritional Status in Regard to Copper

During the last 30 years investigators have used several different variables to monitor the nutritional status of individuals in regard to copper.[13,99,100] The variables most often used are plasma and serum levels of copper.

Hypocupremia is an early and fairly consistent manifestation of experimental copper deficiency in animals.[13] However, as already noted, plasma and serum copper levels are affected by disease states and medications. Women tend to have somewhat higher circulating levels of copper than males.[13,58,99] Diurnal variations in plasma copper have also been noted.[110]

Few investigators have measured plasma or serum copper levels in elderly individuals. In four small surveys investigators found no evidence of low plasma or serum copper or ceruloplasmin levels among elderly subjects. In fact, Garofalo et al.[104] found that 60 females (20 to 45 years of age) had significantly lower levels of copper in their serum than 26 older females (67 to 98 years of age). Sempos et al.[108] observed a weak correlation between age and serum copper levels among 137 women (35 to 67 years of age), but Vir and Love[56] noted no correlation between plasma copper levels and age among 53 elderly subjects.

Serum copper levels were monitored among more than 14,000 individuals participating in the Second Health and Nutrition Examination Survey (HANES II).[58] Data from this survey are reported in Table 5. In the survey, gender, economic status, and race all were found to significantly affect serum copper levels. Age (at least among adults) was not found to significantly affect serum copper levels of participants in this large survey.

Plasma copper levels of <80 $\mu g/d\ell$ are often considered to be an indication of poor nutritional status in regard to copper.[99] Only 1.6 and 0.3% of the elderly men and elderly women (65 to 74 years of age), respectively, participating in HANES II survey had plasma

Table 5
SERUM COPPER LEVELS OF
SUBJECTS PARTICIPATING IN HANES
II[50]

Age	Gender	Status	No.	Mean serum copper level (mg/dℓ)
65—74	M	P[a]	142	129
65—74	M	AP[b]	838	120
65—74	F	P	212	134
65—74	F	AP	917	133
18—44	M	P	269	110
18—44	M	AP	2059	106
18—44	F	P	426	143
18—44	F	AP	2092	138

[a] P = Poverty level.
[b] AP = Above poverty level.

copper levels of <80 μg/dℓ. Thus, the incidence of poor nutritional status in regard to copper among the elderly appears to be minimal. Of course, disease states that elevate circulating copper levels are confounding factors.

Other variables that have been used to assess nutritional status in regard to copper include levels of copper in hair and red blood cells and levels of ceruloplasmin and other copper-containing enzymes in plasma.[13,99,100] Most of these measures have not been used to assess the nutritional status of elderly subjects. Vir and Love[56] measured the levels of copper in hair samples from 103 elderly individuals. They found the values ranged from 6.4 to 22.2 μg copper per gram of hair. They, as had others, concluded that the levels of copper in hair samples were not useful indications of nutritional status.[111,112] Denko and Gabriel[113] monitored serum ceruloplasmin levels in 125 normal adults. Both men and women over the age of 55 had significantly elevated levels of ceruloplasmin in their sera; women had higher serum levels of ceruloplasmin than men at all ages.

E. Significance of Alterations in Copper Metabolism Among the Elderly

Copper is an essential component of many enzymes and proteins, including ceruloplasmin, tyrosinase, uricase, lysyl oxidase, superoxide dismutase, amine oxidases, ascorbic acid oxidase, and cytochrome oxidase.[13,62,114] Symptoms of copper deficiency in laboratory animals include anemia, enlarged hearts, defects in skeleton and vasculature related to faulty cross-linking of collagen and elastin, disorders of the central nervous system, and impaired immune function.[13,62,115,118] These symptoms of copper deficiency are rarely seen in humans.[62] Patients with Menkes kinky hair syndrome and occasionally premature infants who are administered low levels of copper either orally or through parenteral solutions will exhibit some of these symptoms.[101]

Some investigators, notably Klevay, have hypothesized that the long-term consumption of marginally adequate levels (i.e., <1.0 mg/day) of copper will produce hypercholesterolemia.[95,119-124] Klevay and others have demonstrated that rats and monkeys fed low levels of copper or fed marginally adequate levels of copper with other factors that depress the utilization of copper will develop hypercholesterolemia.[95,120-122] The mechanism by which this occurs is not clear, but it may involve plasma lecithin:cholesterol acyltransferase.[123,124] The significance of low copper intakes by young, middle-aged, and elderly individuals to the development of ischemic heart disease has not been assessed. Further work is needed.

In general, it appears that many Americans consume less than recommended levels of copper. The long-term consequences of consumption of marginally adequate levels of copper are not clear. Furthermore, the sensitivity of plasma copper levels to a variety of disease states makes it difficult to monitor the nutritional status in regard to copper of any group of individuals, especially the elderly.

IV. MANGANESE

Only a limited amount of data have been published on manganese intake and metabolism by human subjects; much less has been published on manganese intake and metabolism in elderly individuals.

Adults in Western societies consume 0.9 to 7.0 mg manganese daily; the average intake of manganese is probably about 2.5 to 3.5 mg/day.[77,80,83,125-127] Those individuals who consume liberal amounts of nuts, legumes, whole grain products, and tea consume larger amounts of manganese than those who consume primarily animal products.[128,129] Gibson et al.[77] noted that the manganese intakes of 66 elderly women appeared to be similar to the manganese intakes of young adults.

The Food and Nutrition Board suggested that a safe and adequate intake of manganese for adults would range from 2.5 to 5.0 mg/day.[12] It seems likely that the manganese intake of most elderly individuals would be within this range, provided that their overall energy intake was adequate.

Utilization of manganese by young adults has been monitored in several metabolic balance studies.[21,22,24,130-132] Generally, subjects, when fed 2.5 to 3.5 mg manganese daily, excreted in their feces approximately the amount of manganese that they consumed.[21,22,24,130-132]

Kirchgessner et al.[133] have suggested that rats absorbed manganese less efficiently as they aged. High levels of dietary calcium and iron can depress the absorption of manganese by animals.[13] The significance of these observations to humans is not known.

Currently, scientists find it difficult to monitor the nutritional status of individuals in regard to manganese. Serum levels of manganese are very low (<2 µg/ℓ) and are difficult to measure properly.[134-136] Heparin is contaminated with manganese.[136] Thus, plasma, blood, and red blood cells that have been processed with heparin are contaminated with manganese.

Circulating levels of manganese are also affected by disease states. Sullivan et al.[52] found patients with congestive heart failure, infections, and psychosis had elevated levels of manganese in their sera. Versieck et al.[137] noted patients with liver metastases had elevated levels of manganese in their sera.

Two groups of investigators have attempted to assess the effect of age on the level of manganese in blood.[134,138] One group found no correlations between the ages of subjects and their blood manganese levels.[134] The other group observed that 10 elderly individuals (>70 years of age) had significantly higher concentrations of manganese in their red blood cells than 40 young adults (20 to 40 years of age).[138] Heparin was used to process the samples in both studies. Hence, the significance of these observations is questionable.

The number of manganese metalloenzymes is limited.[13,62,139] Two of these enzymes are manganese-containing superoxide dismutase (located primarily in mitochondria) and pyruvate carboxylase. However, manganese can activate a large number of metal-enzyme complexes involving transferase, hydrolase, lyase, isomerase, and ligase reactions. Thus, it is not surprising that the symptoms of manganese deficiency in animals are diverse and include defective growth, bone abnormalities, reproductive dysfunction, central nervous system manifestations, and disturbances in lipid and carbohydrate metabolism.[13,62,139]

It is reasonable to speculate that nutritional status in regard to manganese could affect the aging process. The activity of manganese-containing superoxide dismutase in some tissues of animals is sensitive to changes in dietary manganese levels.[140,141] However, little has been published on changes in the activity of this enzyme during aging. One group of investigators has observed that the activity of two functionally distinct forms of monoamine oxidase change in the brains, livers, hearts, spleens, and kidneys of rats as they age.[142,143] Many of these changes were prevented when manganese (1 mg $MnCl_2 \cdot 4H_2O/m\ell$ water) was added to the drinking water of rats. The significance of these findings to humans is unknown. However, further work on the relationship of manganese exposure to certain aspects of the aging process seem warranted.

V. CHROMIUM

Many papers have been published on the importance of chromium to carbohydrate and lipid metabolism in humans; some of these studies involved elderly subjects. Unfortunately, the methodology used in many of these studies was very inaccurate. Hence, the conclusions drawn in many of these papers are now questionable.

Estimates of the chromium content of Western-type diets vary greatly.[81,83,125,144,145] Most individuals probably consume between 5 to 100 μg chromium daily. Anderson et al.[146] suggested that the average American consumed 60 μg chromium daily.

Levine et al.[144] observed that the chromium intake of 10 elderly subjects ranged from 5 to 115 μg/day, while the chromium intake of a young adult ranged from 4 to 115 μg/day. This suggests that if elderly individuals consume adequate amounts of energy their chromium intake is apt to be similar to that of young adults. However, food choices are important. Usually, highly refined products that contain high concentrations of sugar, fat, or alcohol contain minimal amounts of chromium. Foods such as Brewer's yeast, egg yolk, cheese, liver, and whole grain products contain significant amounts of chromium.[147,148] Foods, especially acidic foods, accumulate chromium when prepared and stored in stainless steel containers.[149] The significance of this source of dietary chromium has not been assessed.

The Food and Nutrition Board has suggested that safe and adequate intakes of chromium range between 50 and 200 μg/day.[12] At this time it is not possible to assess how many Americans of any age group consume less than the recommended amount of chromium.

A. Chromium Metabolism and Nutritional Status

Many investigators have attempted to study chromium metabolism in human subjects. Much of the work completed before 1978 is now known to be invalid because of methodological problems.[150-153]

Human subjects have been found to absorb only 0.69 and 0.5% of doses of inorganic chromium.[154,155] Anderson et al.[146] estimated that subjects absorbed the chromium naturally present in their diets as efficiently as the chromium in inorganic supplements. However, some "organic" forms of chromium in Brewer's yeast have been hypothesized to be more bioavailable than inorganic chromium.[156,157]

Urine is believed to be the major excretory route for absorbed chromium.[13,158] The best estimates of urinary chromium losses by adults now appear to be between 0.2 and 0.8 μg/day (<1 ng chromium per milliliter of urine).[146,153,159,160] Some scientists have used urinary losses of chromium (both total daily losses and those losses that occurred after a glucose challenge) as indexes of nutritional status in regard to chromium.[162,163] Although daily urinary losses of chromium have been found to be proportional to recent dietary intake of chromium, Anderson and associates[146,161] do not believe that urinary excretion of chromium is a good index of long-term nutritional status in regard to chromium.

Schroeder et al.[149] noted that tissue chromium levels declined with age. Several investigators have observed that patients with diabetes and coronary heart disease had low levels of chromium in their hair.[164-166] The significance of these measurements are now questionable in light of the technical problems observed in the measurement of chromium in urine samples before 1978. Furthermore, Doisy et al.[158] observed that the age of the subjects was not correlated to their ability to absorb inorganic chromium; Anderson et al.[146] observed no correlation between urinary excretion of chromium and the age of subjects.

B. Clinical Significance of Nutritional Status in Regard to Chromium

Chromium deficiency in laboratory animals is characterized by impaired growth and longevity, and disturbances in carbohydrate and lipid metabolism.[13] Chromium deficiency has also been identified in humans who were administered total parenteral nutrition for long periods of time.[167,168] Symptoms included weight loss, impaired glucose tolerance, and neuropathy.

The function of chromium in lipid and carbohydrate metabolism is related to the function of insulin.[13,62] Chromium is believed to facilitate the binding of insulin to cell membranes by forming a "bridge" between the insulin molecule and the membrane.[169] Trivalent chromium as part of an organic complex, called the "glucose tolerance factor", has been found to be more effective than inorganic chromium in potentiating the action of insulin.

The data already cited have naturally caused many scientists to wonder whether low intakes of chromium could be related to the incidence of diabetes, particulary among the elderly. Several clinical trials have been conducted.[144,170-175]

Glinsman and Mertz[170] observed improved glucose tolerance among 3 of 6 diabetics (27 to 48 years of age) who were given oral supplements (150 to 1000 μg trivalent chromium) for 150 to 200 days. Chromium supplements had no effect on individuals with normal glucose tolerance.

Levine et al.[144] administered orally inorganic chromium (150 μg/day) to 10 elderly (74 to 96 years of age) individuals who had abnormal glucose tolerance for 1 to 4 months. Four subjects responded to the chromium with improvements in their response to a glucose load test.

Researchers at Lincoln University administered supplements consisting of 5 g of Brewer's yeast extract that contained 4 μg chromium to 27 women (40 to 75 years of age) for 3 months.[156,171] Prior to the supplementation period, 12 of the women were hyperglycemic as judged by 3-hr glucose tolerance tests. At the end of the supplementation period, the response of the women who initally were hyperglycemic to a glucose load was significantly improved. Their blood insulin and glucose levels rose less in response to a glucose load after the supplementation period. All the women had significantly lower fasting blood cholesterol levels after the supplemental period.

Polansky et al.[172] conducted a double-blind supplementation trial with 76 adults (21 to 69 years of age). Subjects received daily 200 μg chromium as chromium chloride for 3 months and placebo tablets for 3 months in this cross-over design study; 18 subjects had impaired glucose tolerance at the beginning of the trial. Chromium supplements did not affect the glucose concentrations of those subjects. However, the 90-min plasma glucose concentrations of those subjects with impaired glucose tolerance initially were slightly but significantly lower after the chromium supplementation period.

Offenbacher and Pi-Sunyer[175] supplemented the diets of 24 elderly individuals (mean age, 78 years) for 8 weeks with either 9 g/day of Brewer's yeast, which contained a high level of chromium, or 9 g/day of Torula yeast, which contained little chromium. Individuals receiving the Brewer's yeast had significant improvements in their response to glucose load tests and reductions in their serum cholesterol levels.

Riales and Albrink[174] administered either 200 μg trivalent chromium in water or plain water to 23 adult men (31 to 60 years of age) 5 days/week for 12 weeks. The study was a double-blind trial. There was a significant improvement of glucose tolerance due to chromium after 6 weeks but not after 12 weeks of supplementation. Chromium supplementation did not affect serum total cholesterol levels but did result in the elevation of high-density lipoprotein (HDL) cholesterol levels after 12 weeks.

In 1983, Uusitupa et al.[175] conducted a double-blind, cross-over design study with 10 noninsulin-dependent diabetics (37 to 68 years of age). Subjects received 200 μg chromium in solution for 12 weeks and a placebo solution for 12 weeks. Subjects excreted ninefold more chromium in their urine after chromium supplementation. Glucose tolerance, serum lipid levels, and fasting and 2-hr post-glucose serum insulin levels were unaffected by the treatments. Chromium supplementation resulted in significantly lower serum insulin levels 1 hr after a glucose load.

In general, the effects of chromium supplements on glucose tolerance and serum lipid levels are modest. Some elderly individuals may benefit somewhat from additional chromium intake. However, at this time there appears to be no practical way to identify those individuals, except by observing their response to a chromium supplementation trial.

VI. SELENIUM

During the last 10 years a number of studies on the metabolism of selenium by human subjects have been published. Most of these studies have involved children or young adults as subjects. Thus, only a very limited amount of data have been published on the intake and metabolism of selenium by elderly individuals.

Most residents of North America have been observed to consume 50 to 200 μg selenium daily,[77,176-181] while residents of New Zealand have been found to consume less than 70 μg selenium daily.[182] Food choices can affect the selenium content of the diet greatly. Seafood, organ meats, meat, and whole grains (sometimes) are rich sources of selenium.[183,184] The selenium content of grains are dependent on the level and availability of selenium in the soils on which they are grown.[182,183]

Gibson et al.[77] estimated the average selenium intake of 66 elderly women in Canada to be about 110 μg/day. The Food and Nutrition Board estimated that safe and adequate intakes of selenium ranged from 50 to 200 μg/day.[12] Probably most elderly individuals in the U.S. and Canada, provided they consume adequate amounts of energy, consume adequate amounts of selenium. However, those individuals depending on commercial formula diets which have not been fortified with selenium may consume relatively low levels (<50 μg/day) of selenium.[177]

Investigators in New Zealand have observed young adults absorbed 40 to 98% of their selenium intake; diets generally contained 20 to 40 μg selenium daily.[182,185-187] Similarly, young adults in the U.S. who consumed more than 100 μg selenium daily generally were found to absorb 40 to 80% of their selenium intake.[188-191] A variety of dietary factors, including the forms of selenium administered and the levels of protein and tin in the diet, were found to affect selenium utilization in these studies.[186,188,189,192] However, in all of these studies the majority of subjects were in positive balance in regard to selenium. It is interesting to note that 6 young American males, who probably normally consumed more than 100 μg selenium daily, were unable to adjust to a diet that contained 33 to 36 μg selenium daily.[191] They remained in negative balance in regard to selenium throughout a 45-day period. Thus, Levander et al.[191] estimated the selenium requirements of young adult Americans was about 70 μg/day.

There is no evidence that selenium utilization is altered by age among humans. However, Oštádalová et al.[193] observed that mature rats excreted toxic doses of selenium more completely than young animals. Despite their less efficient excretion of selenium, young rats were found to survive exposure to large doses of selenium that were lethal to adult rats.

Nutritional status in regard to selenium has been monitored in human subjects by measuring whole blood and plasma selenium levels and erythrocyte and whole blood glutathione peroxidase activities.[182,194-196] These measures of nutritional status have been used successfully to determine those individuals whose intake of selenium and overall nutritional status in regard to selenium were very low or marginally adequate. They have not always been found to be proportional to selenium intake when dietary levels of selenium were high ($>$100 μg/day).[194]

The impact of age on the tissue levels of selenium and glutathione peroxidase is difficult to assess. Levander[195] observed no consistent differences in selenium levels of tissues that could be attributed to the age of subjects. Robinson et al.[196] noted among New Zealand residents that elderly individuals had lower blood selenium levels than middle-aged and young adults. Pesigehl et al.[197] observed that the selenium content of several organs in humans tended to increase with age. Burch et al.[198] found that the selenium content of kidneys, livers, and lungs tended to increase from day 20 to day 130 in rats. Csallany et al.[199] found that the glutathione peroxidase activities in the testes, hearts, kidneys, and livers of 19-month old rats were greater than the activites of the enzyme tissues of weanling rats.

In recent years a number of investigators have hypothesized that low selenium intakes, poor nutritional status in regard to selenium, and low tissue levels of glutathione peroxidase (a selenium-containing enzyme) were correlated to the incidence of several chronic conditions in residents of Western societies. The interrelationships between selenium metabolism and coronary vascular disease or cancer are complex. Only brief comments are included here.

There are several pieces of evidence that suggest selenium may be related to some forms of cardiovascular disease. In livestock, selenium and vitamin deficiencies are manifested by a variety of symptoms, including cardiovascular failure.[13] Selenium deficiency symptoms, including muscle pain and/or cardiomyopathy, have been identified in several patients in Western societies who subsisted on TPN or very restrictive diets for long periods of time.[200-202] Keshan disease, a cardiomyopathy that affects primarily children, was endemic in certain regions of China.[203] A large-scale intervention trial carried out over 4 years and involving more than 12,000 children demonstrated the efficacy of selenium in preventing this disease.[204]

Several investigators have reported that individuals living in areas with high amounts of selenium in their soils or water had lower age-specific death rates for cardiovascular disease.[205-207] However, other investigators have not observed any differences in the selenium levels in tissues of patients and individuals who died of coronary vascular disease and atherosclerosis vs. those of control subjects.[182,208-211]

Both epidemiological and experimental evidence suggest that selenium may offer some protection against risk of cancer.[212] Shamberger et al.,[213,214] Schrauzer et al.,[215-217] and Jansson et al.[218,219] have noted inverse relationships between the incidence of certain types of cancer and exposure to selenium or blood levels of selenium.

Many investigators have demonstrated that supplementation of the diet or drinking water with selenium protects animals against tumors induced by a variety of chemical carcinogens and at least one viral agent. (The Committee on Diet, Nutrition and Cancer[212] cited 11 papers published prior to 1981 on this topic; Medina et al.[220] in 1983 cited 10 additional reports on this topic.) The mechanisms by which selenium exerts antitumorigenic effects are unknown but experts doubt that the effects of selenium can be totally explained by the action of glutathione peroxidase.[212]

In 1982 the committee from the National Academy of Sciences concluded that the relevance of these experimental data to humans was questionable because the levels of selenium used in these studies far exceeded dietary requirements and were often at nearly toxic levels.[212] However, this is an active area of research that potentially seems promising to many scientists.

VII. SUMMARY

Research on the metabolism of trace elements by elderly individuals is limited. There are many exciting possibilities and hypotheses. Much work needs to be done to assess the nutritional status in regard to trace elements of elderly individuals, including those with chronic diseases; even more work needs to be done before "optimal" nutritional status in regard to trace elements can be defined.

ACKNOWLEDGMENT

The author appreciated the support of the College of Agricultural and Life Sciences, University of Wisconsin, Madison, Project No. 2623.

REFERENCES

1. Science and Education Administration, Food and Nutrient Intakes of Individuals in 1 Day in the United States, Spring 1977. Prelim. Rep. No. 2, U.S. Department of Agriculture, Washington, D.C., 1980. 40, 45, 75.
2. **Welsh, S. O. and Marston, R. M.,** Zinc levels of the U.S. food supply — 1909—1980, *Food Technol.,* 36, 70, 1982.
3. **Abdulla, M., Jägerstad, M., Norden, A., Quist, I., and Svensson, S.,** Dietary intake of electrolytes and trace elements in the elderly, *Nutr. Metab.,* 21(Suppl.1), 41, 1977.
4. **Greger, J. L. and Sciscoe, B. A.,** Zinc nutriture of elderly participants in an urban feeding program, *J. Am. Dietet. Assoc.,* 70, 37, 1977.
5. **Stiedemann, M. and Harrill, I.,** Relation of immunocompetence to selected nutrients in elderly women, *Nutr. Rep. Int.,* 21, 931, 1980.
6. **Flint, D. M., Wahlquist, M. L., Smith, T. J., and Parish, A. E.,** Zinc and protein status in the elderly, *J. Hum. Nutr.,* 35, 287, 1981.
7. **Anderson, B. M., Gibson, R. S., and Sabry, J. H.,** The iron and zinc status of long-term vegetarian women, *Am. J. Clin. Nutr.,* 34, 1042, 1981.
8. **Wagner, P. A., Jernigan, J. A., Bailey, L. B., Nickens, C., and Brazzi, G. A.,** Zinc nutriture and cell mediated immunity in the aged, *Int. J. Vitam. Nutr. Res.,* 53, 94, 1983.
9. **Hutton, C. W. and Hayes-Davis, R. B.,** Assessment of the zinc nutritional status of selected elderly subjects, *J. Am. Dietet. Assoc.,* 82, 148, 1983.
10. **Greger, J. L.,** Dietary intake and nutritional status in regard to zinc of institutionalized aged, *J. Gerontol.,* 32, 549, 1977.
11. **Sempos, C. T., Johnson, N. E., Elmer, P. J., Allington, J. K., and Matthews, M. E.,** A dietary survey of 14 Wisconsin nursing homes, *J. Am. Dietet. Assoc.,* 81, 35, 1982.
12. Food and Nutrition Board, *Recommended Dietary Allowances,* 9th ed., National Academy of Sciences, Washington, D.C., 1980.
13. **Underwood, E. J.,** *Trace Elements in Human and Animal Nutrition,* 4th ed., Academic Press, New York, 1977, 196, 56, 170, 258, 302.
14. **Kirchgessner, M. and Weigand, E.,** Zinc absorption and excretion in relation to nutrition, in *Metals Ions in Biological Systems,* Vol. 15, Sigel, H., Ed., Marcel Dekker, New York, 1983, 319.
15. **Bunker, V. W., Lawson, M. S., Delves, H. T., and Clayton, B. E.,** Metabolic balance studies for zinc and nitrogen in healthy elderly subjects, *Hum. Nutr. Clin. Nutr.,* 36C, 213, 1982.
16. **Burke, D. M., De Micco, F. J., Taper, L. J., and Ritchey, S. J.,** Copper and zinc utilization in elderly adults, *J. Gerontol.,* 36, 558, 1981.
17. **Spencer, H., Osis, D., Kramer, L., and Norris, C.,** Studies of zinc metabolism in man, in *Trace Substance in Environmental Health,* Vol. 5, Hemphill, D. D., Ed., University of Missouri Press, Columbia, 1972, 193.
18. **Spencer, H., Kramer, L., and Osis, D.,** Zinc balance in humans, in *Clinical, Biochemical, and Nutritional Aspects of Trace Elements,* Prasad, A. S., Ed., Alan R. Liss, New York, 1982, 103.
19. **Turnlund, J., Costa, F., and Margen, S.,** Zinc, copper, and iron balance in elderly men, *Am. J. Clin. Nutr.* 34, 2641, 1981.
20. **Crews, M. G., Taper, L. J., and Ritchey, S. J.,** Effects of oral contraceptive agents on copper and zinc balance in young women, *Am. J. Clin. Nutr.,* 33, 1940, 1980.
21. **Greger, J. L. and Baier, M. J.,** Effect of dietary aluminum on mineral metabolism of adult males, *Am. J. Clin. Nutr.,* 38, 411, 1983.

22. **Greger, J. L. and Snedeker, S. M.**, Effect of dietary protein and phosphorus levels on the utilization of zinc, copper and manganese by adult males, *J. Nutr.*, 110, 2243, 1980.
23. **Hess, F. M., King, J. C., and Margen, S.**, Zinc excretion in young women on low zinc intakes and oral contraceptive agents, *J. Nutr.*, 107, 1610, 1977.
24. **Johnson, M. A., Baier, M. J., and Greger, J. L.**, Effects of dietary tin on zinc, copper, iron, manganese and magnesium metabolism of adult males, *Am. J. Clin. Nutr.*, 35, 1332, 1982.
25. **Kelsay, J. L., Jacobs, R., and Prather, E. S.**, Effect of fiber from fruits and vegetables on metabolic responses of human subjects. III. Zinc, copper and phosphorus balances, *Am. J. Clin. Nutr.*, 32, 2307, 1979.
26. **McCance, R. A. and Widdowson, E. M.**, The absorption and excretion of zinc, *Biochem. J.*, 36, 692, 1942.
27. **Reinhold, J. G., Faradji, B., Abadi, P., and Ismail-Beigi, F.**, Decreased absorption of calcium, magnesium, zinc and phosphorus by humans due to increased fiber and phosphorus consumption as whole wheat, *J. Nutr.*, 106, 493, 1976.
28. **Sandstead, H. H.**, Availability of zinc and its requirements in human subjects, in *Clinical, Biochemical, and Nutritional Aspects of Trace Elements*, Prasad, A. S., Ed., Alan R. Liss, New York, 1982, 83.
29. **Sandstead, H. H., Henriksen, L. K., Greger, J. L., Prasad, A. S., and Good, R. S.**, Zinc nutriture in the elderly in relation to taste acuity, immune response and wound healing, *Am. J. Clin. Nutr.*, 36, 1046, 1982.
30. **Sandstead, H. H., Munoz, J. M., Jacob, R. A., Klevay, L. M., Reck, S. J., Logan, G. M., Jr., Dintzis, R. F., Inglett, G. E., and Shuey, W. C.**, Influence of dietary fiber on trace element balance, *Am. J. Clin. Nutr.*, 31, S180, 1978.
31. **Snedeker, S. M., Smith, S. A., and Greger, J. L.**, Effect of dietary calcium and phosphorus levels on the utilization of iron, copper and zinc by adult males, *J. Nutr.*, 112, 136, 1982.
32. **Greger, J. L.**, Effects of phosphorus-containing compounds on iron and zinc utilization: a review of the literature, in *Nutritional Bioavailability of Iron*, Kies, C., Ed., American Chemical Society, Washington, D.C., 1982, 107.
33. **Kies, C., Beshgetoor, D., and Fox, H. M.**, Dietary fiber and zinc bioavailability for humans, in *Antinutrients and Natural Toxicants in Foods*, Ory, R. L., Eds., Food & Nutrition Press, Westport, Conn., 1981, 319.
34. **Solomons, N. W.**, Biological availability of zinc in humans, *Am. J. Clin. Nutr.*, 35, 1046, 1982.
35. **Weigand, E. and Kirchgessner, M.**, Change in apparent and true absorption and retention of dietary zinc with age in rats, *Biol. Trace Elem. Res.*, 1, 347, 1979.
36. **Turnlund, J. R., Michel, M. C., Keyes, W. R., King, J. C., and Margen, S.**, Use of enriched stable isotopes to determine zinc and iron absorption in elderly men, *Am. J. Clin. Nutr.*, 35, 1033, 1982.
37. Committee on Medical and Biologic Effects of Environmental Pollutants, *Zinc*, University Park Press, Baltimore, 1979, 225, 123.
38. **Prasad, A. S., Ed.**, *Zinc in Human Nutrition*, CRC Press, Boca Raton, Fla., 1979, 17.
39. **Askari, A., Long, C. L., and Blakemore, W. S.**, Urinary zinc, copper, nitrogen, and potassium losses in response to trauma, *J. Parenteral Enteral Nutr.*, 3, 151, 1979.
40. **Hallböök, T. and Hedelin, H.**, Zinc metabolism and surgical trauma, *Br. J. Surg.*, 64, 271, 1977.
41. **Cohen, I. K., Schechter, P. J., and Henkin, R. I.**, Hypogeusia, anorexia, and altered zinc metabolism following thermal burn, *JAMA*, 223, 914, 1973.
42. **Henkin, R. I., Patten, B. M., Re, P. K., and Bonzert, D. A.**, A syndrome of acute zinc loss, *Arch. Neurol.*, 32, 745, 1975.
43. **Fell, G. S., Fleck, A., Cuthbertson, D. P., Queen, K., Morrison, C., Bessent, R. G., and Hussain, S. L.**, Urinary zinc levels as an indication of muscle catabolism, *Lancet*, 1, 280, 1973.
44. **Vallee, B. L., Wacker, W. E. C., Bartholomay, A. F., and Hoch, F. L.**, Zinc metabolism in hepatic dysfunction, *Ann. Intern. Med.*, 50, 1077, 1959.
45. **Sullivan, J. F. and Lankford, H. G.**, Zinc metabolism and chronic alcoholism, *Am. J. Clin. Nutr.*, 17, 57, 1965.
46. **Kiilerich, S., Dietrichson, O., Lood, F. B., Naestoft, J., Christofferson, P., Juhl, E., Kjems, G., and Christiansen, C.**, Zinc depletion in alcoholic liver disease, *Scand. J. Gastroenterol.*, 15, 363, 1980.
47. **McDonald, J. T. and Margen, S.**, Wine versus ethanol in human nutrition. IV. Zinc balance, *Am. J. Clin. Nutr.*, 33, 1096, 1980.
48. **Roe, D. A.**, *Drug-Induced Nutritional Deficiencies*, AVI, Westport, Conn., 1976, 145.
49. **Wester, P. O.**, Tissue zinc at autopsy-relation to medication with diuretics, *Acta Med. Scand.*, 208, 269, 1980.
50. Multicentre Trial Group, Controlled trial of D(−) penicillamine in severe rheumatoid arthritis, *Lancet*, 1, 275, 1973.
51. **Falchuk, K. H.**, Effect of acute disease and ACTH on serum zinc proteins, *N. Engl. J. Med.*, 196, 1129, 1977.

52. **Sullivan, J. F., Blotcky, A. J., Jetton, M. M., Hahn, H. D., Jr., and Burch, R. E.,** Serum levels of selenium, calcium, copper, magnesium, manganese and zinc in various human diseases, *J. Nutr.*, 109, 1432, 1979.

53. **Walker, B. E., Hughes, S., Simmons, A. V., and Chandler, G. N.,** Plasma zinc after myocardial infarction, *Eur. J. Clin. Invest.*, 8, 193, 1978.

54. **Husain, S. L. and Bessant, R. G.,** Oral zinc sulfate in the treatment of leg ulcers, in *Clinical Applications of Zinc Metabolism*, Pories, W. J., Strain, W. H., Hsu, J. M., and Woosley, R. L., Eds., Charles C Thomas, Springfield, Ill., 1974, 168.

55. **Fisher, S., Hendricks, D. G., and Mahoney, A. W.,** Nutritional assessment of senior rural Utahns by biochemical and physical measurements, *Am. J. Clin. Nutr.*, 31, 667, 1978.

56. **Vir, S. C. and Love, A. H. G.,** Zinc and copper status of the elderly, *Am. J. Clin. Nutr.*, 32, 1472, 1979.

57. **Wagner, P. A., Krista, M. L., Bailey, L. B., Christakis, G. J., Jernigan, J. A., Araujo, P. E., Appledorf, H., Davis, G. C., and Dinning, J. S.,** Zinc status of elderly black Americans from urban low-income households, *Am. J. Clin. Nutr.*, 33, 1771, 1980.

58. **Fulwood, R., Johnson, C. L., Bryner, J. D., Gunter, E. W., and McGrath, C. R.,** Hematological and Nutritional Biochemistry Reference Data for Persons 6 Months — 74 Years of Age: United States, 1976—80, DHHS Publ. No. (PHS) 83-1682, Department of Health and Human Services, Washington, D.C., 1982.

59. **Lindeman, R. D., Clark, M. L., and Colmore, J. P.,** Influence of age and sex on plasma and red-cell zinc concentrations, *J. Gerontol.*, 26, 358, 1971.

60. **Chooi, M. K., Todd, J. K., and Boyd, N. D.,** Influence of age and sex on plasma zinc levels in normal and diabetic individuals, *Nutr. Metab.*, 20, 135, 1976.

61. **Wagner, P. A., Bailey, L. B., Krista, M. L., Jernigan, J. A., Robinson, J. D., and Cerda, J. J.,** Comparison of zinc and folacin status in elderly women from differing socioeconomic backgrounds, *Nutr. Res.*, 1, 565, 1981.

62. **Li, T. K. and Vallee, B. L.,** The biochemical and nutritional roles of other trace elements, in *Modern Nutrition in Health and Disease*, 6th ed., Goodhart, R. S. and Shils, M. E., Eds., Lea & Febiger, Philadelphia, 1980, 408.

63. **Vreeman, H. J., Venter, C., Leegwater, J., Olivera, C., and Weiner, M. W.,** Taste, smell and zinc metabolism in patients with chronic renal failure, *Nephron*, 26, 163, 1980.

64. **Greger, J. L. and Geissler, A. H.,** Effect of zinc supplementation on taste acuity of the aged, *Am. J. Clin. Nutr.*, 31, 633, 1978.

65. **Seligson, F. H.,** Sodium intake, preference and taste acuity in elderly subjects before and after zinc supplementation, Unpublished manuscript.

66. **Haeger, K., Lanner, E., and Magnusson, P. O.,** Oral zinc sulfate in the treatment of venous leg ulcers, in *Clinical Applications of Zinc Metabolism*, Pories, W. J., Strain, W. H., Hsu, J. M., and Woosley, R. L., Eds., Charles C Thomas, Springfield, Ill., 1974, 158.

67. **Hallbook, T. and Lanner, E.,** Serum-zinc and healing of venous leg ulcers, *Lancet*, 2, 780, 1972.

68. **Weismann, K., Wanscher, B., and Krakaver, R.,** Oral zinc therapy in geriatric patients with selected skin manifestations and a low plasma zinc level, *Acta Dermatatol. (Stockholm)*, 58, 157, 1978.

69. **Fernandes, G., West, A., and Good, R. A.,** Nutrition, immunity and cancer, a review. III. Effects of diet on the diseases of aging, *Clin. Bull.*, 9, 91, 1979.

70. **Duchateau, J., Delepesse, G., Vrijens, R., and Collet, H.,** Beneficial effects of oral zinc supplementation on the immune response of old people, *Am. J. Med.*, 70, 1001, 1981.

71. **Langan, M. J., and Yearick, E. S.,** The effects of improved oral hygiene on taste perception and nutrition of the elderly, *J. Gerontol.*, 31, 413, 1976.

72. **Cooper, R. M., Bilask, M. A., and Zubek, J. P.,** The effect of age on taste sensitivity, *J. Gerontol.*, 14, 56, 1959.

73. **Wacker, W. E. C.,** Role of zinc in wound healing: a critical review, in *Trace Elements in Human Health and Disease I*, Prasad, A. S., Ed., Academic Press, New York, 1976, 107.

74. **Scholoen, L. H., Fernandes, G., Garofalo, J. A., and Good, R. A.,** Nutrition, immunity and cancer. II. Zinc, immune function and cancer, *Clin. Bull.*, 9, 63, 1979.

75. **Holden, J. M., Wolf, W. R., and Mertz, W.,** Zinc and copper in self-selected diets, *J. Am. Dietet. Assoc.*, 75, 23, 1979.

76. Safe Drinking Water Committee, *Drinking Water and Health*, Vol. 3, National Academy of Science, Washington, D.C., 1980, 312.

77. **Gibson, R. A., Anderson, B. M., and Sabry, J. H.,** The trace metal status of a group of post-menopausal vegetarians, *J. Am. Dietet. Assoc.*, 82, 246, 1983.

78. **Tipton, D. H., Stewart, P. L., and Martin, P. G.,** Trace elements in diets and excreta, *Health Phys.*, 12, 1683, 1966.

79. **White, H. S.,** Inorganic elements in weighted diets of girls and young women, *J. Am. Dietet. Assoc.,* 55, 38, 1969.
80. **Gormican, A.,** Inorganic elements in foods used in hospital menus, *J. Am. Dietet. Assoc.,* 56, 397, 1970.
81. **Guthrie, B. E.,** Daily dietary intakes of zinc, copper, manganese, chromium, and cadmium by some New Zealand woman, *Proc. Univ. Otago Med. School,* 51, 47, 1973.
82. **Guthrie, B. and Robinson, M. F.,** Daily intake of manganese, copper, zinc and cadmium by New Zealand women, *Br. J. Nutr.,* 38, 55, 1977.
83. **Walker, M. A. and Page, L.,** Nutritive content of college meals, *J. Am. Dietet. Assoc.,* 70, 260, 1977.
84. **Klevay, L. M., Reck, S. J., and Barcome, D. F.,** Evidence of dietary copper and zinc deficiencies, *JAMA,* 241, 1916, 1979.
85. **Pennington, J. T. and Calloway, D. H.,** Copper content of foods, *J. Am. Dietet. Assoc.,* 63, 143, 1973.
86. **Paul, A. A. and Southgate, D. A. T.,** *McCance and Widdowson's The Composition of Foods,* 4th ed., Elsevier/North-Holland, New York, 1978.
87. **Schletwein-Gsell, D. and Mommsen-Straub, S.,** Übersicht Spurenelemente in Lebensmitteln. VI. Kupfer, *Int. Z. Vit. Ern. Forsch.,* 41, 554, 1971.
88. **Sandstead, H. H., Munoz, J. M., Jacob, R. A., Klevay, L. M., Reck, S., Logan, G. M., Jr., Dintzis, F. R., Inglett, G. E., and Shuey, W. C.,** Influence of dietary fiber on trace element balance, *Am. J. Clin. Nutr.,* 31, S180, 1978.
89. **Klevay, L. M., Reck, S. J., Jacob, R. A., Logan, G. M., Jr., Munoz, J. M., and Sandstead, H. H.,** The human requirement for copper. I. Healthy men fed conventional, American diets, *Am. J. Clin. Nutr.,* 33, 45, 1980.
90. **Jacob, R. A., Sandstead, H. H., Munoz, J. M., Klevay, L. M., and Milne, D. B.,** Whole body surface loss of trace metals in normal males, *Am. J. Clin. Nutr.,* 34, 1379, 1981.
91. **Colin, M. A., Taper, L. J., and Ritchey, S. J.,** Effect of dietary zinc and protein levels on the utilization of zinc and copper by adult females, *J. Nutr.,* 113, 1480, 1983.
92. **Turnlund, J. R., Michel, M. C., Keyes, W. R., Schultz, Y., and Margen, S.,** Copper absorption in elderly men determined by using stable 65Cu, *Am. J. Clin. Nutr.,* 36, 587, 1982.
93. **Sandstead, H. H.,** Copper bioavailability and requirements, *Am. J. Clin. Nutr.,* 35, 809, 1982.
94. **Greger, J. L., Zaikis, S. C., Abernathy, R. P., Bennett, O. A., and Huffman, J.,** Zinc, nitrogen, copper, iron and manganese balance in adolescent females fed two levels of zinc, *J. Nutr.,* 108, 149, 1978.
95. **Milne, D. B., Omaye, S. T., and Amos, W. H., Jr.,** Effect of ascorbic acid on copper and cholesterol in adult cynomolgus monkeys fed a diet marginal in copper, *Am. J. Clin. Nutr.,* 34, 2389, 1981.
96. **Klevay, R. M.,** Hypercholesterolemia due to ascorbic acid, *Proc. Soc. Exp. Biol. Med.,* 151, 579, 1976.
97. **Prasad, A. S., Brewer, C. J., Schoomaker, E. B., and Rabbani, P.,** Hypocupremia induced zinc therapy in adults, *JAMA,* 240, 2166, 1978.
98. **Price, N. O., Bunce, G. E., and Engel, R. W.,** Copper, manganese and zinc balance in preadolescent girls, *Am. J. Clin. Nutr.,* 23, 258, 1970.
99. **Mason, K. E.,** A conspectus of research on copper metabolism and requirements of man, *J. Nutr.,* 109, 1979, 1979.
100. **Solomons, N. W.,** On the assessment of zinc and copper nutriture in man, *Am. J. Clin. Nutr.,* 32, 856, 1979.
101. **Scheinberg, H.,** Humans health effects of copper, in *Copper in the Environoment, Part II: Health Effects,* Nriagu, J. O., Ed., John Wiley & Sons, New York, 1979, 17.
102. **Askari, A., Long, C. L., and Blakemore, W. S.,** Zinc, copper and parenteral nutrition in cancer. A review, *J. Parenteral Enteral Nutr.,* 4, 561, 1980.
103. **Cohen, D. I., Illowsky, B., and Linder, M. C.,** Altered copper absorption in tumor-bearing and estrogen-treated rats, *Am. J. Physiol.,* 2363, E309, 1979.
104. **Garofalo, J. A., Ashikari, H., Lesser, M. L., Menendez-Botet, C., Cunningham-Rundles, S., Schwartz, M. K., and Good, R. A.,** Serum zinc, copper, and the Cu/Zn ratio in patients with benign and malignant breast lesions, *Cancer,* 46, 2682, 1980.
105. **Youssef, A. A. R., Woods, B., and Baron, D. N.,** Serum copper: a marker of disease activity in rheumatoid arthritis, *J. Clin. Pathol.,* 36, 14, 1983.
106. **Versieck, J., Barbier, F., Speecke, A., and Hoste, J.,** Influence of myocardial infarction on serum manganese, copper and zinc concentrations, *Clin. Chem.,* 21, 578, 1975.
107. **Gregoriadis, G. C., Apostolidis, N. S., Romanos, A. N., and Paradellis, T. P.,** Postoperative changes in serum copper value, *Surg. Gynecol. Obstet.,* 154, 217, 1982.
108. **Sempos, C. T., Greger, J. L., Johnson, N. E., Smith, E. L., and Seyedabadi, F. M.,** Levels of serum copper and magnesium in normotensives and untreated and treated hypertensives, *Nutr. Rep. Int.,* 27, 1013, 1983.
109. **Deering, T. B., Dickson, E. R., Fleming, C. R., Geall, M. G., McCall, J. T., and Baggenstoss, A. H.,** Effect of D-penicillamine on copper retention in patients with primary biliary cirrhosis, *Gastroenterology,* 72, 1208, 1977.

110. **Lifschitz, M. D. and Henkin, R. I.,** Circadian variation in copper and zinc in man, *J. Appl. Physiol.,* 31, 88, 1971.

111. **Hambidge, K. M.,** Increase in hair copper concentration with increasing distance from the scalp, *Am. J. Clin. Nutr.,* 26, 1212, 1973.

112. **Klevay, L. M.,** Hair as a biopsy material. II. Assessment of copper nutriture, *Am. J. Clin. Nutr.,* 23, 1194, 1970.

113. **Denko, C. W. and Gabriel, P.,** Age and sex related levels of albumin, ceruloplasmin, α antitrypsin, α acid glycoprotein, and transferrin, *Ann. Clin. Lab. Sci.,* 11, 63, 1981.

114. **Owen, C. A., Jr.,** *Biochemical Aspects of Copper: Copper Proteins, Ceruloplasmin and Copper Protein Binding,* Noyes, Park Ridge, N.J., 1982.

115. **Prohaska, J. R. and Heller, L. J.,** Mechanical properties of the copper-deficient rat heart, *J. Nutr.,* 112, 2142, 1982.

116. **Rucker, R. B. and Murray, J.,** Cross-linking amino acids in collagen and elastin, *Am. J. Clin. Nutr.,* 31, 1221, 1978.

117. **Feller, D. J. and O'Dell, B. L.,** Dopamine and norepinephrine in discrete areas of copper-deficient rat brain, *J. Neurochem.,* 34, 1259, 1980.

118. **Prohaska, J. R. and Lukasewycz, O. A.,** Copper deficiency suppresses the immune responses of mice, *Science,* 213, 559, 1981.

119. **Klevay, L. M.,** Coronary heart disease: the zinc/copper hypothesis, *Am. J. Clin. Nutr.,* 28, 764, 1975.

120. **Petering, H. G., Murthy, L., and O'Flaherty, E.,** Influence of dietary copper and zinc on rat lipid metabolism, *J. Agric. Food Chem.,* 25, 1105, 1977.

121. **Lei, K. Y.,** Oxidation, excretion, and tissue distribution of [26^{14}C]cholesterol in copper-deficient rats, *J. Nutr.,* 108, 232, 1978.

122. **Allen, K. G. D. and Klevay, L. M.,** Hyperlipoproteinemia in rats due to copper deficiency, *Nutr. Rep. Int.,* 22, 295, 1980.

123. **Law, B. W. C. and Klevay, L. M.,** Plasma lecithin: cholesterol acyltransferase in copper-deficient rats, *J. Nutr.,* 11, 1698, 1981.

124. **Harvey, P. W. and Allen, K. G. D.,** Decreased plasma lecithin: cholesterol acyltransferase activity in copper-deficient rats, *J. Nutr.,* 11, 1855, 1981.

125. **Waslein, C. I.,** Human intake of trace elements, in *Trace Elements in Human Health and Disease,* Vol. 2, *Essential and Toxic Elements,* Prasad, A. S. and Oberleas, D., Eds., Academic Press, New York, 1976, 347.

126. **Milne, D. B., Schnakenberg, D. D., Johnson, H. L., and Kuhl, G. L.,** Trace mineral intake of enlisted military personnel, *J. Am. Dietet. Assoc.,* 76, 41, 1980.

127. **Srivastava, U. S., Nadeau, M. H., and Carbonneau, N.,** Mineral intakes of university students: cadmium and manganese content, *Nutr. Rep. Int.,* 18, 375, 1978.

128. Committee on Biologic Effects of Atmospheric Pollutants, *Manganese,* National Academy of Sciences, Washington, D.C., 1973, 77.

129. **Schroeder, H. A., Balassa, J. J., and Tipton, I. H.,** Essential trace metals in man: manganese, *J. Chronic Dis.,* 19, 545, 1966.

130. **North, B. B., Leichsenring, J. M., and Norris, L. M.,** Manganese metabolism in college women, *J. Nutr.,* 72, 217, 1960.

131. **Lang, V. M., North, B. B., and Morse, L. M.,** Manganese metabolism in college men consuming vegetarian diets, *J. Nutr.,* 85, 132, 1965.

132. **McLeod, B. E. and Robinson, M. F.,** Metabolic balance of manganese in young women, *Br. J. Nutr.,* 27, 221, 1972.

133. **Kirchgessner, M., Weigand, E., and Schwarz, F. J.,** Absorption of zinc and manganese in relation to age, in *Trace Element Metabolism in Man and Animals (TEMA-4),* McC.Howell, J., Gawthorn, J. M., and White, C. L., Eds., Australian Academy of Science, Canberra, 1981, 123.

134. **Pleban, P. A. and Pearson, K. H.,** Determination of manganese in whole blood and serum, *Clin. Chem.,* 25, 1915, 1979.

135. **Versieck, J., Cornelis, R., Lemey, G., and DeRudder, J.,** Determination of manganese in whole blood and serum, *Clin. Chem.,* 26, 531, 1980.

136. **Alcock, N. W.,** Serum versus plasma for trace metal analysis, in *Trace Element Metabolism in Man and Animals (TEMA-4),* McC.Howell, J., Gawthorne, J. M., and White, C. L., Eds., Australian Academy of Science, Canberra, 1981, 678.

137. **Versieck, J., Hoste, J., and Barbier, F.,** Determination of manganese, copper and zinc in serum in normal controls and patients with liver metastases, *Acta Gastro-Enterol. Belg.,* 39, 340, 1976.

138. **Hataro, S., Nishi, Y., and Usui, T.,** Erythrocyte manganese concentration in healthy Japanese children, adults and the elderly, and in cord blood, *Am. J. Clin. Nutr.,* 37, 457, 1983.

139. **Leach, R. M., Jr.,** Metabolism and function of manganese, in *Trace Elements in Human Health and Disease*, Vol. 2, *Essential and Toxic Elements*, Prasad, A. S. and Oberleas, D., Eds., Academic Press, New York, 1976, 235.

140. **Paynter, D. I.,** Changes in activity of the manganese superoxide dismutase enzyme in tissues of the rat with changes in dietary manganese, *J. Nutr.*, 110, 437, 1980.

141. **Rosa, G., Keen, C. L., Leach, R. M., and Hurley, L. S.,** Regulation of superoxide dismutase activity by dietary manganese, *J. Nutr.*, 110, 795, 1980.

142. **Leung, T. K. C., Lai, J. C. K., and Lim, L.,** The regional distribution of monoamine oxidase activities toward different substrates; effects in rat brain of chronic administration of manganese chloride and of aging, *J. Neurochem.*, 36, 2037, 1981.

143. **Lai, J. C. K., Leung, T. K. C., and Lim, L.,** Monoamine oxidase activities in liver, heart, spleen and kidney of the rat, *Exp. Gerontol.*, 17, 219, 1982.

144. **Levine, R. A., Streeten, D. H. P., and Doisy, R. J.,** Effects of oral chromium supplementation on the glucose tolerance of elderly human subjects, *Metabolism*, 17, 114, 1968.

145. **Kumpulainen, J. T., Wolf, W. R., Veillon, C., and Mertz, W.,** Determination of chromium in selected United States diets, *J. Agric. Food Chem.*, 27, 490, 1979.

146. **Anderson, R. A., Polansky, M. M., Bryden, N. A., Patterson, K. Y., Veillon, C., and Glinsmann, W. H.,** Effects of chromium supplementation on urinary Cr excretion of human subjects and correlation of Cr excretion with selected clinical parameters, *J. Nutr.*, 113, 276, 1983.

147. **Toepfer, E. W., Mertz, W., Roginski, E. E., and Polansky, M. M.,** Chromium in foods in relation to biological activity, *J. Agric. Food Chem.*, 21, 69, 1973.

148. **Mertz, W.,** Mineral elements: new perspectives, *J. Am. Dietet. Assoc.*, 77, 258, 1980.

149. **Schroeder, H. A., Balassa, J. J., and Tipton, I. H.,** Abnormal trace metals in man-chromium, *J. Chronic Dis.*, 15, 941, 1962.

150. **Guthrie, B. E., Wolf, W. R., and Veillon, C.,** Background correction and related problems in the determination of chromium in urine by graphite furnace atomic absorption spectrometry, *Anal. Chem.*, 50, 1900, 1978.

151. **Veillon, C., Wolf, W. R., and Guthrie, B. E.,** Determination of chromium in biological materials by stable isotope dilution, *Anal. Chem.*, 51, 1022, 1979.

152. **Veillon, C., Guthrie, B. E., and Wolf, W. R.,** Retention of chromium by graphite furnace tubes, *Anal. Chem.*, 52, 457, 1980.

153. **Guthrie, B. E.,** Chromium analysis, in *New Zealand Workshop on Trace Elements in New Zealand*, University of Otago, Dunedin, 1981, 311.

154. **Doisy, R. J., Streeten, D. H. P., Levine, R. A., and Chodos, R. B.,** Effects and metabolism of chromium in normals, elderly subjects, and diabetics, in *Trace Substances in Environmental Health*, Vol. 2, Hemphill, D. D., Ed., University of Missouri Press, Columbia, 1968, 75.

155. **Donaldson, R. M. and Barreras, R. F.,** Intestinal absorption of trace quantities of chromium, *J. Lab. Clin. Med.*, 68, 484, 1966.

156. **Nordstrom, J. W.,** Trace mineral nutrition in the elderly, *Am. J. Clin. Nutr.*, 36, 788, 1982.

157. **Polansky, M. M., Anderson, R. A., Bryden, N. A., and Glinsmann, W. H.,** Chromium (Cr) and brewer's yeast supplementation of human subjects: effect on glucose tolerance, serum glucose insulin and lipid parameters, *Fed. Proc. Fed. Am. Soc. Exp. Biol.*, 41(Abstr.), 709, 1982.

158. **Doisy, P. J., Streeten, D. H. P., Freiberg, J. M., and Schneider, A. J.,** Chromium metabolism in man and biochemical effects, in *Trace Elements in Human Health and Disease*, Vol. 2, *Essential and Toxic Elements*, Prasad, A. S. and Oberteas, D., Academic Press, New York, 1976, 79.

159. **Guthrie, B. E., Wolf, W. R., Veillon, C., and Mertz, W.,** Chromium in urine, in *Trace Substances in Environmental Health*, Vol. 12, Hemphill, D. D., Ed., University of Missouri Press, Columbia, 1978, 490.

160. **Anderson, R. A., Polansky, M. M., Bryden, N. A., Roginski, E. E., Patterson, K. Y., and Reamer, D. C.,** Effect of exercise (running) on serum glucose, insulin, glucagon and chromium excretion, *Diabetes*, 31, 212, 1982.

161. **Anderson, R. A., Polansky, M. M., Bryden, N. A., Roginski, E. E., Patterson, K. Y., Veillon, C., and Glinsmann, W.,** Urinary chromium excretion of human subjects: effects of chromium supplementation and glucose loading, *Am. J. Clin. Nutr.*, 36, 1184, 1982.

162. **Gurson, C. T. and Saner, G.,** The effect of glucose loading on urinary excretion of chromium in normal adults, in individuals from diabetic families and in diabetics, *Am. J. Clin. Nutr.*, 31, 1158, 1978.

163. **Gurson, C. T. and Saner, G.,** Urinary chromium excretion, diurnal changes, and relationship to chromium excretion in healthy and sick individual of different ages, *Am. J. Clin. Nutr.*, 31, 1162, 1978.

164. **Benjanuvatra, N. K. and Bennion, M.,** Hair chromium concentration of Thai subjects with and without diabetes mellitus, *Nutr. Rep. Int.*, 12, 325, 1975.

165. **Hambidge, K. M., Rodgerson, D. O., and O'Brien, D.,** Concentration of chromium in the hair of normal children and children with diabetes mellitus, *Diabetes*, 17, 517, 1968.

166. **Côte, M. and Shapcott, D.,** Hair chromium concentrations and arteriosclerotic heart disease, in *Trace Element Metabolism in Man and Animals (TEMA-4),* McC.Howell, J., Gawthorne, J. M., and White, C. L., Eds., Australian Academy of Science, Canberra, 1981, 521.

167. **Jeejeebhoy, K. N., Chu, R. C., Marliss, E. B., Greenberg, G. R., and Bruce-Robertson, A.,** Chromium deficiency, glucose intolerance and neuropathy reversed by chromium supplementation in a patient receiving long-term total parenteral nutrition, *Am. J. Clin. Nutr.,* 30, 531, 1977.

168. **Freund, H., Atamian, S., and Fischer, J. E.,** Chromium deficiency during total parenteral nutrition, *JAMA,* 241, 496, 1979.

169. **Mertz, W.,** Effects and metabolism of glucose tolerance factor, *Nutr. Rev.,* 33, 129, 1975.

170. **Glinsmann, W. H. and Mertz, W.,** Effect of trivalent chromium on glucose tolerance, *Metabolism,* 15, 510, 1966.

171. **Liu, V. J. K. and Morris, J. S.,** Relative chromium response as an indicator of chromium status, *Am. J. Clin. Nutr.,* 31, 942, 1978.

172. **Polansky, M. M., Anderson, R. A., Bryden, N. A., Roginski, E. E., Mertz, W., and Glinsman, W. H.,** Chromium supplementation of free-living subjects effect on glucose tolerance and insulin, *Fed. Proc. Fed. Am. Soc. Exp. Biol.,* 40, 3721(abstract), 1981.

173. **Offenbacher, E. G. and Pi-Sunyer, F. X.,** Beneficial effects of chromium-rich yeast on glucose tolerance and blood lipids in elderly subjects, *Diabetes,* 29, 219, 1980.

174. **Riales, R. and Albrink, M. J.,** Effect of chromium chloride supplementation on glucose tolerance and serum lipids including high-density lipoprotein of adult men, *Am. J. Clin. Nutr.,* 34, 2670, 1981.

175. **Uusitupa, M. I. J., Kumpulainen, J. T., Voutilainen, E., Hersio, K., Sarlund, H., Pyorälä, K. P., Koivistoinen, P. E., and Lehto, J. T.,** Effect of inorganic chromium supplementation on glucose tolerance, insulin response, and serum lipids in noninsulin-dependent diabetics, *Am. J. Clin. Nutr.,* 38, 404, 1983.

176. **Thompson, J. N., Erdody, P., and Smith, D. C.,** Selenium content of food consumed by Canadians, *J. Nutr.,* 105, 274, 1975.

177. **Zabel, N. L., Harland, J., Gormican, A. T., and Ganther, H. E.,** Selenium content of commercial formula diets, *Am. J. Clin. Nutr.,* 31, 850, 1978.

178. **Schrauzer, G. N. and White, D. A.,** Selenium in human nutrition; dietary intakes and effects of supplementation, *Bioinorg. Chem.,* 8, 303, 1978.

179. **McConnell, K. P., Smith, J. C., Jr., Higgins, P. G., and Blotcky, A. J.,** Selenium content of selected hospital diets, *Nutr. Res.,* 1, 235, 1981.

180. **Welsh, S. O., Holden, J. M., Wolf, W. R., and Levander, O. A.,** Selenium in self-selected diets of Maryland residents, *J. Am. Dietet. Assoc.,* 79, 277, 1981.

181. **Palmer, I. S., Olson, O. E., Ketterling, L. M., and Shank, C. E.,** Selenium intake and urinary excretion in persons living near a high selenium area, *J. Am. Dietet. Assoc.,* 82, 511, 1983.

182. **Thomson, C. D. and Robinson, M. F.,** Selenium in human health and disease with emphasis on those aspects peculiar to New Zealand, *Am. J. Clin. Nutr.,* 33, 303, 1980.

183. **Morris, V. C. and Levander, O. A.,** Selenium content of foods, *J. Nutr.,* 100, 1383, 1970.

184. **Lane, H. W., Taylor, B. J., Stool, E., Servance, D., and Warren, D. C.,** Selenium content of selected foods, *J. Am. Dietet. Assoc.,* 82, 24, 1983.

185. **Robinson, M. F., McKenzie, J. M., Thomson, C. D., and van Rij, A. L.,** Metabolic balance of zinc, copper, cadmium, iron, molybdenum and selenium in young New Zealand women, *Br. J. Nutr.,* 30, 195, 1973.

186. **Griffiths, N. M., Stewart, R. D. H., and Robinson, M. F.,** The metabolism of [Se^{75}]selenomethionine in four women, *Br. J. Nutr.,* 35, 373, 1976.

187. **Stewart, R. D. H., Griffiths, N. M., Thomson, C. D., and Robinson, M. F.,** Quantitative selenium metabolism in normal New Zealand women, *Br. J. Nutr.,* 40, 45, 1978.

188. **Greger, J. L. and Marcus, R. E.,** Effect of dietary protein, phosphorus, and sulfur amino acids on selenium metabolism of adult males, *Ann. Nutr. Metab.,* 25, 97, 1981.

189. **Greger, J. L., Smith, S. A., Johnson, M. A., and Baier, M. J.,** Effects of dietary tin and aluminum on selenium utilization by adult males, *Biol. Trace Elem. Res.,* 4, 269, 1982.

190. **Janghorbani, M., Christensen, M. J., Nahapetian, A., and Young, V. R.,** Selenium metabolism in healthy adults: quantitative aspects using the stable isotope $^{75}SeO_3^{2-}$, *Am. J. Clin. Nutr.,* 35, 647, 1982.

191. **Levander, O. A., Sutherland, B., Morris, V. C., and King, J. C.,** Selenium balance in young men during selenium depletion and repletion, *Am. J. Clin. Nutr.,* 34, 2262, 1981.

192. **Young, V. R., Nahapetian, A., and Janghorbani, M.,** Selenium bioavailability with reference to human nutrition, *Am. J. Clin. Nutr.,* 35, 1076, 1982.

193. **Ostádalová, I., Babicky, A., and Kopoldova, A.,** Ontogenic changes in selenite metabolism in rats, *Arch. Toxicol.,* 49, 247, 1982.

194. **Burk, R. F.,** Selenium in nutrition, *World Rev. Nutr. Diet,* 30, 83, 1978.

195. **Levander, O. A.,** Selenium and chromium in human nutrition, *J. Am. Dietet. Assoc.,* 66, 338, 1975.

196. **Robinson, M. F., Godfrey, P. J., Thomson, C. D., Rea, H. M., and van Rij, A. M.,** Blood selenium and glutathione peroxidase activity in normal subjects and in surgical patients with and without cancer in New Zealand, *Am. J. Clin. Nutr.,* 32, 1477, 1979.

197. **Persigehl, M., Schicha, H., Kasperek, K., and Klein, H. J.,** Trace element concentration in human organs in dependence of age, *Beitr. Pathol.,* 161, 209, 1977.

198. **Burch, R. E., Sullivan, J. F., Jetton, M. M., and Hahn, H. K. J.,** The effect of aging on trace element content of various rat tissues. I. Early stages of aging, *Age,* 2, 103, 1979.

199. **Csallany, A. S., Zaspel, B. J., and Ayaz, K. L.,** Selenium and aging, in *Selenium in Biology and Medicine,* Spallholz, J. E., Martin, J. L., and Ganther, H. E., Eds., AVI, Westport, Conn., 1981, 118.

200. **Van Rij, A. M., Thomson, C. D., McKenzie, J. M., and Robinson, M. F.,** Selenium deficiency in total parenteral nutrition, *Am. J. Clin. Nutr.,* 32, 2076, 1979.

201. **Johnson, R. S., Baker, S. G., Fallon, J. T., Maynard, E. P., Ruskin, J. N., Wen, Z., Ge, K., and Cohen, H. J.,** An occidental case of cardiomyopathy and selenium deficiency, *N. Engl. J. Med.,* 304, 1210, 1981.

202. **Collipp, P. J. and Chen, S. Y.,** Cardiomyopathy and selenium deficiency in a two-year-old girl, *N. Engl. J. Med.,* 304, 1304, 1981.

203. Keshan Disease Research Group of the Chinese Academy of Medical Sciences, Epidemiologic studies on the etiologic relationship of selenium and Keshan disease, *Chin. Med. J.,* 92, 477, 1979.

204. Keshan Disease Research Group of the Chinese Academy of Medical Sciences, Observations on effect of sodium selenite in prevention of Keshan disease, *Chin. Med. J.,* 92, 471, 1979.

205. **Frost, D. V.,** The two faces of selenium — can selenophobia be cured? *CRC Crit. Rev. Toxicol.,* 1, 467, 1972.

206. **Shamberger, R. J., Tytko, S. A., and Willis, C. E.,** Selenium and heart disease, in *Trace Substances in Environmental Health,* Vol. 9, Hemphill, D. D., Ed., University of Missouri Press, Columbia, 1975, 15.

207. **Bostrom, H. and Wester, P. O.,** Trace elements in drinking water and death rates of cardiovascular disease, *Acta Med. Scand.,* 181, 465, 1967.

208. **Masironi, R. and Parr, R.,** Selenium and cardiovascular disease, in *Proc. Symp. Selenium-Tellurium in the Environment,* University of Notre Dame, South Bend, Ind., 1976, 316.

209. **Wester, P. O.,** Trace elements in human myocardial infarction determined by neutron activation analysis, *Acta Med. Scand.,* 178, 765, 1965.

210. **Wester, P. O.,** Trace elements in coronary arteries in the presence and absence of atherosclerosis, *Atherosclerosis,* 13, 395, 1971.

211. **Wester, P. O.,** Trace elements in serum and urine from hypertensive patients treated for six months with chlorthalidone, *Acta Med. Scand.,* 196, 489, 1974.

212. Committee on Diet, Nutrition and Cancer, Minerals, in *Diet, Nutrition and Cancer,* National Academy Press, Washington, D.C., 1982.

213. **Shamberger, R. J. and Willis, C. E.,** Selenium distribution and human cancer mortality, *CRC Crit. Rev. Clin. Lab. Sci.,* 2, 211, 1971.

214. **Shamberger, R. J., Tytko, S. A., and Willis, C. E.,** Antioxidants and cancer. VI. Selenium and age-adjusted human cancer mortality, *Arch. Environ. Health,* 31, 231, 1976.

215. **Schrauzer, G. N.,** Selenium and cancer: a review, *Bioinorg. Chem.,* 5, 275, 1976.

216. **Schrauzer, G. N., White, D. A., and Schneider, C. J.,** Cancer mortality correlation studies. III. Statistical associations with dietary selenium intakes, *Bioinorg. Chem.,* 7, 23, 1977.

217. **Schrauzer, G. N., White, D. A., and Schneider, C. J.,** Cancer mortality correlation studies. IV. Associations with dietary intakes and blood levels of certain trace elements, notably Se-antagonists, *Bioinorg. Chem.,* 7, 35, 1977.

218. **Jansson, B., Seibert, G. B., and Speir, J. F.,** Gastrointestinal cancer: its geographic distribution and correlation to breast cancer, *Cancer,* 36, 2373, 1975.

219. **Jansson, B., Jacobs, M. M., and Griffin, A. C.,** Gastrointestinal cancer: epidemiology and experimental studies, *Adv. Exp. Med. Biol.,* 91, 305, 1978.

220. **Medina, D., Lane, H. W., and Tracey, C. M.,** Selenium and mouse mammary tumorigenesis: an investigation of possible mechanisms, *Cancer Res.,* 43, 2460s, 1983.

Index

INDEX

S

T

Printed and bound by CPI Group (UK) Ltd, Croydon, CR0 4YY

22/10/2024

01777632-0009